Brazilian Marine Biodiversity

Series editor
Alexander Turra
São Paulo, Brazil

This book series provides an in-depth analysis of Brazilian marine biodiversity for a broad and international readership.

The diversity of marine habitats found in Brazil is astonishing and includes estuaries, coral reefs, rocky shores, sandy beaches, rhodolith beds, mangroves, salt marshes, deep-sea habitats, vegetated bottoms, and continental shelf. These habitats are addressed from an ecosystem perspective across the series, and characterized in terms of distribution along the Brazilian coast, records of relevant species, and information on the prevailing structuring ecological and oceanographic processes.

The series presents an analysis of the role of biodiversity and the importance of ecosystem services, and discusses it in relation to the threats to each habitat, such as pollution, invasive species, and global environmental changes. It also introduces educational and outreach strategies to address the Brazilian population's environmental questions related to the coasts and oceans, which could inspire educational projects on marine biodiversity worldwide.

This series is an initiative of the Brazilian Network for Monitoring Coastal Benthic Habitats (ReBentos). Supported by the Brazilian National Council for Scientific and Technological Development (CNPq) and the Research Program on Biodiversity Characterization, Conservation, Restoration and Sustainable Use of the São Paulo Research Foundation (BIOTA-FAPESP), ReBentos is also part of the Brazilian Network on Global Climate Change Research (Rede Clima) at the Ministry of Science, Technology and Innovation (MCTI).

More information about this series at http://www.springer.com/series/15050

Paulo da Cunha Lana
Angelo Fraga Bernardino
Editors

Brazilian Estuaries

A Benthic Perspective

Springer

Editors
Paulo da Cunha Lana
Universidade Federal do Paraná
Centro de Estudos do Mar
Pontal do Sul, Paraná, Brazil

Angelo Fraga Bernardino
Departamento de Oceanografia e Ecologia
Universidade Federal do Espirito Santo
Vitória, Espírito Santo, Brazil

ISSN 2520-1077 ISSN 2520-1085 (electronic)
Brazilian Marine Biodiversity
ISBN 978-3-030-08537-7 ISBN 978-3-319-77779-5 (eBook)
https://doi.org/10.1007/978-3-319-77779-5

Softcover re-print of the Hardcover 1st edition 2018

Cover illustration: Rhodolit close-up - *João Paulo Krajewski*; Fish school at a coral reef, Northeastern Brazil - *Ruy Kenji P. Kikuchi*; Rocky shore intertidal and subtidal fauna, Southeastern Brazil - *Áthila Bertoncini*; Canoes in the estuary of Jaguaribe River, Northeastern Brazil - *Francisco Carlos Rocha de Barros Junior*; Paranaguá estuary, Southern Brazil - *Gisele Morais*; Mangrove at Iriri River, Bertioga, Southeastern Brazil - *Guilherme Moraes de Oliveira Abuchahla*

Printed on acid-free paper

This Springer imprint is published by the registered company Springer International Publishing AG part of Springer Nature.
The registered company address is: Gewerbestrasse 11, 6330 Cham, Switzerland

Coordination

Support

Foreword

The knowledge of marine biodiversity, which has become a global field of research, asks for an integrative view of social, ecological, and physical interactions. This perception holds especially true for the coastal regions worldwide, where most of the human populations are currently located. Current trends in climate and in land usage pose severe threats to coastal and estuarine ecosystems especially in developing countries. Among coastal areas, estuaries are unique in sustaining and providing essential goods and services to mankind. They include a variety of habitats with their own structure and dynamics, mostly already altered during the last few decades. Recent evidence suggests that estuaries may be more relevant in providing ecological services than previously believed, but their social relevance remains essentially unknown to much of the population.

Estuaries are preeminent ecosystems along the 9200 km of the Brazilian coastal margin, widely differ in their geology, and are under variable meteorological and oceanographic conditions. These variable conditions are expected to influence the biology and ecology of estuarine ecosystems, resulting in spatial and temporal differences in their assemblages. In order to understand the expected changes in estuarine dynamics and its influence on benthic assemblages, these ecosystems can be spatially divided into three major biogeographic provinces known as Marine Ecoregions (Spalding et al. 2007). The Marine Ecoregions' approach was applied in a global scale in order to advance conservation of marine ecosystems, and therefore, it is expected that some regional subdivisions may not fully represent spatial differences in biological assemblages. The congruence of climatic and environmental differences along the Brazilian coast has yet to be empirically tested for most coastal ecosystems, including estuaries. However, at this initial stage, the spatial classification and ordination of Marine Ecoregions can support spatial conservation planning along broad areas with a high climatic and geomorphic variability, which is the case for estuaries along the Brazilian coast. Although this is the case for the estuarine benthic assemblages in Brazil, we will here follow the regional subdivision proposed by Spalding et al. (2007) and present general biodiversity, ecological processes, and major threats to benthic assemblages in estuaries from the five Brazilian Marine Ecoregions.

There are no extensive reviews about estuarine conservation status in Brazil or about their current changes that can be used by researchers and by the private and public sectors. The last comprehensive books on South-American estuaries were published by Springer almost two decades ago (Perillo et al. 1999; Seelinger and Kjerfve 2001). Brazilian estuaries, mainly from the Northern and Northeastern sectors, are poorly represented in both books. The *Encyclopedia of Estuaries* (Kennish 2016) did not focus on specific countries or regions.

This book presents the main drivers of structure and processes in the benthic assemblages from estuaries along the Brazilian coast, assesses the influence of natural and human disturbance, and discusses their relevance to management needs. Information is organized in an introduction and five chapters dedicated to each of the five ecoregions, from the tropical Amazonian to the warm temperate Rio Grande. Our compilation and synthesis will enable readers to compare broad biodiversity patterns of Brazil's estuaries as well as will serve researchers, students, and policy makers on their own work in the fields of marine sciences and conservation.

Universidade Federal do Espirito Santo Angelo Fraga Bernardino
Vitória, Brazil
Universidade Federal do Paraná Paulo da Cunha Lana
Pontal do Sul, Brazil

References

Kennish MJ (ed) (2016) Encyclopedia of estuaries. Springer, Dordrecht
Perillo GME, Piccolo MC, Pino-Quivira M (eds) (1999) Estuaries of South America. Springer, Berlin Heidelberg
Seeliger U, Kjerfve B (eds) (2001) Coastal marine ecosystems of Latin America. Springer, Berlin Heidelberg
Spalding MD, Fox HE, Allen GR et al (2007) Marine ecoregions of the world: a bioregionalization of coastal and shelf areas. Bioscience 57:573–583. doi:10.1641/B570707

Acknowledgements

This book is dedicated to the many students, technicians, and researchers who have contributed to the study of estuarine ecology and biodiversity in Brazil. The editors thank Prof. Alexander Turra for the research initiative that led to funding of the ReBentos, the Brazilian Network for the Monitoring of Coastal Benthic Biodiversity (SISBIOTA-CNPq/FAPESP). We also deeply thank all authors who contributed to this book. Brazil's research agencies, including federal (CNPq, Capes), state (FAPs), and private, are also acknowledged for the support of a number of research groups that have led to essential scientific assessments of estuaries.

Contents

Contributors

Cristina de Almeida Rocha-Barreira Divisão de Oceanografia Biótica, Instituto de Ciências do Mar Labomar, Universidade Federal do Ceará, Fortaleza, Ceará, Brazil

Francisco Carlos Rocha de Barros Jr Laboratório de Ecologia Bentônica, Programa de Pós-Graduação em Ecologia e Biomonitoramento, Instituto de Biologia, Centro Interdisciplinar de Energia e Ambiente, Universidade Federal da Bahia, Salvador, Bahia, Brazil

Thayanne Lima Barros Universidade Federal do Paraná, Centro de Estudos do Mar, Pontal do Sul, Paraná, Brazil

Colin Robert Beasley Laboratório de Conservação da Biodiversidade e das Águas, Instituto de Estudos Costeiros, Universidade Federal do Pará, Bragança, Pará, Brazil

Angelo Fraga Bernardino Grupo de Ecologia Bêntica, Departamento de Oceanografia e Ecologia, CCHN, Universidade Federal do Espírito Santo, Vitória, Espírito Santo, Brazil

Lorena Bonno Bissoli Grupo de Ecologia Bêntica, Departamento de Oceanografia e Ecologia, CCHN, Universidade Federal do Espírito Santo, Vitória, Espírito Santo, Brazil

César França Braga Laboratório de Ecologia e Conservação Ambiental, Universidade Federal Rural da Amazônia, Capitão Poço, Pará, Brazil

Kalina M. Brauko Laboratório de Biodiversidade e Conservação Marinha, Núcleo de Estudos do Mar, Universidade Federal de Santa Catarina, Campus Universitário Trindade, Florianópolis, Santa Catarina, Brazil

Ronaldo Christofoletti Instituto do Mar, Universidade Federal de São Paulo, Santos, SP, Brazil

André Colling Laboratório de Ecologia de Invertebrados Bentônicos, Instituto de Oceanografia, Universidade Federal do Rio Grande, Rio Grande, Rio Grande do Sul, Brazil

Luiz César Corrêa-Gomes Instituto de Geociências, Universidade Federal da Bahia, Salvador, Bahia, Brazil

Rauquírio Albuquerque Marinho da Costa Laboratório de Plâncton e Cultivo de Microalgas, Instituto de Estudos Costeiros, Universidade Federal do Pará, Bragança, Pará, Brazil

Tânia Marcia Costa Universidade Estadual Paulista, Instituto de Biociências, São Vicente, São Paulo, Brazil

Paulo da Cunha Lana Universidade Federal do Paraná, Centro de Estudos do Mar, Pontal do Sul, Paraná, Brazil

Heleny Noronha David Divisão de Oceanografia Biótica, Universidade Federal do Ceará, Instituto de Ciências do Mar Labomar, Fortaleza, Ceará, Brazil

Antônio Carlos Dórea Pereira Filho Laboratório de Ecologia Bentônica, Programa de Pós-Graduação em Ecologia e Biomonitoramento, Instituto de Biologia, Centro Interdisciplinar de Energia e Ambiente, Universidade Federal da Bahia, Salvador, Bahia, Brazil

José Souto Rosa Filho Departamento de Oceanografia, Centro de Tecnologia, Universidade Federal de Pernambuco, Recife, Pernambuco, Brazil

Alessandra L. Fonseca Laboratório de Biodiversidade e Conservação Marinha, Núcleo de Estudos do Mar, Universidade Federal de Santa Catarina, Campus Universitário Trindade, Florianópolis, Santa Catarina, Brazil

João Bosco Gusmão Jr Departamento de Biología Marina, Facultad de Ciencias del Mar, Millennium Nucleus for Ecology and Sustainable Management of Oceanic Islands (ESMOI), Coquimbo, Chile

Guilherme C. Lessa Instituto de Geociências, Universidade Federal da Bahia, Salvador, Bahia, Brazil

Rafaela Camargo Maia Instituto Federal de Educação, Ciência e Tecnologia do Ceará, Avenida Desembargador Armando de Sales Louzada, Acaraú, CE, Brazil

Alisson Sousa Matos Departamento de Biologia, Universidade Federal do Ceará, Centro de Ciências, Fortaleza, Ceará, Brazil

Helena Matthews-Cascon Departamento de Biologia, Centro de Ciências, Universidade Federal do Ceará, Fortaleza, Ceará, Brazil

Marcela Carneiro Monteiro Laboratório de Oceanografia Costeira e Estuarina, Instituto de Estudos Costeiros, Universidade Federal do Pará, Bragança, Pará, Brazil

Nils Edvin Asp Instituto de Estudos Costeiros, Universidade Federal do Pará, Bragança, Pará, Brazil

Sérgio Antônio Netto Laboratório de Ciências Marinhas, Universidade do Sul de Santa Catarina, Tubarão, SC, Brazil

Luiz Eduardo de Oliveira Gomes Grupo de Ecologia Bêntica, Departamento de Oceanografia e Ecologia, CCHN, Universidade Federal do Espírito Santo, Vitória, Espírito Santo, Brazil

Paulo Roberto Pagliosa Laboratório de Biodiversidade e Conservação Marinha, Núcleo de Estudos do Mar, Universidade Federal de Santa Catarina, Campus Universitário Trindade, Florianópolis, SC, Brazil

Luci Cajueiro Carneiro Pereira Laboratório de Oceanografia Costeira e Estuarina, Instituto de Estudos Costeiros, Universidade Federal do Pará, Bragança, Pará, Brazil

Alice Reis Laboratório de Ecologia Bentônica, Programa de Pós-Graduação em Ecologia e Biomonitoramento, Instituto de Biologia, Centro Interdisciplinar de Energia e Ambiente, Universidade Federal da Bahia, Salvador, BA, Brazil

Cinthya Simone Gomes dos Santos Departamento de Biologia Marinha, Universidade Federal Fluminense, Niterói, Rio de Janeiro, Brazil

Emanuel Soares dos Santos Instituto Federal de Educação, Ciência e Tecnologia do Ceará. Avenida Desembargador Armando de Sales Louzada, Acaraú, Ceará, Brazil

Felipe M. Santos Grupo de Oceanografia Tropical, Universidade Federal da Bahia, Salvador, Bahia, Brazil

Daiane Aviz Laboratório de Oceanografia Biológica, Instituto de Geociências, Universidade Federal do Pará, Belém, Pará, Brazil

Abílio Soares Gomes Departamento de Biologia Marinha, Universidade Federal Fluminense, Niterói, Brazil

Pedro W. Souza Filho Instituto de Geociências, Universidade Federal do Pará, Belém, Pará, Brazil

Daphne Spier Centro de Estudos do Mar, Universidade Federal do Paraná, Pontal do Sul, Paraná, Brazil

Abbreviations

AVISO	Satellite Altimetry Data
BNC	Brazilian Northern Coast
CONAMA	National Environmental Council
DHN	Brazilian Hydrographic Authority
DTM	Digital Terrain Model
EME	Eastern Brazil Marine Ecoregion
ENSO	El-Nino Southern Oscillation
GPR	Ground Penetrating Radar
INMET	Brazilian National Institute of Meteorology
ITCZ	Intertropical Convergence Zone
MPA	Marine Protected Areas
REBIO	Brazilian Federal Biological Reserve
SBE	Southern Brazil Marine Ecoregion
SLR	Sea-level rise
SRTM	Shuttle Radar Topography Mission

Project Acronyms

GERCO	Brazilian National Coastal Management Project
GLOBESAR	NE Para and coastal plain Bragança project
MADAM	Mangrove dynamics and management project
PNGC	Brazilian National Plan of Coastal Management
PIATAM	Potential Environmental Impacts of Petroleum project
PROMAR	Research Program in Marine Sciences
RADAM	Amazonas Radar Project
RECOS	Millennium Institute Project
ReBentos	National benthic habitats network (Sisbiota-CNPq)

Chapter 1
Brazilian Estuaries: A Geomorphologic and Oceanographic Perspective

Guilherme C. Lessa, Felipe M. Santos, Pedro W. Souza Filho, and Luiz César Corrêa-Gomes

Abstract Estuaries are primarily sites where rivers meet the sea, where fresh water is mixed with sea water by the tidal flow and where salinity gradients are set up along a mixing zone. In this chapter, we consider that the estuary is an accommodation space landward of the coastline, where fluvial sediments are largely retained with no appreciable effect on the sedimentation processes of the open coast. Here we recognize 41 large estuaries along the Brazilian coast, ranging in size from 40 km^2 to more than 10,000 km^2, encompassing areas with different topography, climate, geology, tidal ranges, and alongshore drift. The majority of these estuaries are located along fault lines or within grabens, and although scattered, structural and stratigraphic evidences suggest that land subsidence has caused sea level to locally rise, which has been an important factor in the preservation of these coastal features. A larger number of estuaries exist in the macrotidal and humid north coastal sector where large-scale coastal subsidence occurs. The estuary sizes along the coast bear no relation with tidal ranges, but have a positive linear correlation with river discharge in the east-southeast and northeast coastal sectors. Based on the existing hydrographic data, it could be stated that the large Brazilian estuaries are

G. C. Lessa (✉)
Instituto de Geociências, Universidade Federal da Bahia, Campus Ondina, Salvador, Bahia, Brazil

Grupo de Oceanografia Tropical, Universidade Federal da Bahia, Campus Ondina, Salvador, Bahia, Brazil
e-mail: gclessa@ufba.br

F. M. Santos
Grupo de Oceanografia Tropical, Universidade Federal da Bahia, Campus Ondina, Salvador, Bahia, Brazil

P. W. Souza Filho
Instituto de Geociências, Universidade Federal do Pará, Belém, Pará, Brazil

Instituto Tecnológico Vale, Belém, Pará, Brazil
e-mail: walfir@ufpa.br

L. C. Corrêa-Gomes
Instituto de Geociências, Universidade Federal da Bahia, Campus Ondina, Salvador, Bahia, Brazil

P. C. Lana, A. F. Bernardino (eds.), *Brazilian Estuaries*, Brazilian Marine Biodiversity, https://doi.org/10.1007/978-3-319-77779-5_1

hypersynchronous, ebb-dominated, and well mixed. In the deeper estuaries of the east-southeast sector, the longitudinal density gradient is an important driver for water circulation, causing vertically stratified residual flows that end up retaining suspended sediments within the estuary and promoting the accumulation of organic, muddy sediments.

Keywords Neotectonic · Estuarine circulation · Coastal geomorphology

1.1 Introduction

Various definitions of estuaries exist in the literature, reflecting the diversity of scientific interests in coastal areas. For most (Pritchard 1952; Cameron and Pritchard 1963; Fairbridge 1980; Lincoln et al. 1982; Perillo 1995; Allaby and Allaby 1999; Potter et al. 2010), estuaries are primarily sites where rivers meet the sea, where fresh water is mixed with sea water by the tidal flow and where salinity gradients are set up along a mixing zone. Also included is the existence of a partially enclosed body of water, or inlet of the sea. Thus, estuaries may encompass several coastal landforms such as coastal bays, lower river courses (permanently or seasonally opened to the ocean), lagoon entrances, and even distributary channels in a river delta. This interpretation is physically and biochemically biased and represents a problem for geologists more interested in the transport and fate of sediments. The geological incoherence resides on the fact that those premises can consider fluvial depositional systems as transitional environments. From a geological perspective (Dalrymple et al. 1992; Boyd et al. 1992), estuaries are associated with long-term coastal flooding and recession, and not with coastal emergence and progradation (coastline advancing upon the sea), such as the case of river deltas. The definition developed by Dalrymple et al. (1992) states that estuaries are "*the seaward portion of a drowned valley system that receives sediment from both fluvial and marine sources, and contains facies influenced by tide, wave and fluvial processes. The estuary is considered to extend from the inner limit of tidal facies at its head to the outer limit of coastal facies at its mouth.*" This implies the existence of gradients in sediment size and composition, and acknowledges the estuary as an accommodation space, or sediment trapping area, landward of the coastline. It does not exclude the existence of a mixing zone, and as a matter of fact this must exist if a river (perennial or seasonal) brings sediments to the estuary.

In this chapter, we adopt a physiographically biased definition, borrowing concepts of both geological and physical–chemical perspectives in defining estuaries. We consider estuaries as a coastal indentation where fluvial sediments are largely retained, with no appreciable effect on the sedimentation processes of the open coast, and where biogeochemical gradients may exist. As such, our definition includes within the estuary realm the classical types of estuaries (coastal plain, bar-built, tectonic, fjords—as in Perillo 1995), lagoons and salt marsh estuaries (as in

Kjerfve and Magill 1989). The estuary in this sense is a geologically ephemeral feature, and ceases to exist when completely filled-in with sediments. At this point the estuary is extinct and converted into a fluvial/tidal plain with a relatively small drainage channel where an equilibrium exists between sediment yield and transport capacity, as proposed by Roy (1984) in his model of estuarine evolution. Therefore, the preservation of an estuary is viable in submerging coasts such as in most of North America and NW Europe where sea level has been rising for millennia, counteracting the effect of sediment shoaling (Nichols 1989; Spencer et al. 1998; Jaeger et al. 2009). In coasts where relative sea level has fallen, such as along the Brazilian and Australia coasts (Angulo et al. 2006; Lewis et al. 2013), estuaries tend to disappear due to the additive effects of sedimentation (Lessa and Masselink 2006). A large section of the Brazilian coast has undergone a relative fall of sea level of about 3.5 m in the last 6000 years (Angulo et al. 2006), and as a result large-scale coastal progradation and infilling of estuaries have occurred (Dominguez et al. 1987; Martin et al. 2003).

The largest estuaries in East North America and West and Northwest Europe, where sea level rising rate was about 1 mm/year during the Late Holocene (Woodworth et al. 2009; Engelhart et al. 2011), are classified as coastal-plain estuaries (Nienhuis and Smaal 1994; Aubry and Elliott 2006). These type of estuaries arise from marine flooding of the lower river course, commonly generating funnel-shaped morphologies as the drowned valley becomes more incised upstream. River deltas in those regions do not exist, in accordance with the World Delta Database (Hart and Coleman 2009). On the other hand, the large majority of important rivers along the Brazilian coast exit strait to the sea, affect sedimentary processes on the open coast and are generally associated with sizeable sandy coastal plains that were built under the effect of the fluvial discharge (Dominguez et al. 1983, 1987). Figure 1.1 illustrates the contrasting coastal geomorphology of eastern United States and eastern Brazil. The images show the north-american coast fully indented by large coastal-plain estuaries, barrier-islands, lagoons, and salt marsh, whereas the Brazilian coast presents three wave-dominated deltas (Jequitinhonha, Doce, and Paraíba do Sul river deltas) projecting the coastline close to the river mouths. Out of the 97 Brazilian rivers with available discharge data (Oliveira et al. 2018), only the equatorial rivers on the subsiding north coast (Souza Filho 2000; Souza Filho et al. 2009; Rossetti et al. 2008) can be classified as coastal-plain estuaries (Fig. 1.2). In the remainder of the country, to the east of Turiaçu River, only 6 sizeable rivers ($Q > 100\ m^3/s$) exit inside estuaries, namely Mearim River (São Marcos Bay—Maranhão state), Itapecuru and Munin rivers (São José Bay—Maranhão state), Tubarão River (Laguna estuary—Santa Catarina state), and Jacuí, Camaquã and Piratini rivers (Lagoa dos Patos—Rio Grande do Sul state).

In spite of this regressive scenario and large-scale coastal progradation, large estuaries are still observed along the Brazilian coast, within which are located the most important harbors in the country. An investigation of the coast with satellite images (1:10,000) allowed for the identification of 41 estuaries larger than 40 km^2 (including intertidal areas, Fig. 1.2). Taking political boundaries into consideration, 18 estuaries are located in the Brazilian North Region, 13 in the Northeast Region,

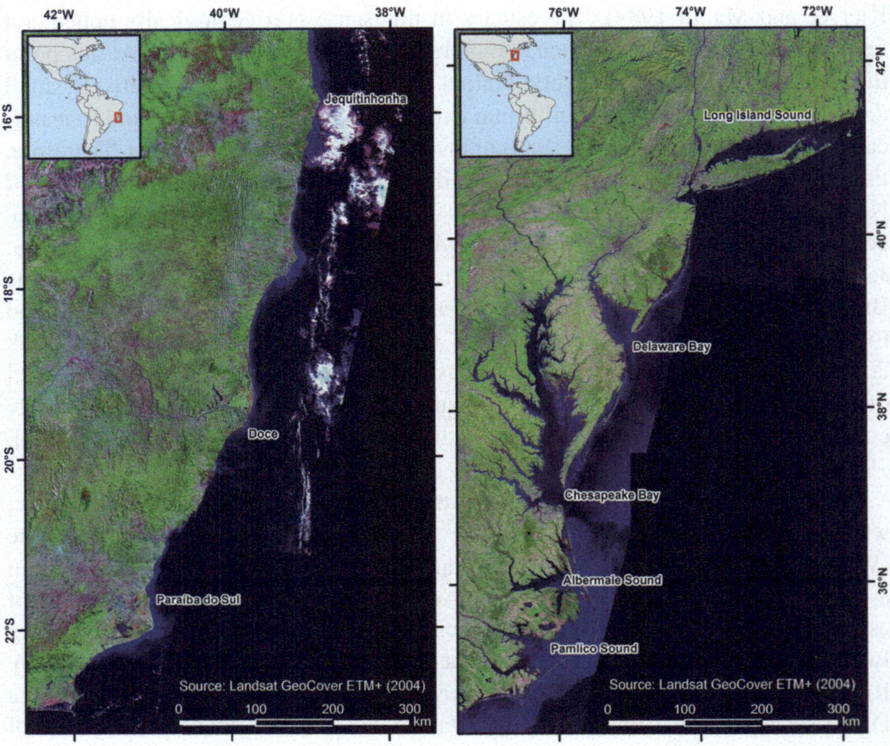

Fig. 1.1 Contrasting coastal geomorphologies of two wave-dominated coastal settings: (1) eastern Brazil, with deltaic forms associated with Paraiba do Sul, Doce, and Jequitinhonha rivers, and (2) eastern United States, with large coastal-plain, drowned-river valley estuaries

5 in the Southeast Region, and 5 in the South Region. These regions present a diversity of climatic (from temperate to equatorial), geologic (from Precambrian high-relief, metamorphic rocks to Tertiary sedimentary tablelands), and oceanographic (microtides and high wave energy to macrotides and moderate wave energy) conditions (Dominguez 2009). Amongst these estuaries only 6 present drainage basins larger than 10,000 km^2, a fact that is partly explained by the control exerted by the high-relief coastal massifs on the establishment of the continental drainage along the eastern border. Although most of the estuaries bear importance for the country's economy, scientific information on their geological and physical characteristics is not abundant, and only a few of them can be treated in some detail. The geology and water circulation within estuaries are important constraints for biological and chemical processes that will end up determining ecological gradients. Therefore, this work aims to make a general assessment of the physical and geological characteristics of the large Brazilian estuaries, highlighting differences and similarities that are important for identifying primary geochemical and biological processes that are key to ecological studies, support important ecosystem services and may be highly influenced by climate changes.

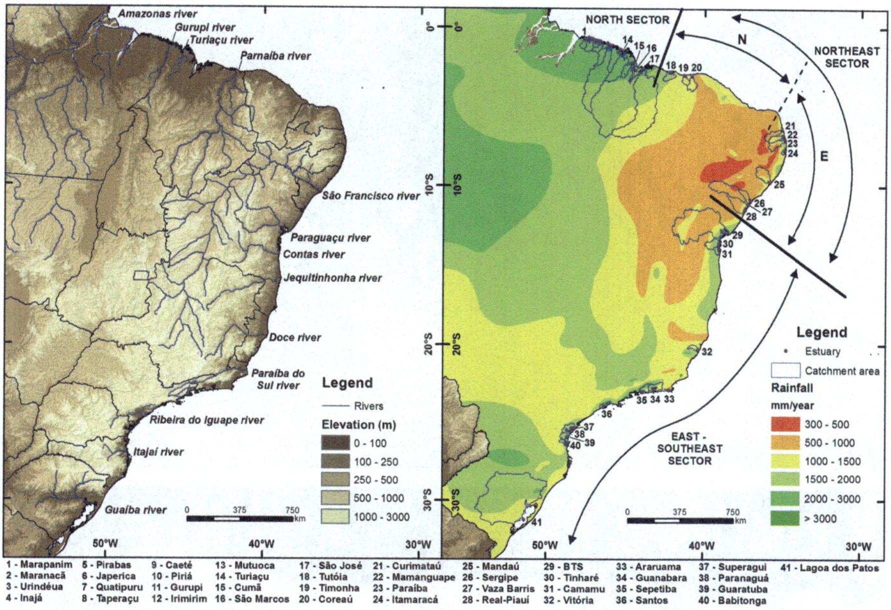

Fig. 1.2 (Left) Elevation map and the location of the Brazilian rivers with mean discharge is higher than 100 m^3/s. (Right) Distribution of the mean annual rainfall in the continent, location of the estuaries (>40 km^2) and their respective catchment areas, and the identification of the coastal sectors described in the text. Estuaries number 2–13 (not identified because of their closeness) form a continuum from left to right between estuaries 1 and 14

The Brazilian coast will be subdivided into three large sectors, North, Northeast, and East-Southeast, on the basis of climate, geomorphology, and mainly structural geology (Fig. 1.2). It is a simplified model of the coastal and shelf subdivisions proposed by Knoppers et al. (1999) and Dominguez (2009). The northern sector (with 17 estuaries), in the Amazonian coast, is about 580 km long and runs from Marajó Bay in the west to São José in the east. It has a humid climate and an overall subsiding trend that adds transgressive characteristics to the coastal geomorphology (Souza Filho et al. 2009) (Fig. 1.3a). The northeast coastal sector (with 11 estuaries) fringes a semi-arid region, extending for 1800 km from Tutóia in the north to Real-Piauí in the south. Sedimentary plateaus (tablelands) are common all along the coast, forming continuous sections of both active and inactive cliffs. The north-facing segment of this coastal sector (Fig. 1.2) is marked by intense east–west littoral drift (Hesp et al. 2009) and abundant deposition of sand at the mouth of estuaries (Fig. 1.3b). The East-Southeast Sector (with 13 estuaries), is about 3000 km long and located in a humid setting, has a higher coastal relief with granitic rocks at most of the estuary entrances (Fig. 1.3c), and because of relatively small fluvial discharge (smaller catchment areas) and alongshore oriented grabens, presents the least evidence of fluvial incision in the geomorphology of the estuaries. The 620 km long section of extensive sand barriers in South Brazil, that encompass Lagoa dos Patos, is also included in this sector.

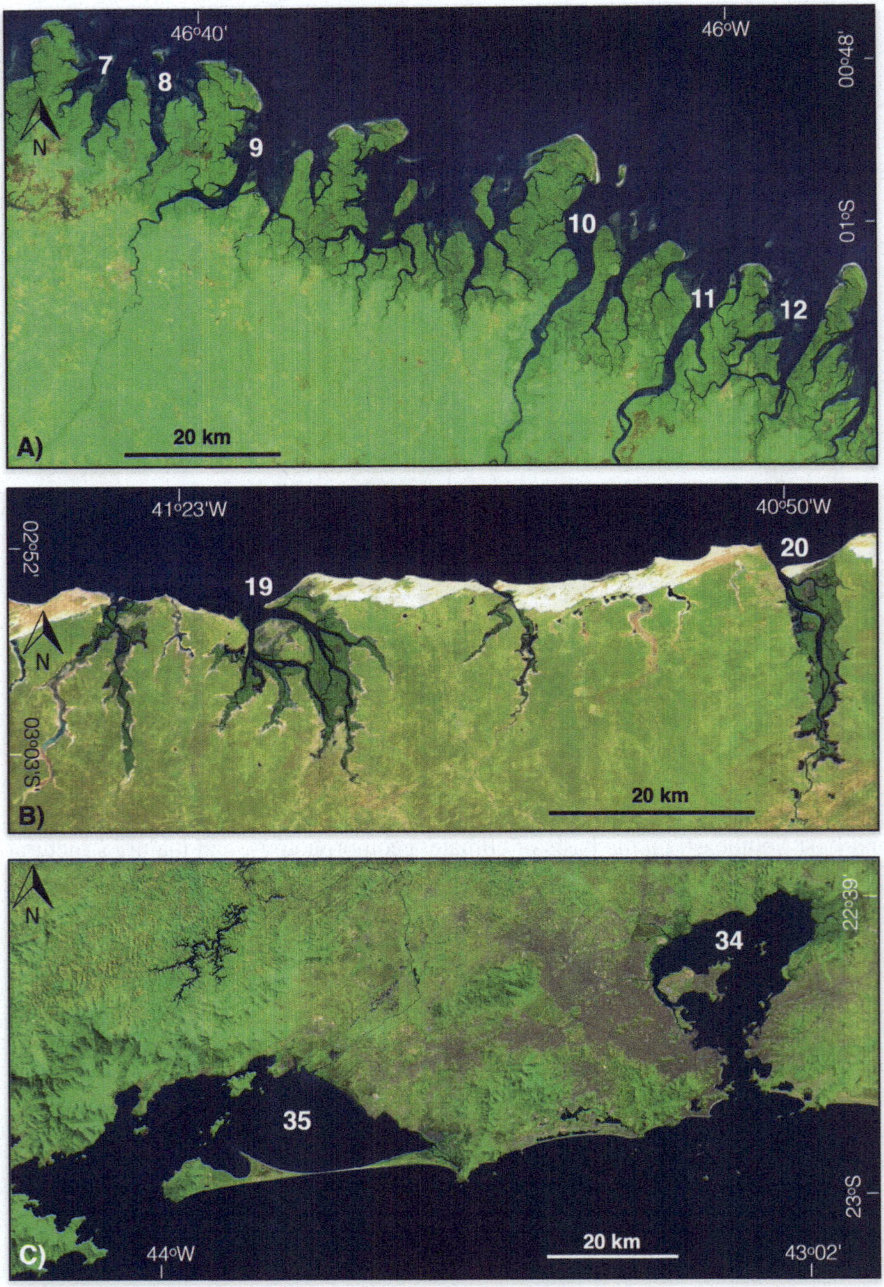

Fig. 1.3 Characteristic geomorphology of coastal sectors with a few estuaries: North Sector (**a**), north-facing NE Sector (**b**), East-Southeast Sector (**c**). See Fig. 1.2 for location of coastal sectors

1.2 Geomorphology

1.2.1 Structural Control of the Estuaries

The estuaries identified in Fig. 1.2 have a common strong structural control, i.e., geological faults either influence the overall shape of the estuary or define a limit to its longitudinal or lateral extension. Graben structures, related to the downward movement of a crustal block between two normal-fault lines, have been mapped in all coastal sectors where the estuaries are located. The tectonic bearing on the geomorphology obscures the evidence of fluvial erosion that has likely occurred during successive lowstand periods in the Late Quaternary, as claimed by Lessa et al. (2000) in relation to Todos os Santos. We argue that Holocene subsidence within these grabens has locally offset the sea level fall of the last few thousand years, creating conditions for either the preservation or regeneration of accommodation spaces, thus allowing for the existence of the current estuaries.

In the North Sector, the continental margin is controlled by E–W transcurrent faults and several structural compartments (Souza Filho 2000; Souza Filho et al. 2009). Most of the coastal plain east of the Amazon River runs along a paleo-cliff 1 m to 3 m high and parallel to the coast, that apparently marks an active fault associated with the downward movement of a block that supports the coastal plain (Souza Filho et al. 2009). Neotectonic movements in this region have been indicated by ground-penetrating radar (GPR) images showing a myriad of small-scale normal faults with strike angles parallel to coastline and fault blocks intersecting the paleo-cliff (Rossetti 2003). Also, Ferreira Jr et al. (1996) have shown sets of strike-slip faults with meter scale throws ascribed to four tectonic pulses that occurred between the Middle Pleistocene and Holocene. The orientation of these fault lines is similar to the faults measured at the mouth of the Amazon River by Costa et al. (2002) and Rossetti et al. (2008). Rossetti et al. (2008, 2012) described numerous structures related to seismogenic deformation during or shortly after deposition of Pleistocene and Holocene sediments. The downward movement of this large coastal sector has given rise to numerous coastal-plain estuaries (Souza Filho et al. 2009), harboring the largest mangrove area in the world (7600 km^2) (Souza Filho 2005; Nascimento Jr et al. 2013). Souza Filho et al. (2009) have shown that the coastal evolution during the Mid- to Late Holocene has been modulated by small-scale subsidence events, with the onset of three phases of barrier development. There is no record of any higher sea level position in the last 6000 years, and transgression is apparently an ongoing process. Cohen et al. (2005), based on pollen analysis, also suggest that a recent sea level rise is pushing the mangrove forest to higher elevation zones.

In the Northeast Sector, the sedimentary plateaus are intersected by numerous faults reaching Holocene sequences, which generated a series of horst and graben structures along the coastal plain. Incised river valleys are organized along the graben structures oriented normal to the coastline (Bezerra et al. 2001). The Mamanguape, Paraíba, and Itamaracá estuaries are located in these grabens (Lima et al. 2016; Rossetti et al. 2011), with sedimentological and chronostratigraphic evidence of Holocene subsidence existing for Mamanguape and Paraíba estuaries (Rossetti et al. 2011; Alvez 2015). The local relative sea level rise that promoted the inundation of these two estuaries appears to have occurred in the last 6000 years, and more abruptly within the last 1000 years, as indicated by radiocarbon dates of samples within mud deposits published by Alvez (2015).

The East-Southeast Sector shows larger graben structures that tend to be aligned with the coastline. The Todos os Santos, Tinharé-Boipeba, and Camamu estuaries are located within the continuous Recôncavo and Camamu sedimentary basins, large hemi-grabens that are limited in the west by the Maragojipe Fault. These grabens are highly fractured and subdivided in several sub-grabens (Magnavita et al. 2005; Born et al. 2011) that apparently underwent distinct relative vertical movements (Martin et al. 1986). The Maragogipe fault forms a scarp several tens of meters high delimiting the western margin of the graben and, as pointed out by Martin et al. (1986) and Suguio and Martin (1996), some rivers have not yet had time to carve valleys down this scarp, and reach the coast as waterfalls. Carvalho (2000) observed frequent intercalations of colluvium and estuarine sediments in sediment cores obtained at the base of the fault, that were ascribed to debris-flow events triggered by small seismic episodes. Earthquakes around Todos os Santos have been recorded in nine occasions in the historic time (Brazilian Seismic Catalog 2009; Bezerra et al. 2006). Carvalho (2000) also showed that no radiocarbon-dated samples indicate sea levels higher than today's in the past 9200 years in the western extreme of Todos os Santos, nor there is geomorphologic evidence that a higher sea level existed in the past. In addition, soil fractures close to Camamu were radiocarbon-dated at 27,000 cal years B.P. by Corrêa-Gomes et al. (2012). Martin et al. (1986) and Suguio and Martin (1996) suggested that Todos os Santos was formed by differential subsidence of a series of faulted blocks. Several geomorphological and geochronological evidences for tectonic control on the Quaternary sedimentation inside Todos os Santos exist, including the presence of Pleistocene marine terraces only at the southern side of the bay, suggesting that the bay is tilted to the north where submergence prevails. In the northern bay margin, paleo-sea level indicators are vertically offset in relation to a reference paleo-sea level curve established for the open coast (Martin et al. 1986), and indicate varying degrees of submergence in different sectors of the bay.

Further south, between the States of Rio de Janeiro and Santa Catarina, seven estuaries (Fig. 1.2) are located in grabens that bear their same name (Guanabara graben, for instance), and evidence of Quaternary tectonic movements exist in four of these grabens: Guanabara, Santos-Bertioga, Cananéia, and Paranaguá. In the Guanabara graben, Ferrari (2001) identified tectonic events in the Pleistocene and Holocene, which caused deformation of sediment fill and segmentation of sedimentary deposits. In the Santos-Bertioga estuary, Barbosa et al. (2012) and Souza (2015) have pointed to several seismic and morphological evidences that indicate tectonic pulses at the end of the Pleistocene and during the Middle to Late Holocene, amongst them: drainage anomalies, sets of systemic fractures in marine terraces, awkward arrangement of Quaternary sedimentary units on the coastal plain, and the presence of guide layers in Pleistocene and Holocene deposits with throws up to 7.5 m. Further south the Cananéia hemi-graben, studied by Souza et al. (1996), is tilted to the NW in association with Cenozoic reactivations of old fault lines (Riccomini 1992). The Cananéia estuary, located between Santos and Superagui (not mapped in this study for being smaller than 40 km^2), has the longest tide gauge record in Brazil, which shows a mean RSL rise rate (average of 50 years) of 5.6 mm $year^{-1}$ (Ducarme et al. 2006). Concurrent GPS measurements at the tide gauge site between 2002 and 2005 show that 67% of the RSL long-term rise would be due to a land subsidence of 3.8 mm $year^{-1}$, therefore indicating the tectonic instability of the graben.

Fewer and less compelling evidence exist for neotectonic movements in Paranaguá and Babitonga-Guaratuba grabens. The geomorphology suggests non-coaxial offsets within the tectonic blocks. Paleo-mangrove plains abound in the southern Guaratuba margin, whereas present mangrove vegetation thrives in the northern side, suggesting emergence of the southern and submergence of the northern side, a situation comparable to that identified in Todos os Santos. In the Paranaguá graben, Nascimento et al. (2013) has pointed to considerable morphological evidence for Late Quaternary tectonism, such as alluvium deposits segmented by knickpoints, anomalous drainage lines, and uplifted catchments. In Lagoa dos Patos, the southernmost estuary, sedimentological studies have postulated that sea level has been rising in the last millennium (Tomazelli et al. 1998; Lorscheitter and Dillenburg 1998). Saadi (1993, 2002) has mapped the Pelotas Fault, extending alongside Patos and Mirim lagoons and delimiting Mirim and Porto Alegre graben in the west, as an active fault in the Quaternary. Additionally, Fonseca (2005) has reported on the existence of deformational structures within Pleistocene marine terraces in the northern margin of Lagoa dos Patos.

1.2.2 Dimensions

Orthorectified Landsat 7 ETM images (15 m resolution) were used as a base map to digitize the contour of the estuarine water body and vegetated intertidal and supratidal areas, at a 1:25,000 scale. Digital elevation models generated from Shuttle Radar Topography Mission (SRTM) with 90 m spatial resolution were used to map the catchment areas and extract the drainage lines. Nautical charts produced by the Brazilian Hydrographic Authority (CHM) are only available for 8 estuaries, namely Paraíba, Itamaracá, Todos os Santos, Camamu, Guanabara, Vitória, Sepetiba, and Paranaguá. The charts were digitized and a digital terrain model (DTM) produced. Hypsometric curves and volumes of these six estuaries were then calculated. Volumes for Araruama and Mundaú-Manguaba lagoons were obtained from Kjerfve et al. (1996) and Oliveira and Kjerfve (1993), respectively. Because the maximum extent of the intertidal area could not be determined systematically for all estuaries, we will define "intertidal" henceforth as the area encompassing the vegetated intertidal and the supratidal areas. The "water surface" area incorporates both subtidal (below the hydrographic zero line) and non-vegetated intertidal. For consistency, this norm will also apply to those six charted estuaries, unless stated otherwise. Intertidal vegetation includes mangrove and salt marshes (only in Lagoa dos Patos where mangroves are non-existent) and does not incorporate the freshwater vegetation that occupies the intertidal area beyond the limit of salt intrusion or in supratidal zones in regions with higher rainfall rates. The outer and inner estuary limits were defined by the narrowest cross section close to the mouth and the disappearance of inundation zones, respectively. In the coastal-plain estuaries of the north region, this limit was rather difficult to pinpoint because both water level oscillations and reversing currents can be observed much farther upstream than the innermost mangrove forest or the estuary funnel zone. In estuaries with more than one inlet (Santos, Todos os Santos, Mundaú-Manguaba, Itamaracá, and Tutóia), the estuary length was measured taking the largest, or main, entrance as a reference starting point. The estuary length, as well as mangrove/salt marsh excursion into the estuary, was measured following the sinuous path of the drainage channels, since this is the distance that must be covered by the water flow.

Table 1.1 presents the spatial dimensions of all 41 estuaries. The largest Brazilian estuary is Lagoa dos Patos (31.77°S), with an area of 10,360 km^2. It is followed by 4 three-digits-area estuaries, which are São Marcos (2.39°S–2568 km^2), São José (2.39°S–1309 km^2), Todos os Santos (13°S–1233 km^2), and Turiaçu (2.45°S–1171 km^2). Four estuaries have areas between 500 km^2 and 1000 km^2 (Cumã, Tutóia, Sepetiba, and Paranaguá), 23 estuaries (40% of the total) between 100 km^2 and 500 km^2, while 9 estuaries (30% of the total) are smaller than 100 km^2. Figure 1.4a shows the north-south distribution of estuarine areas, showing that most of the smaller estuaries are located in the Northeast Sector (blue-shaded region), where the average area is 102 km^2 (±45 km^2; n = 11). Larger estuaries are observed in the North Sector (pink-shaded region) with an average area of 529 km^2 (±610 km^2; n = 18) and where an eastward-trend of increasing estuary size is observed. The estuary sizes in the East-Southeast Sector are highly variable, with an average area of 371 km^2 (±337 km^2; n = 13) not including Lagoa dos Patos. The intertidal areas of the estuaries in the North and Northeast sectors take up most of the estuary expense (Fig. 1.4b), with the exception of Unindéua, São Marcos, São José and Mundaú-Manguaba. Larger water surface areas prevail in the East-Southeast Sector, except for Tinharé-Boipeba, Vitória, Araruama, and Guaratuba where intertidal areas are slightly larger. The largest relative water surface area occurs in Araruama (99.9%) and Lagoa dos Patos (99.2%).

With the exception of Lagoa dos Patos, São Marcos, and Turiaçu, which are, respectively, 300 km, 176 km, and 107 km long, all estuary channels (or the main estuary axis in lagoons) are shorter than 100 km (Fig. 1.4c). Twenty-five estuaries (61% of the total) are shorter than 50 km, and most of the shortest estuaries are in the northeast region, where the average length is 29 km (±10 km). Itamaracá, 13.4 km long between the northern inlet and Itapissuma, is the shortest estuary. The average estuary lengths in the North and East-Southeast sectors are 59 km (±38 km) and 46 km (±16 km), respectively.

Intertidal, brackish vegetation is observed along the whole extent of the estuaries in all three geographic regions (Fig. 1.4c). It suggests a rather ample estuarine

Table 1.1 List of estuaries, from north to south, with their number, coastal sector, areas, along-channel length, fluvial catchment area, along-channel mangrove extension of vegetation, and mean annual river discharge

			Area (km²)				Length	Catchment area	Mangrove extension	River discharge
Estuary	Estuary number	Sector	Intertidal vegetated	Supratidal	Water surface + non-vegetated intertidal	Total	(km)	(km²)	(km)	(m³/s)
Marapanim	1	N	126.0	*	124.7	251.0	62.7	2323	61.3	64.0
Maracanã	2	N	181.2	*	156.4	338.0	71.7	3560	62.2	98.0
Unindéua	3	N	43.0	*	57.8	101.0	25.9	26	25.9	**
Inaja	4	N	35.0	*	28.4	64.0	17.6	35	17.6	**
Pirabas	5	N	57.7	*	52.0	111.0	23.7	137	23.7	**
Japerica	6	N	148.1	*	142.2	292.0	32.7	206	32.7	**
Quatipuru	7	N	166.0	*	143.0	312.0	46.4	1097	44.1	**
Taperaçu	8	N	103.0	*	77.7	182.0	27.5	40	27.5	**
Caete	9	N	218.1	*	191.2	412.0	54.3	2195	39.7	60.4
Piriá	10	N	209.7	*	84.5	294.0	60.2	5358	56.1	**
Gurupi	11	N	154.5	*	122.9	277.0	57.7	35,053	31.1	**
Irimirim	12	N	145.0	*	135.1	280.0	33.3	206	33.3	**
Mutuoca	13	N	197.3	*	163.5	361.0	41	277	41	**
Turiaçu	14	N	534.3	*	637.0	1171.0	106.8	13,957	91.5	290.1
Cumã	15	N	292.0	*	265.6	558.0	56.9	4077	56.9	**
São Marcos	16	N	669.1	*	1898.4	2568.0	176.4	100,394	145.9	436.8
São José	17	N	466.0	*	843.4	1309.0	84.8	68,577	84.8	335.9
Tutóia	18	NE	274.2	103.1	265.2	642.5	75.2	3046	61.4	**
Timonha	19	NE	54.2	30.2	32.9	117.0	27.6	2165	26.1	14.5
Coreaú	20	NE	37.8	22.7	15.1	75.6	29.2	4185	28.6	29.8
Curimataú	21	NE	41.8	7.8	9.1	58.8	22.5	4103	22.5	2.5

(continued)

Table 1.1 (continued)

Estuary	Estuary number	Sector	Area (km^2) Intertidal vegetated	Supratidal	Water surface + non-vegetated intertidal	Total	Length (km)	Catchment area (km^2)	Mangrove extension (km)	River discharge (m^3/s)
Mamanguape	22	NE	44.2	5.9	10.0	60.2	23.8	3282	23.6	13.3
Paraíba	23	NE	45.4	8.2	31.2	84.7	25.6	19,938	24.7	27.4
Itamaracá	24	NE	43.0	10.4	35.9	89.2	13.4	692	13.2	25.9
Mundaú-Manguaba	25	NE	14.9	0.4	80.1	95.0	28.4	7670	17.9	64.0
Sergipe	26	NE	69.4	18.9	43.5	131.7	44.5	3540	41.4	13.7
Vaza Barris	27	NE	63.4	12.3	44.9	120.6	28.8	16,267	26.4	11.1
Real-Piauí	28	NE	114.7	21.9	67.0	203.6	44.4	8776	44.4	25.6
Todos os Santos	29	SE	155.6	17.2	1060.2	1233.0	82.4	58,927	74.8	113.4
Tinharé-Boipeba	30	E-SE	134.3	2.8	100.6	237.7	41.6	4713	38.5	68.1
Camamu	31	E-SE	179.5	15.9	204.7	400.1	51.1	1374	51.1	53.4
Vitória	32	E-SE	22.6	3.8	18.2	44.6	24.3	1728	24.3	18.7
Araruama	33	E-SE	0.1	0.0	222.9	223.0	50.6	350	6.1	2.3
Guanabara	34	E-SE	84.9	1.3	362.6	448.8	48.2	3768	47.1	105.8
Sepetiba	35	E-SE	30.5	10.1	507.1	547.7	48.0	2388	48.0	32.2
Santos	36	E-SE	85.5	2.0	59.5	147.0	23.6	1052	21.1	42.8
Superagui	37	E-SE	SE 31.5	3.0	55.6	90.0	35.2	127	35.2	**
Paranaguá	38	E-SE	159.5	12.3	535.8	707.5	56.2	3145	54.5	230.4
Guaratuba	39	E-SE	59.2	4.6	54.5	118.3	30.6	1810	21.8	80.0
São Francisco	40	E-SE	68.2	12.0	177.4	257.6	55.0	1198	54.4	22.2
Lagoa dos Patos	41	E-SE	72.0	0.0	9028.0	9100.0	300.0	201,626	43	2400.0

See Fig. 1.2 for the geographic location of these estuaries. River discharge data source: National Water Agency
*not yet possible to assess, **not measured

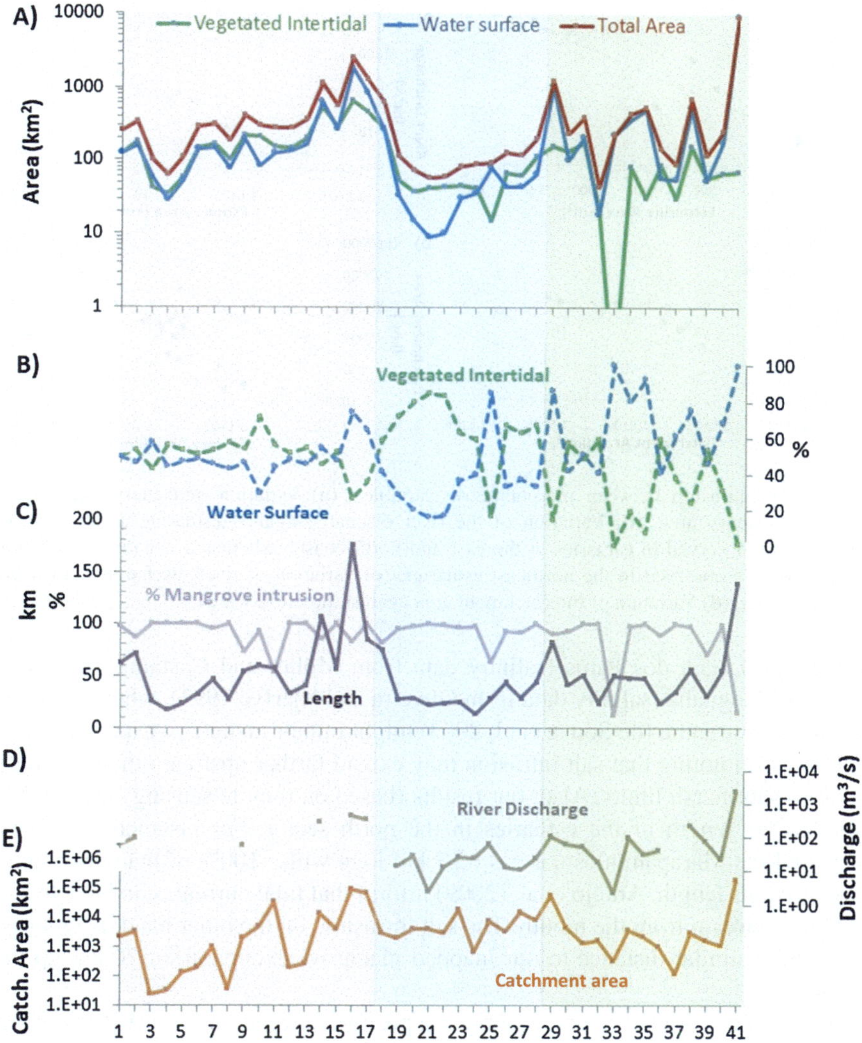

Fig. 1.4 North-South distribution of estuarine areas (**a**), relative extent of intertidal vegetation (mangroves except for Lagoa dos Patos—41) and high-tide water surface area (**b**), estuary channel length (km) and distance of mangrove (salt marsh in 41) intrusion relative to channel length (%) (**c**), river discharge (**d**), and catchment area (**e**). Dashed vertical lines separate the north equatorial region (left section of the graphs), the northeast (mostly) semi-arid region (center section), and the southeastern and south humid regions (right section). Numbers 1–41 in *x* axis refer to estuary number presented in Fig. 1.2 and Table 1.1

mixing zone and also means that tidal range does not modulate the excursion of brackish vegetation amongst these estuaries, i.e., relatively longer saline intrusion is not restricted to the macrotidal North Sector. An exception exists for the lagoonal estuaries of Mundaú-Manguaba, Araruama, and Lagoa dos Patos, and also for the Gurupi estuary. Fresh water limits the intrusion of brackish vegetation in the mainly

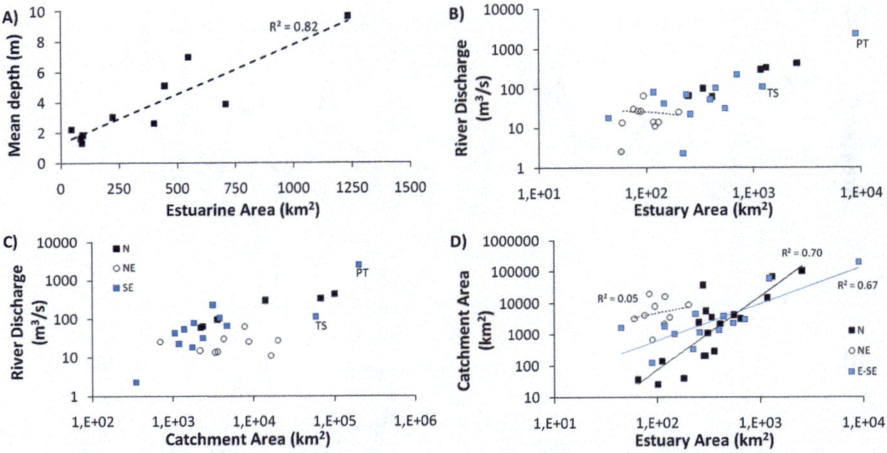

Fig. 1.5 Relationship between morphometric variables. (**a**) Variation of estuary-mean depth relative to estuary area. (**b**) Variation of the river discharge against estuarine areas. Positive correlation is observed in estuaries at the east and north coasts, whereas a negative correlation (dashed line) is observed in the northeast estuaries. (**c**) Variation of river discharge against the catchment area. (**d**) Variation of the catchment area against the estuary area

oligohaline Lagoa dos Patos (salinity data from Möller and Castaing 1999) and Mundaú-Manguaba (salinity data from Oliveira and Kjerfve 1993), whereas hypersalinity in Araruama (de Souza et al. 2003) might impair mangrove expansion.

It is worth noting that salt intrusion may extend farther upstream than the mangrove or salt marsh limits. Also, our results (based on remote sensing) may underestimate the length of the estuaries in the north sector. For instance, while we measured the Marapanim estuary as 62.7 km long with ~100% of mangrove presence along its length, Araújo et al. (2009) inform that tidal currents can be detected 150 km upstream from the mouth. The salt intrusion, on the other hand, is reported to extend a similar distance to our mapped mangrove excursion, or 70 km up the estuary.

Amongst the charted estuaries, there is a positive correlation ($R^2 = 0.86$) between mean estuarine depth and estuarine area (Fig. 1.5a). Mean estuarine depth varies from 1.3 m in Itamaracá to 9.6 m in Todos os Santos, whereas mean channel depth varies from 10 m to 30 m (in Todos os Santos) (Table 1.2). The few data points available indicate that in smaller estuaries the difference between mean estuarine depth and channel depth is larger.

1.2.3 *River Discharges*

The river discharge data used was obtained from the National Water Agency website (farthest downstream gauging stations in the catchment), and technical and scientific literature. The level of detail given to the calculation of annual averages is

Table 1.2 Tidal ranges, tidal prisms, and geometric characteristics of charted estuaries

Estuary	SR (m)	STP (× 10^9 m^3)	MD/CD (m)	V (× 10^9 m^3)	Q (m^3/s)	FR (× 10^{-3})	$\delta s/\delta x$ (psu/km)	U_e (m/s)	Tf (h)
Coreaú	2.96	0.224	–	–	29.8	5	–	–	–
Paraíba	2.18	0.184	1.6/–	0.13	27.4	7	–	–	8.9
Itamaracá[1]	2.27	0.202	1.3/10	0.12	25.9	7	0.30	0.03	8.5
Mundau-Manguaba[2]	0.14	0.014	1.8/–	0.13	64.0	205	–	–	93.1
Sergipe	2.01	0.265	–/–	–	13.7	2	–	–	–
BTS[3]	2.50	3.078	9.6/30	11.84	113.4	2	0.09	0.06	47.6
Camamu[4]	2.01	0.803	2.6/10	1.05	53.4	3	0.42	0.04	16.1
Vitória[5]	1.39	0.062	2.2/10	0.10	18.7	13	0.79	0.08	19.6
Araruama[6]	0.01	0.001	3.0/–	0.62	2.3	98	−1.83	0.01	6653
Guanabara	1.05	0.471	5.1/20	2.28	105.8	10	0.19	0.07	59.4
Sepetiba	1.15	0.632	7.0/18	3.84	32.2	2	0.30	0.15	75.2
Paranaguá	1.44	1.017	3.9/15	2.75	230.4	10	0.80	0.17	33.2
Guaratuba	2.27	0.269	–/–	–	80.0	13	0.56	0.04	–

Morphology data from: 1—Medeiros and Kjerfve (1993); 2—Oliveira and Kjerfve (1993); 3—Lessa et al. (2009); 4—Amorim (2005); 5—Rigo (2004); 6—Kjerfve et al. (1996)

SR spring tidal range, *STP* spring tidal prism, *MD* mean depth, *CD* channel depth, *V* volume, *Q* freshwater discharge, *FR* flow ratio (*R*/P), $\delta s/\delta x$ longitudinal salinity gradient, *Tf* flushing time

asymmetrical amongst the estuaries, with some, such as Lagoa dos Patos, Paranaguá, Guanabara, Araruama, Camamu, Todos os Santos, and Itamaracá, being more thoroughly investigated than others. The average annual river discharges are overall positively correlated with both the catchment areas and the estuarine areas (Fig. 1.5b, c). A closer look at the data spread for the Northeast Sector, however, shows a negative correlation, with estuarine and catchment areas increasing with a decrease in river discharges areas (Fig. 1.5b).

The catchment areas vary five orders of magnitude amongst the estuaries, from 26 km^2 in Unindéua to 202 × 10^3 km^2 in Lagoa dos Patos (Fig. 1.4e, Table 1.1). The smallest catchments are in the East-Southeast sector, where the average is 6.7 × 10^3 km^2 (± 16.5 × 10^3 km^2—not including Lagoa dos Patos), followed by the Northeast and North sectors with averages of 7.3 × 10^3 km^2 (± 6.5 × 10^3 km^2) and 13.4 × 10^3 km^2 (± 27.7 × 10^3 km^2), respectively. It is worth noting that the catchment areas in the North and in East-Southeast sectors are positively correlated with the estuary area (Fig. 1.5d), with coefficients of determination (R^2) of 0.70 and 0.67. No relationship, however, can be observed for these two attributes in the Northeast Sector (R^2 = 0.05), which could be ascribed to a lesser scouring power of river discharges and/or stronger structural control of the estuary size.

Half of the estuaries with data availability have mean annual freshwater discharges smaller than 50 m^3/s (Table 1.1; Fig. 1.4d). The largest freshwater inputs occur in the North Sector (214 ± 161 m^3/s; n = 18), and the smallest in the Northeast Sector (24 ± 18 m^3/s; n = 11). Mean freshwater discharge in the East-Southeast

Sector is almost three times larger than in the Northeast Sector, or 70 m^3/s, but with large spatial variability (± 64 m^3/s; $n = 13$). These values do not include Lagoa dos Patos which, with an average annual discharge of 2400 m^3/s, is the freshest of the estuaries and where salt marshes have one of the shortest excursions (14 km), second only to Araruama Lagoon (6 km). The lowest mean annual discharge is found in Araruama Lagoon, with only 2.3 m^3/s considering river flow and water pumped for human consumption (Kjerfve et al. 1996).

1.3 Tides and Hydrography

1.3.1 Tidal Ranges, Tidal Prisms, and Tidal Distortions

Tidal ranges were estimated on the basis of the main lunar and solar semi-diurnal tidal constituents, respectively, M_2 and S_2. The mean tidal range is given by $2.M_2$ and the mean spring tidal range by $2(M_2 + S_2)$. The large majority of the tidal harmonic analysis was performed by the Brazilian Hydrographic Authority (CHM) on time series of water elevation ranging between 15 days and 1 year (modal length = 30 days). We used tidal analysis from 163 stations, split between open coast and estuarine stations, encompassing data from CHM as well as results from several other publications on estuarine circulation and hydrology. Tide constituents from AVISO Global Tide Model (FES_2014, 1/16° resolution—http://www.aviso.altimetry.fr) were also utilized to allow for a more detailed and complete coverage of the tidal ranges along the open coast (using the grid cells closest to the shore).

Overall, the tidal range on the open coast increases toward the north (Fig. 1.6a, b). Spring tidal ranges vary from 0.15 m close to Lagoa dos Patos to about 8 m in the north extreme, whereas mean tidal ranges vary from 0.1 m in the south to 6.6 m in the north (Fig. 1.6b). Therefore, the coast can be classified, according to Hayes (1979), as microtidal (0 m–1 m tidal range) between 34°S and 19.5°S (about Doce River), as low-mesotidal (1 m–2 m tidal range) between 19.5°S and 3°S (close to Tinhonha estuary), then high-mesotidal (2 m–3.5 m tidal range) between 3°S and 1.3°S, and finally as macrotidal (tidal range >3.5 m) in the North Sector between 1.3°S and 3.4°N. The northbound tidal amplification is uneven, however, with 3 coastal regions presenting localized amplification, namely: (i) Paraná-Rio de Janeiro Bight (26°S–23°S), where the continental shelf widens; (ii) the internal part of Abrolhos Bank (19°S), also within a wider shelf, and (iii) in the initial E–W section of the Northeastern coast. There is a close agreement between observed (CHM) and modeled (AVISO/FES-2014) tidal ranges (red line in Fig. 1.6b). For the North Sector with no tide gauge station, the model results show varying tidal ranges apparently caused by a complex bathymetry, very indented costal contour, and strong influence of the Amazon River discharge.

Larger tidal ranges in the Northeast and North sectors can potentially cause bigger tidal prisms (volume of ocean water flows into the estuary during the rising tide). However, while the average spring tidal range gets bigger by a factor of 2, the

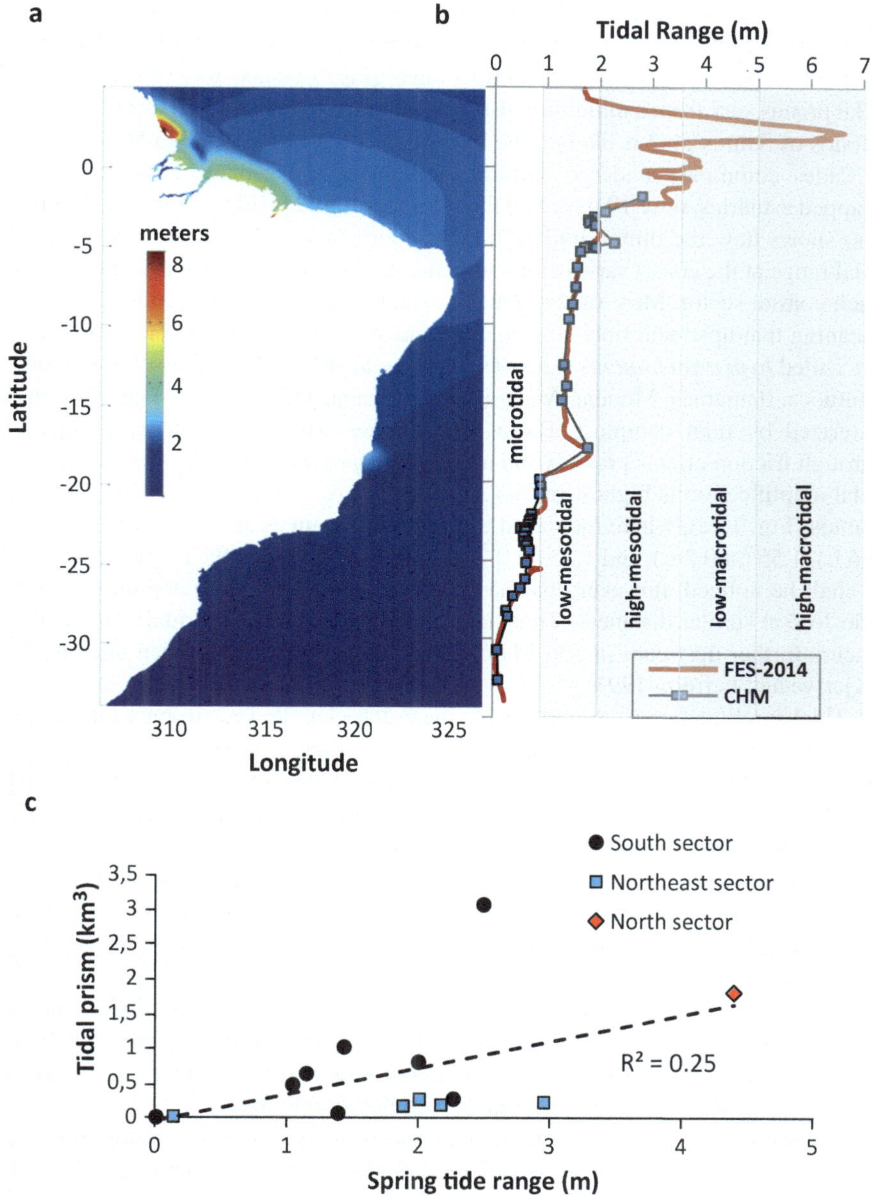

Fig. 1.6 (**a**) Spatial variation of the mean tidal range (estimated by $2.M_2$) based on AVISO Global Tide Model (FES_2014); (**b**) Latitudinal changes of mean tide ranges from observed open coast data (obtained by the Brazilian Hydrographic Authority—CHM) and their corresponding AVISO (FES_2014) ranges. Tidal range classification based on Hayes (1979). (**c**) Variation of the estuarine tidal prism against the spring tide range (estimated by $2[M_2 + S_2]$) on the open coast

average estuary size gets smaller by a factor of 3.5 between the south and northeast sector (Fig. 1.4a). Hence, tidal prisms also diminish producing a very weak positive correlation between tidal prism and tidal range (Fig. 1.6c and Table 1.2). Measured tidal prisms vary from a minimum of 1×10^6 m^3 (Araruama Lagoon) to 3.08×10^9 m^3 (Todos os Santos), but is likely to be larger than 1.5×10^{10} m^3 in São Marcos.

Tides commonly undergo amplification inside the estuaries. Out of the 41 mapped estuaries, only 19 have had the tides monitored inside. Figure 1.7 (left panels) shows how the dimensionless tidal range (estuary tidal range divided by the tidal range at the coast) varies along increments (x) of the total estuary length (L) in each coastal sector. Most of the estuaries show tidal ranges increasing landwards, meaning that upstream tidal convergence and shoaling prevail over friction. These are called *hypersynchronous* estuaries (Allen et al. 1980). Only 5 estuaries, namely Mutuóca, Itamaracá, Mundaú-Manguaba, Araruama, and Lagoa dos Patos, are characterized by tidal damping. These are estuaries where tidal-energy dissipation through friction effects prevails and are called *hyposynchronous* estuaries. Relative tidal amplification is highest in São José, São Marcos (Fig. 1.7a) and in Todos os Santos (Fig. 1.7c), where measured estuarine tidal ranges are, respectively, 1.6 (at 0.6 L), 1.55 (at 0.2 L), and 1.53 (at 0.7 L) times larger than in the neighboring shelf. It shall be noticed, however, that São Marcos has greater tidal amplification than São José at similar distances. Thus, it is possible that additional tidal amplification occurs further upstream in São Marcos, where tidal bores have been documented (Kjerfve and Ferreira 1993).

Tidal amplification is proportional to the estuary length, i.e., stronger tidal amplification is observed in longer estuaries (Fig. 1.7—inset) as well as at stations farther upstream. This positive correlation is apparently (due to small number of data points) stronger in the north sector ($R^2 = 0.85$) possibly due to (i) lesser sediment deposition (formation of transverse sand bars) by wave action at the mouth of the tide-dominated estuaries (Hayes 1979) and (ii) less structural control of the inlet morphology in a predominantly prograding coastal plain (Souza Filho et al. 2009). Both factors contribute for an unrestricted control of tidal power on the estuarine morphology.

Estuaries at the northeast sector are affected by strong sediment drift, which in some cases completely block tidal inlets, whereas in the southeast sector rock headlands can partially (at Todos os Santos, Paranaguá, and Guaratuba) or entirely (at Vitória, Guanabara, and Santos) control the inlet morphology.

Amongst the five hyposynchronous estuaries, tidal damping is more intense in the three lagoonal-type estuaries of Mundaú-Manguaba, Araruama, and Patos. In these locations, tidal range falls to about 40% of the ocean range in less than 10% of the total estuarine length (<0.1 L), and tidal oscillations cease to exist in the first half of the estuaries. In Patos Lagoon, the tide influence dies out at about 0.1 L in accordance with Möller et al. (2001), and diurnal oscillations in water level beyond this point is ascribed to the wind only.

The few estuaries with tidal information in the Northeast Sector (Fig. 1.7b) show either a strong tidal damping or a small amplification, but given the small number of estuaries, this shall not be taken as a norm for the estuaries in this sector. In the

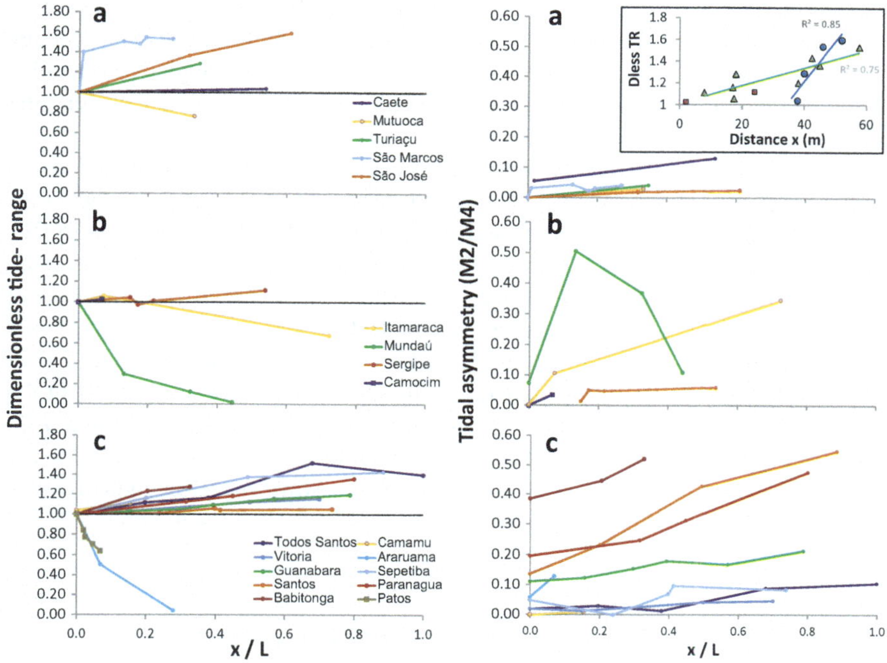

Fig. 1.7 Changes in tidal range (left) and degree of tidal distortion (or tidal asymmetry; right) in different stations along the estuaries in the north sector (**a**), northeast sector (**b**), and east-southeast sector (**c**). Distances inside the estuary (*x*, where $x = 0$ km at the entrance) are fractions of the estuary length (*L*, where $L = x$ at the estuary head). Tidal range is expressed as the ratio of the tidal range at each station and the tidal range outside the estuary; ratios larger (smaller) than 1 mean tidal amplification (dampening). Tidal asymmetry is expressed as the ratio of M_4 (first M_2 overtide) and M_2 amplitudes; larger ratios mean stronger tidal distortions. INSET: Degree of tidal amplification against distance of the station up the estuary. Green triangles = North sector, blue circles = Southeast sector, and red squares = Northeast sector

estuaries of the East-Southeast Sector (Fig. 1.7c), the weakest relative tidal amplification occurs in Santos (0.7 at 0.7 L), which ranks amongst the shortest estuaries (23.6 km long).

Tidal distortion, or a loss of symmetry in the shape of the tidal wave, is a common phenomenon within estuaries. This distortion is translated either into a short-lasting rising (and long-lasting falling) or a short-lasting falling (long-lasting rising) tide. Due to continuity issues, shorter rising tides quite commonly cause stronger flood-tidal flows, whereas shorter falling tides are associated with faster ebb flows (Dronkers 1986). Because sediment transport as bedload is proportional to tidal velocities to the cube ($\sim u^3$), tidal distortion can be seen as a proxy for net-bedload sediment transport (Fry and Aubrey 1990). The intensity of the distortion is given, in semi-diurnal tides, by the ratio of M_4/M_2, where M_4 is M_2's first sub-harmonic. This ratio is plotted against the dimensionless estuary length (*x*/L) in Fig. 1.7 (right panels). Tidal distortion grows stronger inside all estuaries here investigated, with the exception of São Marcos and Mundaú-Manguaba where the degree of distortion is greater closer to the inlet. The highest degree of distortion, where the amplitude of M_4 is about half of M_2 amplitude, occurs in Mundaú-Manguaba, Sepetiba, Babitonga, and Paranaguá. Overall, the estuaries in the North Sector have less distorted tides.

The tidal distortion in most of the estuaries causes shorter ebbing tides and favors bedload sediment transport toward the sea. Stronger ebb-tidal flows, especially during spring tides, are well documented in Paranaguá (Mantovanelli et al. 2004), Santos (Miranda et al. 2012; Fiedler 2015), Guanabara (Sampaio 2003; Bérgamo 2006; Diretoria de Hidrografia e Navegação 2012), Vitória (Rigo 2004; Veronez Jr et al. 2009), Camamu (Amorim et al. 2016), Todos os Santos (Cirano and Lessa 2007), Curimataú (Andutta 2006) and São Marcos (González-Gorbeña et al. 2015). Ebb-directed, net-bedload transport favors the development of sand shoals in the seaward side of tidal inlets, mainly as a result of the obstruction of the littoral drift. These features are called ebb-tidal deltas, common features fronting estuaries in the east-southeast and northeast sectors (Fig. 1.8), where wave energy is higher relative to tidal range and sand bars are easier to form outside the estuary mouths.

1.3.2 Estuarine Mixing, Gradients, and Circulation

Estuarine mixing is a function of the stirring power of the tidal currents. Stronger tidal flow enhances turbulence and generates better-mixed water columns. In the absence of strong tidal flow, the less-dense fresh water entering the estuary would remain undisturbed and give rise to strong vertical gradients of several scalars such as salinity. A simple way to assess the stirring power of an estuary is to compare the tidal prism (P) with the volume of riverine water (R) that flows out during a tidal cycle. Following the convention proposed by Simmons (1955—cited by Dyer 1997), if $R/P \geq 1.0$ the estuary is highly stratified; for R/P about 0.25 the estuary is partially mixed and for $R/P < 0.1$ the estuary is well mixed. As shown in Table 1.2, flow ratios (FR) are smaller than 0.1 in all estuaries where tidal information exists, except for Mundaú-Manguaba Lagoon. Although spring tidal prisms were employed in these calculations, the use of mean tidal ranges does not alter the results. It is thus anticipated that the majority of the estuaries tend to be well- to possibly mildly partially-mixed (during stronger-than-average river flow), i.e., vertical salinity differences tend to be small.

Proper thermohaline characterization of the Brazilian estuaries is rare. Average vertical salinity differences measured in dry and wet seasons are smaller than 5 psu in all of the documented estuaries, namely Paranaguá (Mantovanelli 1999), Guaratuba (Mizerkowski et al. 2012), Santos (Miranda et al. 2012), Guanabara (Kjerfve et al. 1997; Ribeiro and Kjerfve 2002; Bérgamo 2006), Vitória (Nalesso et al. 2005), Camamu (Amorim 2005; Santos 2009; Amorim et al. 2016), Todos os Santos (Cirano and Lessa 2007), and Itamaracá (Medeiros and Kjerfve 1993). The only exception is Lagoa dos Patos. Because it has the largest drainage basin amongst

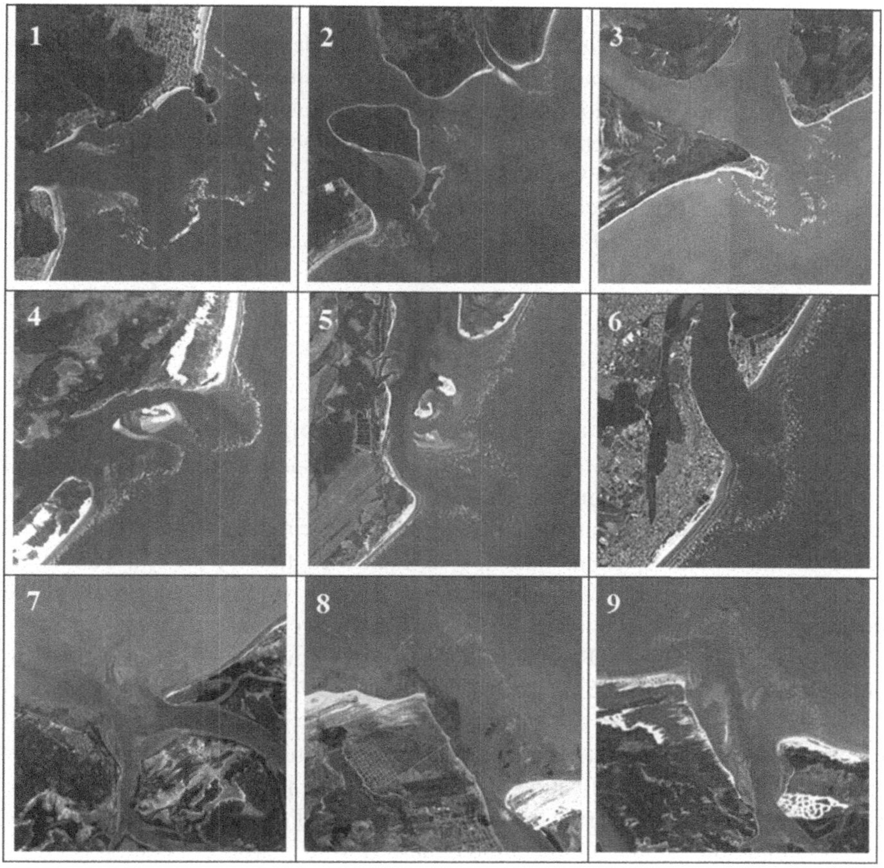

Fig. 1.8 Shallows, highlighted by breaking waves, associated with ebb-tidal deltas fronting some of the Brazilian estuaries: 1—Guaratuba; 2—Paranaguá and Superagui; 3—Todos os Santos (south channel); 4—Real-Piauí; 5—Vaza Barris; 6—Sergipe; 7—Timonha; 8—Coreaú; 9—Tutóia (Images from Google Earth)

the estuaries and receives the biggest fluvial discharge, estuarine mixing conditions easily vary between hydrographic extremes, i.e., from well mixed to salt wedge, and may even become completely fresh (Hartmann and Schettini 1991; Möller and Castaing 1999).

Salinity differences along these estuaries have a similar magnitude ($<10^1$ psu). Average salinity gradients ($\delta s/\delta x$) can be as small as 9×10^{-2} psu/km in the largest sector of Todos os Santos, to as high as ~8.0×10^{-2} psu/km in Vitoria and Paranaguá, and up to 1.83 psu/km along the inlet at Araruama (Table 1.2). Inside Araruama, however, the gradient has an opposite direction due to the lagoon's hypersalinity. The horizontal salinity gradient, by setting up longitudinal density differences, is the master variable controlling the residual estuarine circulation. Both the horizontal density gradient and the channel depth directly interfere with the structure and magnitude of the mean flow (U_e), or tidally averaged flow, by establishing a two layered, stratified flow called estuarine circulation. Hence, deeper estuaries with larger horizontal density gradients develop stronger estuarine circulation, whereas shallower estuaries with more homogeneous water quality present unidirectional tidally averaged flow. Current magnitudes associated with the estuarine circulation are one order of magnitude weaker than the tidal flow (10^{-2} m/s *vs* 10^{-1} m/s). The magnitude of the estuarine circulation (U_e) can be assessed by the equation (Geyer 2010)

$$U_e = a_o \frac{\beta g \frac{\partial s}{\partial x} h^2}{C_d U_t}$$

where β is the coefficient of saline contraction (~7.8×10^{-4} kg m^{-3} psu^{-1}), g is gravity, C_d is the drag coefficient (~3×10^{-3}), a_o is a dimensionless constant (~0.3), and U_t is the tidal velocity magnitude (maximum current velocities reported for each estuary were used). The strength of the estuarine circulation may vary 5-fold amongst the estuaries, from 0.03 m/s in Itamaracá to 0.17 m/s in Paranaguá (Table 1.2). The estuarine circulation is by definition stratified, and in all of the documented estuaries (Itamaracá—Medeiros e Kjerfve 1993; Todos os Santos—Cirano and Lessa 2007 and Pereira and Lessa 2009; Camamu—Amorim et al. 2016; Guanabara—Bérgamo 2006; Paranaguá—Mantovanelli 1999), a primarily vertical shear is identified, with the ocean denser water flowing in closer to the bottom and the less-dense estuarine water flowing out through more surface layers. Figure 1.9 shows the circulation-stratification diagram of Hansen and Rattray (1966) with the varying dynamic condition of 7 estuaries. Zones 1a and 1b in the diagram are associated with unidirectional, non-stratified, mean flow; estuarine circulation is established in zones 2a,b and becomes stronger in zones 3a,b. Zone 4 is associated with salt-wedge estuaries, where no mixing between fresh and salt water exists. The data are limited to a few surveys in all seven estuaries, and in Santos and Todos os Santos they are restricted to one of their most internal sections. It is observed that the dynamic condition in each estuary transits between zones 1 and 2 as a result of changing tidal ranges (spring-neap) and freshwater flow (rainy and dry season, wetter and dryer years), but most of the data points fall within zone 2.

Although the mean estuarine flow, or estuarine circulation, is much weaker than the tidal flow, it is crucial in establishing exchanges between the estuary and the neighboring shelf. Hence, its dynamic is inversely proportional to estuarine time scales such as residence and flushing times (Geyer 2010), i.e., stronger estuarine circulation causes shorter residence and flushing times. The flushing time (ideally the time needed to reduce the concentration of a given substance inside the estuary to zero) can be assessed by considering the tidal prism (P) and fresh water (R) volumes in relation to the estuary volume (V), or $(T.V)/(P + R)$ (Monsen et al. 2002; Miranda et al. 2003), where T is the tidal period (12.4 h). Even though this method underestimates the flushing time by not considering the full complexity of the

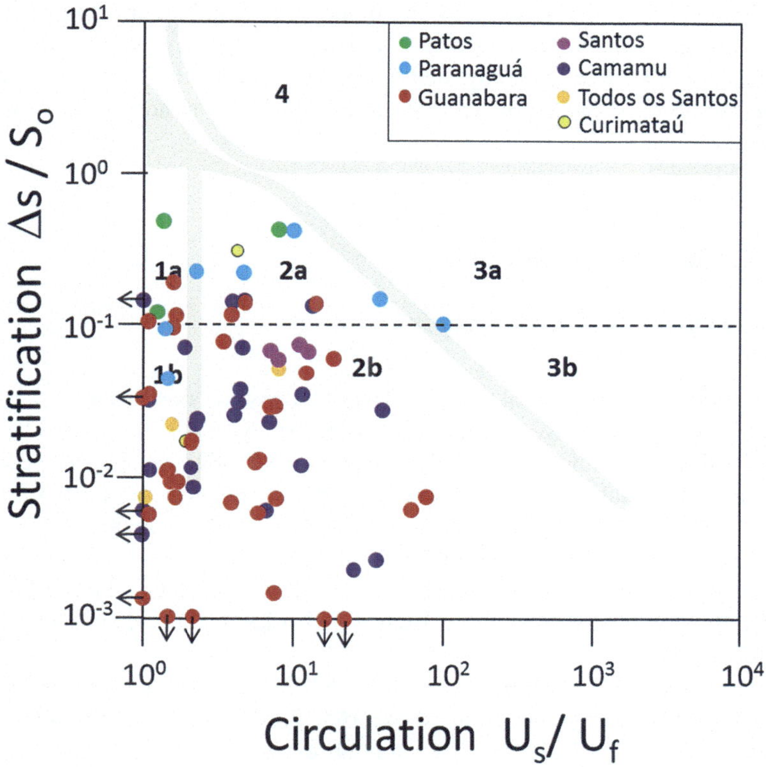

Fig. 1.9 Hansen and Rattray's (1966) circulation-stratification diagram based on current and water quality measurements at Curimataú (Andutta 2006), Camamu (Amorim et al. 2015), Guanabara (Kjerfve et al. 1996; Bérgamo 2006), Santos (Miranda et al. 2012), Lagoa dos Patos (Möller and Castaing 1999), Paranaguá (Mantovanelli 1999; Noernberg et al. 2014). The Todos os Santos data from Genz (2006) is restricted to the fresher, most internal sector of the bay. ΔS mean vertical salinity difference, S_0 mean vertical salinity, U_s mean vertical velocity, U_f mean surface velocity

estuarine circulation, it serves as an initial reference for comparisons and assessment of the slowest and fastest flushed estuaries. The estimated flushing times (Tf) listed in Table 1.2 shows that estuaries with smaller volumes are easier to be flushed. Paraíba and Itamaracá estuaries, as well as Camamu and Vitória, would be flushed in less than a day. The more voluminous Todos os Santos, Guanabara, and Sepetiba would be flushed within 2–3 days, and Araruama Lagoon, given its very small freshwater discharge relative to its volume, would be flushed after 9 months.

Only three, amongst the 41 estuaries, have had their flushing time assessed in the literature. Santana et al. (2018) calculated the dilution rate of a conservative substance in Todos os Santos by means of a 3D baroclinic numerical model forced with climatological means, and the $Tf_{50\%}$, or the time taken to achieve one half of the original concentration, was calculated at around 70 days. The $Tf_{50\%}$ was also calculated for Guanabara (Kjerfve et al. 1997) and Araruama Lagoon (Kjerfve et al.

1996) on the basis of the sum of all water inputs (river, rainfall, gravitational circulation, transverse circulation) to the system divided by the bay volume. The calculated flushing times were 11.4 days and 83.5 days, respectively. These three values, as expected, differ considerably from those in Table 1.2, but are, however, ranked the same.

1.4 Estuary Sedimentation

1.4.1 Fluvial Sediment Supply

Fluvial sediment yields vary at least 3 orders of magnitude amongst the few estuaries where measurements were made, from 10^3 to 10^6 t/year (Table 1.3), but because of the limited data set no spatial analysis can be made. Mean suspended sediment discharge from Guaíba River into Patos Lagoon was calculated as 1.1×10^6 t/year between 2003 and 2006 by Andrade Neto et al. (2012). Because the Guaíba River catchment area (84,760 km^2) corresponds to 42% of the total catchment to the lagoon, the total sediment discharge to the estuary is certainly larger, but it is unlikely to be doubled given the existence of harder, metamorphic rocks in the southern catchments.

In Babitonga, a sediment discharge of 7.6×10^3 t/year was gauged by Oliveira (2006). This discharge comes from an area (684.7 km^2) encompassing 60% of the catchment of the bay. Because little variation exists in rock hardness within the catchment, inserted into a granolithic complex, total sediment yield to the bay may amount to around 0.02×10^6 t/year.

Suspended sediment discharge to the southern sector of Paranaguá (1700 km^2 catchment) was measured by Mantovanelli (1999) as 355 t/day in the summer, and 88 t/day in the winter, which gives an approximate year-average of 221 t/day or 0.081×10^6 t/year. Assuming that the specific sediment discharge is the same for all other sub-catchments (1445 km^2) with similar geology (highly metamorphic granite rocks), the sediment yield for the whole bay would be approximately 0.17×10^6 t/year.

Sediment yield in Sepetiba was calculated by Molisani et al. (2006) as 0.85×10^6 t/year after performing detailed analysis of the water and sediment discharge of Guandú River, the largest sub-catchment making up 93% of the bay's drainage basin (2387 km^2). This sediment discharge is overloaded due to water diversion from a larger neighboring catchment (Paraiba do Sul River) that began in the 1950s. Pre-diversion sediment discharge was 0.58×10^6 t/year (Molisani et al. 2006), which, if extrapolated to the remaining catchment area with similar geology, gives a total discharge of 0.89×10^6 t/year.

Sediment mass accumulation, associated with recent mud deposits, was measured in Todos os Santos by Argollo (2001) by means of ^{210}Pb activity profiles in sediment cores. The calculated accumulation varied between 3.8 and 7.6 kg/m^2/year. Considering that this mud facies occupies an area of 460 km^2 (Lessa and Dias

2009), it is suggested that suspend sediment yield to the bay may vary from 1.75×10^6 t/year to 3.5×10^6 t/year.

The calculated specific fluvial sediment discharges vary considerably amongst the estuaries (Table 1.3), from 5.4 t/km^2/year in Paranaguá to 259 t/km^2/year in Sepetiba (in pre-diversion conditions). Similar broad variations occur for the specific sediment yield to the estuaries, which cannot explain variations observed in measured sedimentation rates. For instance, it is difficult to explain these large differences in sediment discharge when there are similarities in the geology and rainfall amongst the catchment areas. Hence, measurement errors can be much higher than expected. Santos (2009) calculated the sediment discharge for 6 estuaries (Todos os Santos, Camamu, Vitoria, Guanabara, Sepetiba, and Paranaguá) by means of the BQART model (Syvitski and Milliman 2007), which takes into consideration, amongst other things, altimetry, pluviosity, and lithology of the catchment area. These results show specific sediment discharges varying within the same order of magnitude, from 102 t/km^2/year in Todos os Santos to 626 t/km^2/year in Paranaguá.

1.4.2 Surface Sediment Cover and Sediment Deposition Rates

The existence of an estuarine circulation favors the retention of suspended particles inside the estuaries. Flocculation of suspended sediment particles within the mixing zone generates larger aggregates that are heavier and tend to be closer to the bottom, where they are swept by mean currents oriented upstream (the deeper layer in the estuarine circulation). Therefore, it is no surprise to observe that the majority of the estuaries, where the surface sediment cover was mapped, have an inner muddy-sediment zone flanked by sandy sediments of both fluvial origin (in the estuary head) and marine origin (close to the mouth) (Fig. 1.10).

The formation of this muddy lithofacies is clearer in the deeper and wider estuaries, namely Guaratuba, Paranaguá, Sepetiba, Guanabara, and Todos os Santos, where (i) estuarine circulation can be better developed given lesser depth constraints, (ii) larger accommodation space exists (underwater volume available for sediment retention) and (iii) less confined tidal flow favors deposition instead of erosion. Therefore, a less obvious mud-trapping zone exists at the shallower estuaries of Vitória (Veronez et al. 2009), Camamu (dos Santos 2016; Hatje et al. 2008), and Itamaracá (da Silva 2004) (Table 1.3), where the flow is more channelized and erosion of the estuary channel can locally expose relict mud sediments in places that would otherwise be covered with marine sand, as reported in Vitória (Veronez et al. 2009). In Lagoa dos Patos, a muddy bottom occurs all over except for the shallow margins exposed to wave action. The amount of suspended sediment delivered to this estuary is large enough to cause intense sediment flux toward the sea, where it impacts the sediment texture of the coastal zone and occasionally form a field of fluid mud on the inner shelf (Marques et al. 2010).

The estuarine deposition of muddy sediments through the Holocene has apparently caused shoaling of the bays and created sediment packages thicker than 10 m

Table 1.3 Rates of fluvial sediment discharge and muddy sediment deposition in subtidal areas, fluvial catchment size and maximum thickness of the muddy Holocene sedimentation within a few estuaries

	Estuary area (km^2)	Catchment (km^2)	Sediment discharge ($\times 10^6$ t/year)	Specific sediment yield to the estuary (t/year/km^2)	Catchment-specific sediment discharge (t/year/km^2)	Maximum thickness of modern sedimentation (m)	Holocene sedimentation rate (mm/year)	Modern sedimentation rates in subtidal areas (mm/year)
Todos os Santos	1233	58,926	1.75 to 3.5[a]	1419 to 2838	29.7 to 59.4	20		3.8 to 15.0
Camamu	400	1374						
Vitória	44	1728				5	0.2 to 0.5	
Guanabara	449	3768				10	0.2 to 0.8	2.8 to 26.0
Sepetiba	548	2388	0.89 (0.62[b])	1624 (1,1310[b])	372 (259[b])	11		3.0 to 12.0
Santos	147	1052						20.0 to 76.0
Paranaguá	707	3145	0.017[c]	24	5.4	>10		
Guaratuba	118	1810						5.2 to 6.1
Babitonga	258	1198	0.02[c]	77	16.7			
Lagoa dos Patos	9100	201,626	1.1 to <2.2	120 to <242	5.45 to <10.9	6	0.52	3.5 to 8.3

[a]Indirect assessment
[b]Before water diversion into Guandú River
[c]Extrapolated

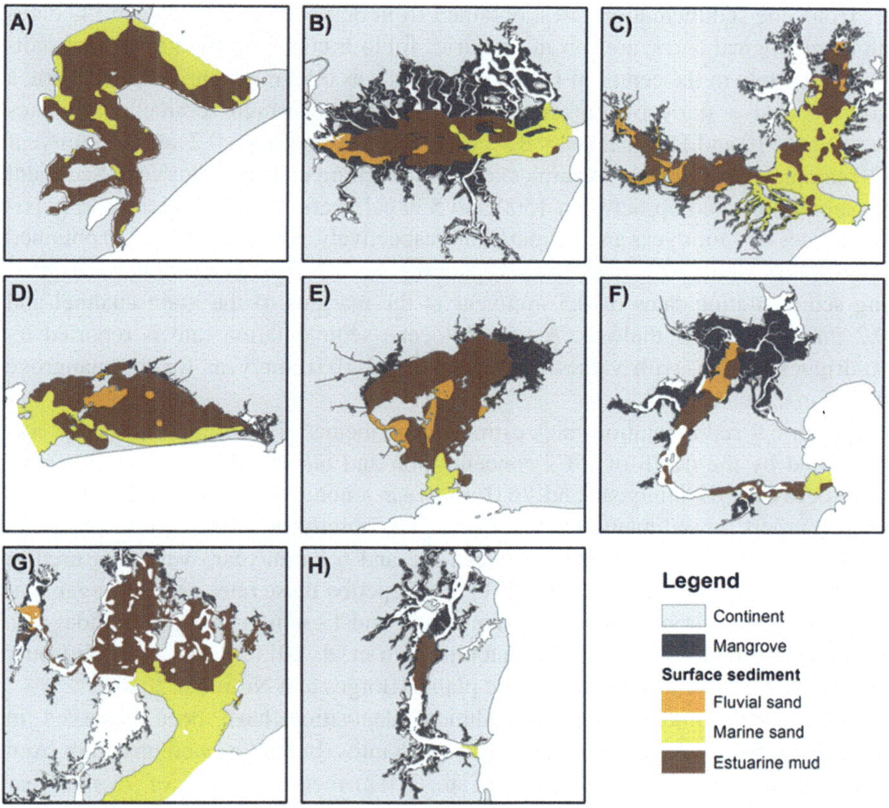

Fig. 1.10 Surface sediment cover of the Brazilian estuaries with available data. (**a**) Patos Lagoon (Demore 2001). (**b**) Guaratuba (Barbosa and Suguio 1999; Zem 2005). (**c**) Paranaguá (Lessa et al. 1998; Lamour et al. 2004). (**d**) Sepetiba (Nogueira Jr 1992; FEEMA 1997—in Montezuma 2007; Villena et al. 2003). (**e**) Guanabara (Kjerfve et al. 1997; Catanzaro et al. 2004; Marino 2011). (**f**) Vitória (Veronez et al. 2009). (**g**) Todos os Santos (Lessa et al. 2000; Lessa and Dias 2009). (**h**) Itamaracá (Da Silva 2004)

in Paranaguá (Lessa et al. 1998), Sepetiba (Borges and Nittrouer 2015), and Todos os Santos (Lessa et al. 2000; Dominguez and Bittencourt 2009). Thinner sediment packages are reported for Guanabara (5–10 m thick—Marino 2011), Vitória (2–5 m thick—Bastos et al. 2010), and Lagoa dos Patos (6 m—Toldo Jr et al. 2000).

It is a premise that estuaries act as sediment traps, holding the entire coarse (sand sized) sediment load and great part of the suspended load. Export of suspended sediment has only been documented to date in Lagoa dos Patos, which discharges around 1.37×10^7 t/year (Marques et al. 2010). The retained fine sediments within the estuaries cause rates of bed accretion that vary between several millimeters to a few centimeters per year, with considerable spatial variation occurring within each estuary. Also, the existing data show that rates of sediment accumulation have increased in historical times in all estuaries, by up to 1 order of magnitude, when compared to bed accretion rates during the Holocene (two last columns in Table 1.3).

Holocene sedimentation rates, obtained from depth intervals between ^{14}C dates along geological cores, are sub-millimetric. Toldo Jr et al. (2006) indicate that sedimentation rate in the center of Lagoa dos Patos was 0.52 mm/year. In Guanabara, a comprehensive study by Figueiredo et al. (2014) measured sedimentation rates between 5910 and 1500 cal years B.P., varying mostly between 0.2 and 0.8 mm/year (10 samples), with a few (3 samples) outliers reaching up to 8.7 mm/year. Sediment accumulation sped up between 1500 and 550 cal years B.P., with mean and maximum rates of 4 mm/year and 16 mm/year, respectively. Bastos et al. (2010) obtained a few radiocarbon dates from Vitoria spanning the last 7240 cal years B.P., suggesting sedimentation rates of 0.5 mm/year at the margins of the main channel and 0.2 mm/year in the thalweg. A last Holocene sedimentation rate is reported by Rodrigues (2014), with values between 3.3 and 10.0 mm/year for the mangrove plains in Coreaú.

Historical sedimentation rates estimated by means of ^{210}Pb, and normally corroborated by the depth of ^{137}Cs concentration and other anthropogenic elements, vary between 3.5 mm/year and 76.0 mm/year amongst estuaries. Toldo Jr et al. (2006) report on sedimentation rates of 3.5–8.3 mm/year in the last 150 years in Lagoa dos Patos. Similar rates, between 5.2 and 6.1 mm/year, were measured in Guaratuba by Patchineelam et al. (2011). In Sepetiba these rates present larger spatial differences, varying between 3.0 mm/year and 12 mm/year in the subtidal area (Forte 1996; Gomes et al. 2009; Patchineelam et al. 2011; Borges and Nittrouer 2016) to 22 mm/year in the mangrove plains (Borges and Nittrouer 2016).

Significantly higher rates of sediment deposition have been observed in Paranaguá, Santos, Guanabara, and Todos os Santos. In Santos, well upstream from port and dredged areas, Luiz-Silva et al. (2012) measured rates of 76 mm/year between the 1950s and 1970s, and 66 mm/y between 1976 and 1984. Since then, the average sedimentation rate has been around 20 mm/year. Noteworthy changes in modern sedimentation rates were also identified by Godoy et al. (1998), who calculated rates between 1.2 mm/year and 2.8 mm/year prior to 1950, and 6.1 mm/year to 26.0 mm/year since then in Guanabara. In Todos os Santos, Argollo (2001) and Andrade et al. (2017) measured rates of deposition varying from 3.8 mm/year to 15.0 mm/year in nine sediment cores, with limited indication of changes in the historical record.

1.5 Final Considerations

Forty-one large estuaries were identified along the Brazilian coast, where significant changes in coastal relief, pluviosity, geology, tidal ranges, and alongshore drift exist. The coast was subdivided in three sectors on the basis of pluviosity (wet and arid coast), costal geomorphology (tide and wave dominated), and structural geology (grabens running along or cross-shore). The North Sector has the highest number of large estuaries, with an average of 1 estuary every 34 km. It encompasses a long subsiding coastal zone, scantily studied, with tide-dominated, funnel-shaped, coastal-plain estuaries. The Northeast Sector, with north and an east facing sub-sectors, is an arid coast, characterized by small mesotidal estuaries within grabens that tend to be transversal to the coast and undergoing strong influence of littoral drift in the north-facing sub-sector. It has the second largest number of estuaries, but 72% of those are located in the East sub-sector. The East-Southeast Sector is the longest coastal span but has the lowest number of large estuaries. This sector is again humid, where wider and deeper estuaries, with generally small catchment areas, are fit within grabens that run parallel to the coast.

The estuary sizes are not positively correlated with the tidal range. Linear positive correlations exist, however, with river discharge in the East-Southeast and North sectors, and may indicate that fluvial incision was an important factor for the development of the estuaries in these zones. This is quite possibly the case in the North Sector, where large-scale coastal subsidence is identified and classic coastal-plain estuaries have evolved. Indeed, a positive linear trend is also observed between catchment area and estuarine area. Weaker positive correlation in the East-Southeast sector is apparently associated with varying tectonic control of the estuaries and their catchment area.

Data on fluvial sediment discharge is still sparse and apparently with large errors involved. Given the significant changes in the historical rates of sediment deposition, sediment yield must be changing considerably on decadal time scales. Sedimentation rates reflect the magnitude of the suspended sediment discharge as well as the area available for deposition. Larger (smaller) estuaries have lower (higher) sedimentation rates for a same sediment discharge. For instance, highest sedimentation rates in Santos are at least three times higher than in Todos os Santos or six times higher than Lagoa dos Patos. Despite having fluvial discharges 2–20 times higher than Santos, these latter estuaries are 10 and 60 times larger than the former. To date, there is no record of classical turbidity maximum zones in the investigated Brazilian estuaries cited herein, albeit zones of maximum turbidity have been reported in places of stronger tidal flows (Mantovanelli 1999).

Estuarine circulation apparently has a relevant role in trapping fine sediments inside the deeper and wider estuaries, where a muddy lithofacies has been mapped. In the German Wadden Sea, estuarine circulation has been found to finally offer a more generic explanation for its extensive mud accumulation, after years of debate on whether scour and settling lags, Stokes drift or asymmetrical tidal currents were key driving processes for the landward mud transport (Floser et al. 2011). The presence of this mud facies in the mixing zone of the estuaries, always flanked by sandy lithofacies in the sea- and landward sides, creates some similarity amongst quite a few estuaries in respect to potential zonation of the benthic fauna, regardless of differences in the thermohaline fields.

The amount of existing information on the estuaries is extremely skewed, with estuaries in the East-Southeast Sector more thoroughly, albeit unevenly, investigated than in the Northeast and North sectors. An exception is the knowledge on their structural control, with Northeast estuaries better investigated in this respect. Although scattered, evidence of tectonic subsidence for several estuaries mapped in this study are compelling enough to encourage a series of chronostratigraphic studies to test this hypothesis. The possibility that local subsidence is accentuating ongoing sea level rising rates (Church et al. 2013) deserves due consideration, as long-term planning for urban and industrial accommodation of the predicted eustatic sea level rise is generally based on open coast sea level data.

Acknowledgments We would like to thank Osmar Möller (FURG) for his help with literature on Patos Lagoon, and Mauro Cirano for sharing his large river flow data set. Bjorn Kjerfve, several years ago, was an important motivator of this work.

References

Allaby A, Allaby M (1999) Dictionary of earth sciences, 2nd edn. Oxford University Press, Oxford, 661 p

Allen GP, Salomon JC, Bassoullet P et al (1980) Effects of tides on mixing and suspended sediment transport in macrotidal estuaries. Sediment Geol 26(1–3):69–90

Alvez FC (2015) Análise morfoestrutural das bacias dos rios Paraíba e Mamanguape (PB) com base em MDE-SRTM e imagem PALSAR/ALOS-1. MSc Dissertation, Instituto Nacional de Pesquisas Espaciais, INPE, 151p. http://sid.inpe.br/mtc-m19/2015/02.26.13.32-TDI. Accessed 20 Nov 2016

Amorim FN (2005) Caracterização oceanográfica da Baía de Camamu e adjacências e mapeamento das áreas de risco à derrames de óleo. MSc Dissertation, Programa de Pós-Graduação em Geologia, UFBA, 191p. http://www.goat.fis.ufba.br/uploads/userfiles/32.pdf. Accessed June 01 2013

Amorim FN, Rezende LF, Lessa GC et al (2016) Oceanography of Camamu Bay (14S, Brasil) during dry and wet conditions. Revista Brazileira de Geofísica 33(4)

Andrade Neto JSR, Thiesen L, Toldo EE Jr et al (2012) Descarga sólida em suspensão do sistema fluvial do Guaíba, RS, e sua variabilidade temporal. Pesquisas em Geociências 39(2):161–171

Andrade RLB, Hatje V, Masqué P, Zurbrick CM, Boyle EA, Santos WPC (2017) Chronology of anthropogenic impacts reconstructed from sediment records of trace metals and Pb isotopes in Todos os Santos Bay (NE Brazil). Mar Pollut Bull 125(1–2):459–471

Andutta FP (2006) Experimentos e Modelagem Numérica com Aplicação ao Estuário Tropical do Rio Curimataú, RN. MSc Dissertation, Instituto Oceanográfico, USP, 122 p. http://www.teses.usp.br/teses/disponiveis/21/21132/tde-11102006-110223/pt-br.php. Accessed 01 May 2017

Angulo RJ, Lessa GC, Souza MC (2006) A critical review of mid- to late-Holocene sea-level fluctuations on the eastern Brazilian coasline. Quat Sci Rev 25:486–506

Araújo CS, Souza Filho PWM, Rodrigues SWP (2009) Morphology and modern sedimentary deposits of the macrotidal Marapanim Estuary (Amazon, Brazil). Cont Shelf Res 29:619–631

Argollo RM (2001) Cronologias de sedimentação recente e de deposição de metais pesados na Baía de Todos os Santos usando Pb210 e Cs137. PhD Thesis, Programa de Pós-Graduação em Geologia, UFBA, 119 p

Aubry A, Elliott M (2006) The use of environmental integrative indicators to assess seabed disturbance in estuaries and coasts: application to the Humber Estuary, UK. Mar Pollut Bull 53:175–185

Barbosa CF, Suguio K (1999) Biosedimentary facies of a subtropical microtidal estuary – an example from Southern Brazil. J Sediment Res 69(3):576–587

Barbosa EM, Prado RL, Sobreira MCA et al (2012) Seismic investigation in the Bertioga coastal plain, Brazil. Revista Brasileira de Geofísica 30:159–167

Bastos AC, Vilela CG, Quaresma VS et al (2010) Mid- to Late-Holocene estuarine infilling processes studied by radiocarbon dates, high resolution seismic and biofacies at Vitoria Bay, Espirito Santo, Southeastern Brazil. An Acad Bras Cienc 82(3):761–770

Bérgamo AL (2006) Características hidrográficas, da circulação e dos transportes de volume e sal na Baía de Guanabara (RJ): variações sazonais e moduladas pela maré. PhD Thesis, Instituto Oceanográfico, USP, 169p. http://www.teses.usp.br/teses/disponiveis/21/21132/tde-22052007-161032/pt-br.php. Accessed 18 Nov 2016

Bezerra FHR, Amaro VE, Vita-Finzi C et al (2001) Pliocene-Quaternary fault control of sedimentation and coastal plain morphology in NE Brazil. J S Am Earth Sci 14(1):61–75

Bezerra FH, Ferreira JM, Souza MOM (2006) Review of seismicity and neogene tectonics in Northeastern Brazil. Rev Asoc Geol Argent 61(4):525–535

Borges HV, Nittrouer CA (2015) The paleo-environmental setting of Sepetiba Bay, Rio de Janeiro, Brazil, in the Late Pleistocene: interpretations from high-resolution seismic stratigraphy. Revista Brasileira de Geofísica 33(4). http://sbgfisica.org/revista/index.php/rbgf/article/viewFile/762/292. Accessed 15 Nov 2016

Borges HV, Nittrouer CA (2016) Sediment accumulation in Sepetiba Bay (Brazil) during the Holocene: a reflex of the human influence. J Sediment Environ 1(1):96–112

Born CC, Scherer CMS, Adegas F et al (2011) Estratigrafia e tectônica da seção rifte no Gráben de Camamu, porção emersa da Bacia de Camamu, Bahia. Revista Brasileira de Geociências 41(2):334–354

Boyd R, Dalrymple RW, Zaitlin BA (1992) Estuarine and incised-valley facies models. In Posamentier HW, Walker RG (eds) Facies models revisited. SEPM Special Publication No. 84, pp 171–235

Brazilian Seismic Catalog (2009) Instituto de Astronomia, Geofísica e Ciências Atmosféricas, Universidade de São Paulo. http://moho.iag.usp.br/sismologia/boletim.php. Accessed 01 Apr 2009

Cameron WM, Pritchard DW (1963) Estuaries. In: Hill MN (ed) The Sea: Ideas and Observations on Progress in the Study of the Seas, vol 2. The composition of sea-water, comparative and descriptive oceanography. Wiley, New York, pp 306–324

Carvalho JSB (2000) Caracterização morfoestratigráfica do preenchimento sedimentar da Baía de Iguape, Bahia: influência das variações eustáticas do nível do mar e atividades tectônicas recentes. MSc Dissertation, Programa de Pós-Graduação em Geologia, UFBA, 125p

Catanzaro LF, Baptista Neto JA, Guimarães MSD et al (2004) Distinctive sedimentary processes in Guanabara Bay – SE/Brazil. based on the analysis of echo-character (7.0 kHz). Braz J Geophys 22(1):69–83

Church JA, Clark PU, Cazenave A et al (2013) Sea level change. In: Stocker TF, Qin D, Plattner GK et al (eds) Climate change 2013: The physical science basis. Contribution of Working Group I to the Fifth Assessment Report of the Intergovernmental Panel on Climate Change. Cambridge University Press, Cambridge, pp 1137–1216

Cirano M, Lessa GC (2007) Oceanographic characteristics of Baía de Todos os Santos, Brazil. Rev Bras Geof 25(4):363–387

Cohen MCL, Souza Filho PWM, Lara RJ et al (2005) A model of Holocene mangrove development and relative sea-level changes on the Braganca Peninsula (Northern Brazil). Wetl Ecol Manag 13:433–443

Corrêa-Gomes LC, Ucha JM, Silva IC (2012) Neotectônica. In: Barbosa JSF et al (eds) Geologia da Bahia – Pesquisa e Atualização. UFBA-CBPM, Salvador, pp 497–515

Costa JBS, Hasui Y, Bemerguy RL et al (2002) Tectonics and paleogeography of the Marajó Basin, northern Brazil. An Acad Bras Cienc 74:519–531

Da Silva LA (2004) Sedimentologia do Canal de Santa Cruz, Ilha de Itamaracá – PE. MSc Dissertation, Geoscience Institute, UFPE, Recife, 131 p. http://repositorio,ufpe,br/bitstream/handle/123456789/6600/arquivo6875_1,pdf?sequence=1&isAllowed=y. Accessed 01 Jun 2017

Dalrymple RW, Zaitlin BA, Boyd R (1992) Estuary facies models: conceptual basis and stratigraphic implications. J Sediment Petrol 62:1130–1146

de Souza WFL, Kjerfve B, Knoppers B et al (2003) Nutrient budgets and trophic state in a hypersaline coastal lagoon: Lagoa de Araruama, Brazil. Estuar Coast Shelf Sci 57:843–858

Demore JP (2001) Aspectos sedimentares do estuário da Lagoa dos Patos e sua interação com a poluição por petróleo: subsídios para um plano de contingência. BSc Monograph, Oceanology, FURG, 30 p. http://saopelotas,furg,br/images/stories/documentosdereferencia/aspectos%20sedimentares%20do%20estuario%20da%20lagoa%20dos%20patos%20_subsidiosparaumplanodecontingencia,pdf. Accessed 01 Oct 2016

Diretoria de Hidrografia e Navegação (2012) Carta de correntes de maré: Baía de Guanabara, Centro de Hidrografia da Marinha, Brasil, 5a, reimpressão. https://www,mar,mil,br/dhn/chm/box-publicacoes/publicacoes/ccm/CCM-Baia-de- Guanabara,pdf. Accessed 23 Sep 2014

Dominguez JML (2009) The coastal zone of Brazil. In: Dillemburg S, Hesp P (eds) Geology and geomorphology of holocene coastal barriers of Brazil, vol 107. Springer, Berlin, pp 17–51

Dominguez JML, Bittencourt ACSP (2009) Geologia. In: Hatje V, Andrade JB (eds) Baía de Todos os Santos: aspectos oceanograficos. EDUFBA, Salvador, pp 29–66

Dominguez JML, Bittencourt ACSP, Martin L (1983) O papel da deriva litorânea de sedimentos arenosos na construcão das planícies costeiras associadas às desembocaduras dos rios São Francisco, Jequitinhonha, Doce e Paraíba do Sul. Revista Brasileira de Geociências 13(2):98–105

Dominguez JML, Martin L, Bittencourt ACSP (1987) Sea-level history and quaternary evolution of river mouth-associated beachridge plains along the east-southeast Brazilian coast: a summary. In: Nummedal D, Pilkey OH, Howard JD (eds) Sea-level fluctuation and coastal evolution. SEPM Special Pub No. 41, pp 115–127

dos Santos DG (2016) Interpretação de processos hidrossedimentológicos nos estuários Serinhaém, Maraú e Sorojó (Baía de Camamu) a partir do estudo de bioclastos recentes. MSc Dissertation, Geoscience Institute, UFBA, Salvador, 155 p. https://repositorio,ufba,br/ri/handle/ri/22527. Accessed 01 Jun 2017

Dronkers J (1986) Tidal asymmetry and estuarine morphology. Neth J Sea Res 20(2/3):117–131

Ducarme B, Venedikov AP, de Mesquita AR et al (2006) New analysis of a 50 years tide gauge record at Cananéia (SP-Brazil) with the VAV tidal analysis program. Monitoring and understanding a dynamic planet with geodetic and oceanographic tools IAG symposium, Cairns, Australia, August, 2005, pp 453–460

Dyer KR (1997) Estuaries, a physical introduction, 2nd edn. Wiley, Chichester, p 198

Engelhart SE, Horton BP, Kemp AC (2011) Holocene sea level changes along the United States' Atlantic Coast. Oceanography 24(2):70–79

Fairbridge RW (1980) The estuary: its definition and geochemical role. In: Olausson E, Cato I (eds) Chemistry and geochemistry of estuaries. Wiley, New York, pp 1–35

Ferrari AL (2001) Evolução Tectônica do Graben da Guanabara. PhD Thesis, Instituto de Geociências, USP, 412 p

Ferreira CRP Jr, Costa JBS, Bermerguy RL et al (1996) Neotectônica na área da Bacia de São Luís. Revista Geociências 15:185–208

Fiedler MFM (2015) Dinâmica estuarina em cenários de aumento do nível do mar: estuário de Santos (SP). MSc Dissertation, Oceanographic Institute, USP, São Paulo, 130 p. http://www,teses,usp,br/teses/disponiveis/21/21136/tde-26022016-144937/pt-br,php. Accessed 01 Apr 2017

Figueiredo AG Jr, Toledo MB, Cordeiro RC et al (2014) Linked variations in sediment accumulation rates and sea-level in Guanabara Bay, Brazil, over the last 6000 years. Palaeogeogr Palaeoclimatol Palaeoecol 415:83–90

Floser G, Burchard H, Riethmuller R (2011) Observational evidence for estuarine circulation in the German Wadden Sea. Cont Shelf Res 31:1633–1639

Fonseca VP (2005) Estudos morfotectônicos aplicados à planície costeira do Rio Grande do Sul e adjacências. PhD Dissertation, Instituto de Geociências, UFRGS, 309 p. http://www,lume,ufrgs,br/handle/10183/6173. Accessed 01 Jun 2017

Forte CMS (1996) Determinação das taxas de sedimentação na porção nordeste da baía de Sepetiba utilizando datação com o radioisótopo 210Pb em excesso. MSc Dissertation,

Programa de Pós-Graduação em Geociências, UFF, 93 p. http://www,iaea,org/inis/collection/NCLCollectionStore/_Public/32/005/32005640,pdf. Accessed 01 Oct 2014
Fry A, Aubrey DG (1990) Tidal transport velocity asymmetries and bedload transport in shallow embayments. Estuar Coast Shelf Sci 30:453–473
Genz F (2006) Avaliação dos efeitos da barragem Pedra do Cavalo sobre a circulação estuarina do rio Paraguaçu e Baía de Iguape. PhD Thesis, Instituto de Geociências, UFBA, 245 p. http://www,pggeologia,ufba,br/v1/index,php/biblioteca-digital/teses/marinha-costeira-e-sedimentar/22-fernando-genz-2006. Accessed 01 Jun 2017
Geyer WR (2010) Estuarine salinity structure and circulation. In Valle-Levinson A (ed) Contemporary issues in estuarine physics. Cambridge University Press, Cambridge, Chap 2, 19 p
Godoy JM, Moreira J, Bragança MJ et al (1998) A study of Guanabara Bay sedimentation rates. J Radioanal Nucl Chem 227(1–2):157–160
Gomes FC, Godoy JM, Godoy MLDP et al (2009) Heavy metals in sediment profiles from Sepetiba and Ribeira Bays: a comparative study. Radioprotection 44(5):243–250
González-Gorbeña E, Rosman PCC, Qassim RY (2015) Assessment of the tidal current energy resource in São Marcos Bay, Brazil. J Ocean Eng Mar Energy 1:421–433
Hansen DV, Rattray M (1966) New dimensions in estuary classification. Limnol Oceanogr 11:319–326
Hart GF, Coleman J (2009) The world deltas database framework. Louisiana State University. http:/www,geol,lsu,edu/WDD. Accessed 31 Mar 2013
Hartmann C, Schettini CAF (1991) Aspectos hidrológicos na desembocadura da Laguna dos Patos, RS. Revista Brasileira de Geociências 21(4):371–377
Hatje V, Barros F, Magalhães W et al (2008) Trace metals and benthic macrofauna distributions in Camamu Bay, Brazil: sediment quality prior oil and gas exploration. Mar Pollut Bull 56:348–379
Hayes MO (1979) Barrier island morphology as a function of wave and tide regime. In: Leatherman SP (ed) Barrier islands from the Gulf of St, Lawrence to the Gulf of Mexico. Academic Press, New York, pp 1–29
Hesp PA, Maia LP, Claudino-Sales V (2009) The Holocene Barriers of Maranhão, Piauí and Ceará States, Northeastern Brazil. In: Dillemburg S, Hesp P (eds) Geology and geomorphology of holocene coastal barriers of Brazil, vol 107. Springer, New York, pp 325–345
Jaeger JM, Mehta A, Faas R, Grella M (2009) Anthropogenic impacts on sedimentary sources and processes in a small urbanized subtropical estuary, Florida. J Coast Res 25(1):30–47
Kjerfve B, Ferreira HO (1993) Tidal bores: first ever measurements. Ciência e Cultura 45(2):135–138
Kjerfve B, Magill KE (1989) Geographic and hydrodynamic characteristics of shallow coastal lagoons. Mar Geol 88:187–199
Kjerfve B, Schettini CAF, Knoppers B et al (1996) Hydrology and salt balance in a large, hypersaline coastal lagoon: Lagoa de Araruama, Brazil. Estuar Coast Shelf Sci 42:701–725
Kjerfve B, Ribeiro CHA, Dias GTM et al (1997) Oceanographic characteristics of an impacted coastal bay: Baía de Guanabara, Rio de Janeiro, Brazil. Cont Shelf Res 17:1609–1643
Knoppers B, Ekau W, Figueiredo AG (1999) The coast and shelf of east and northeast Brazil and material transport. Geo-Mar Lett 19(3):171–178
Lamour MR, Soares CR, Carrilho JC (2004) Mapas de parâmetros texturais de sedimentos de fundo do complexo estuarino de Paranaguá – PR. Boletim Paranaense de Geociências 55:77–82
Lessa GC, Dias K (2009) Distribuição espacial das litofácies de fundo da Baía de Todos os Santos. Quat Environ Geosci 1(2):84–97
Lessa GC, Masselink G (2006) Evidence of a mid-Holocene sea level highstand from the sedimentary record of a macrotidal barrier and paleoestuary system in Northwestern Australia. J Coast Res 22(1):100–112
Lessa GC, Meyers SR, Marone E (1998) Holocene stratigraphy in the Paranagua Bay estuary, southern Brazil. J Sediment Res 68(6):1060–1076

Lessa GC, Bittencourt ACSP, Brichta A et al (2000) Late quaternary sedimentation in Todos os Santos Bay. An Acad Bras Cienc 72(4):573–590

Lessa GC, Cirano M, Tanajura CAS et al (2009) Oceanografia Física. In: Hatje V, Andrade JB (eds) Baía de Todos os Santos: aspectos oceanograficos. EDUFBA, Salvador, pp 68–11

Lewis SE, Sloss CR, Murray-Wallace CV et al (2013) Post-glacial sea-level changes around the Australian margin: a review. Quat Sci Rev 74(15):115–138

Lima JCF, Bezerra FHR, Rossetti DF et al (2016) Neogene-Quaternary fault reactivation influences coastal basin sedimentation and landform in the continental margin of NE Brazil. Quat Int 438(Part A):92–105

Lincoln RJ, Boxshall GA, Clark PF (1982) A dictionary of ecology, evolution and systematics. Cambridge University Press, Cambridge, 371 p

Lorscheitter ML, Dillenburg SR (1998) Holocene palaeoenvironments of the northern coastal plain of Rio Grande do Sul, Brazil, reconstructed from palynology of Tramandai lagoon sediments. Quat South Am Antarct Pen 11:73–97

Luiz-Silva W, Matos RHR, Machado W et al (2012) Taxas de sedimentação em estuário do sudeste do Brasil, baseadas em assinatura geoquímica e histórico industrial. Geociências 31(1):69–78

Magnavita LP, Silva RR, Sanches CP (2005) Guia de Campo da Bacia do Recôncavo, NE do Brasil. Boletim de Geociências da Petrobras 13(2):301–334

Mantovanelli A (1999) Caracterização da Dinâmica Hídrica e do Material Particulado em Suspensão na Baía de Paranaguá e em sua Bacia de Drenagem. MSc Dissertation, Programa de Pós-Graduação em Geologia, UFPR, Brasil, 154 p. http://acervodigital,ufpr,br/bitstream/handle/1884/41425/D%20-%20ALESSANDRA%20MANTOVANELLI,pdf?sequence=2&isAllowed=y. Accessed 01 Apr 2017

Mantovanelli A, Marone E, da Silva ET et al (2004) Combined tidal velocity and duration asymmetries as a determinant of water transport and residual flow in Paranagua Bay estuary. Estuar Coast Shelf Sci 59:523–537

Marino IC (2011) Caracterização do Registro Sedimentar Quaternário da Baía de Guanabara. BSc Monograph, Geophysics, UFF, Brasil, 75 p. http://www,geofisica,uff,br/sites/default/files/projetofinal/monografia_igor,pdf. Accessed 27 Jun 2012

Marques WC, Fernandes EHL, Moraes BC et al (2010) The dynamics of the Patos Lagoon coastal plume and its contribution to the deposition pattern of the southern Brazilian inner shelf. J Geophys Res 115:C10045

Martin L, Flexor J-M, Bittencourt ACSP et al (1986) Neotectonic movements on a passive continental margin: Salvador region, Brazil. Neotectonics 1:87–103

Martin L, Dominguez JML, Bittencourt ACSP (2003) Fluctuating Holocene sea levels in eastern and southeastern Brazil: evidence from multiple fossil and geometric indicators. J Coast Res 19(1):101–124

Medeiros C, Kjerfve B (1993) Hydrology of a tropical estuarine system: Itamaracá, Brazil. Estiuar Coast Shelf Sci 36:494–515

Miranda LB, Castro BM, Kjerfve B (2003) Principios de Ocenografia Fisica de Estuários. EDUSP, São Paulo, 411 p

Miranda LB, Olle ED, Bérgamo AL et al (2012) Circulation and salt intrusion in the Piaçaguera Channel, Santos (SP). Braz J Oceanogr 60(1):11–23

Mizerkowski BD, Machado lEC, Brandini N et al (2012) Environmental water quality assessment in Guaratuba Bay, State of Paraná, Southern Brazil. Braz J Oceanogr 60(2):109–115

Molisani MM, Kjerfve B, Silva AP et al (2006) Water discharge and sediment load to Sepetiba Bay from an anthropogenically-altered drainage basin, SE Brazil. J Hydrol 331:425–433

Möller O, Castaing P (1999) Hydrological characteristics of the estuarine area of Patos Lagoon (30,1S, Brazil). In: GME P, Piccolo MC (eds) Estuaries of South America (their geomorphology and dynamics) – environmental science. Springer, Berlin, pp 83–100

Möller OO, Castaing P, Salomon JC et al (2001) The influence of local and non-local forcing effects on the subtidal circulation of Patos Lagoon. Estuaries 24:275–289

Monsen NE, Cloern JE, Lucas LV et al (2002) A comment on the use of flushing time, residence time and age as transport time scales. Limnol Oceanogr 47(5):1545–1553
Montezuma PN (2007) Impactos nos processos de assoreamento na Baía de Sepetiba-RJ, de sedimentos oriundos da bacia contribuinte e de dragagens portuárias. MSc Dissertation, Pós-Graduação em Engenharia, UFRJ, 94 p. http://www,oceanica,ufrj,br/intranet/teses/2007_mestrado_patricia_ney_de_montezuma,pdf. Accessed 06 Jun 2010
Nalesso R, Joyeux J, Quintana C (2005) Soft-bottom macrobenthic communities of the Vitoria Bay estuarine system, south-eastern Brazil. Braz J Oceanogr 53:23–38
Nascimento WR Jr, Souza-Filho PWM, Proisy C et al (2013) Mapping changes in the largest continuous Amazonian mangrove belt using object-based classification of multisensor satellite imagery. Estuar Coast Shelf Sci 117:83–93
Nascimento ER, Salamuni E, Queiroz GL et al (2013) Morphotectonic and neotectonic evidence in the Serra do Mar relief in the State of Paraná. Revista Brasileira de Geomorfologia 14(3):287–299
Nichols M (1989) Sediment accumulation rates and relative sea-level rise in lagoons. Mar Geol 88:201–219
Nienhuis PH, Smaal AC (1994) The Oosterschelde Estuary (The Netherlands): a case-study of a changing ecosystem. Hydrobiologia 282(283):1–14
Noernberg MA, Mizerkowski BD, Paloschi NG et al (2014) Hydrodynamics and bio-optical assessment of two pristine subtropical estuaries in southern Brazil. Braz J Oceanogr 62(4):265–278
Nogueira J Jr (1992) Estudo sedimentológico da Baía de Sepetiba (RJ) – Sintese dos trabalhos existentes e sua aplicação para fins de geologia de engenharia. In: Proceedings 34th Brazilian congress of geology, vol 4. SBG, Salvador, pp 1676–1688
Oliveira FA (2006) Estudo do aporte sedimentar em suspensão na Baía de Babitonga sob a ótica da geomorfologia. PhD Thesis, Pós-Graduação em Geografia, USP, 285 p
Oliveira AM, Kjerfve B (1993) Environmental response of a tropical coastal lagoon system to hydrological variability: Mundau-Manguaba, Brazil. Estuar Coast Shelf Sci 37:575–591
Oliveira JC, Aguiar W, Cirano M, Genz F, Amorim FN (2018) A climatology of the annual cycle of river discharges into the Brazilian continental shelves: from seasonal to interannual variability. Environ Earth Sci 77:192. 17 p
Patchineelam SM, Sanders CJ, Smoak JM et al (2011) A historical evaluation of anthropogenic impact in coastal ecosystems by geochemical signatures. Braz J Chem Soc 22(1):120–125
Pereira MAG, Lessa GC (2009) Varying Patterns of water circulation in Canal de Cotegipe, Baía de Todos os Santos. Revista Brasileira de Geofísica 27:1–17
Perillo GME (1995) Definitions and geomorphologic classifications of estuaries. In: GME P (ed) Geomorphology and sedimentology of estuaries, developments in sedimentology, vol 53. Elsevier, Berlin, pp 12–47
Potter IC, Chuwen BM, Hoeksema SD et al (2010) The concept of an estuary: a definition that incorporates systems which can become closed to the ocean and hypersaline. Estuar Coast Shelf Sci 87(3):497–500
Pritchard DW (1952) Estuarine hydrography. Adv Geophys I:243–280
Ribeiro CHA, Kjerfve B (2002) Anthropogenic influence on the water quality in Guanabara Bay, Rio de Janeiro, Brazil. Reg Environ Chang 3:13–19
Riccomini C (1992) Evidências geológicas da atividade neotectônica no Estado de São Paulo e regiĺes adjacentes. Boletim IG-USP, Publ, Esp (12):95–96
Rigo D (2004) Análise do escoamento em regiões estuarinas com manguezais – medições e modelagem na Baía de Vitória, ES. PhD Thesis, Programa de Pós-Graduação em Engenharia, UFRJ, 156 p. http://www,teses,oceanica,ufrj,br/pdfs/2004_doutorado_daniel_rigo,pdf. Accessed 01 Apr 2017
Rodrigues SWP (2014) Detecção de mudança e sedimentação no estuário do rio Coreaú. PhD Thesis, Programa de Pós-Graduação em Geologia e Geoquímica, UFPA, 108 p. http://repositorio,ufpa,br/jspui/handle/2011/6350). Accessed 20 Jan 2016

Rossetti DF (2003) Delineating shallow Neogene deformation structures in northeastern Pará State using Ground Penetrating Radar. An Acad Bras Cienc 75:235–248
Rossetti DF, Goes AM, Valeriano MM et al (2008) Quaternary tectonics in a passive margin: Marajó Island, northern Brazil. J Quat Sci 23(2):121–135
Rossetti DF, Bezerra FHR, Góes AM et al (2011) Late Quaternary sedimentation in the Paraíba Basin, Northeastern Brazil: Landform, sea level and tectonics in Eastern South America passive margin. Palaeogeogr Palaeoclimatol Palaeoecol 300:191–204
Rossetti DF, Souza LSB, Prado R et al (2012) Neotectonics in the northern equatorial Brazilian margin. J S Am Earth Sci 37:175–190
Roy PS (1984) New South Wales estuaries – their origin and evolution. In: Thom BG (ed) Developments in Coastal geomorphology in Australia. Academic Press, Sydney, pp 99–121
Saadi A (1993) Neotectônica da plataforma Brasileira: esboço e interpretação preliminares. Geonomos 1(1–2):1–15
Saadi A, Machette MN, Haller KM et al (2002) Map of quaternary faults and lineaments of Brazil, International Lithosphere Program Task Group II-2, Major Active Faults of the World, U.S. Geological Survey Open-File Report 02–230
Sampaio M (2003) Estudo da circulação hidrodinâmica 3D e trocas de massa d'água na Baía de Guanabara, RJ. MSc Dissertation, Programa de Pós-Graduação em Engenharia, UFRJ, 213 p. http://www,coc,ufrj,br/index,php?option=com_content&view=article&id=1830:marcelle-sampaio&catid=103&Itemid=153&lang=pt-br. Accessed 19 Apr 2010
Santana R, Teixeira CEP, Lessa GC (2018) The impact of different forcing agents on the residual circulation in a tropical estuary (Baía de Todos os Santos, Brazil). J Coast Res 34(3) (in press)
Santos VG (2009) Distribuição espaço-temporal do zooplâncton no estuário do rio Maraú, Baía de Camamu – BA. MSc Dissertation, Departamento de Ciências Biológicas da Universidade Estadual de Santa Cruz, Brasil, 71 p
Souza CRG (2015) The Bertioga coastal plain: an example of morphotectonic evolution. In: Vieira BC, Salgado AAR, Santos LJC (eds) World geomorphological landscapes. Springer, Berlin, pp 115–134
Souza Filho PWM (2000) Tectonic control on the coastal zone geomorphology of the northeastern Pará State. Revista Brasileira de Geociências 30(3):527–530
Souza Filho PWM (2005) Costa de manguezais de macromaré da Amazônia: cenários morfológicos, mapeamento e quantificação de áreas usando dados de senores remotos. Revista Brasileira de Geofísica 23(4):427–435
Souza Filho PW, Lessa GC, Cohen MCL et al (2009) Trangressive barrier estuarine systems of the Eastern Amazon coast, Northern Brazil. In: Dillemburg S, Hesp P (eds) Geology and geomorphology of holocene coastal barriers of Brazil, vol 107. Springer, New York, pp 347–375
Souza LAP, Tessler MG, Galli VL (1996) O graben de Cananéia. Revista Brasileira de Geociências 26(3):139–150
Spencer CD, Plater AD, Long AJ (1998) Rapid coastal change during the mid- to late Holocene: the record of barrier estuary sedimentation in the Romney Marsh region, southeast England. The Holocene 8(2):143–163
Suguio K, Martin L (1996) The role of neotectonics in the evolution of the Brazilian coast. Geonomos 4(2):45–53
Syvitski JPM, Milliman JD (2007) Geology, geography, and human battle for dominance over the delivery of fluvial sediment to the coastal ocean. J Geol 115:1–19
Toldo EE Jr, Dillenburg SR, Corrêa ICS et al (2000) Holocene sedimentation in Lagoa dos Patos Lagoon, Rio Grande do Sul, Brazil. J Coast Res 16(3):816–822
Toldo EE Jr, Dillenburg SR, Corrêa ICS et al (2006) Sedimentação de Longo e Curto Período na Lagoa dos Patos, Sul do Brasil. Pesquisas em Geociências 33(2):79–86
Tomazelli LJ, Villwock JA, Dillenburg SR et al (1998) Significance of present-day coastal erosion and marine transgression, Rio Grande do Sul, Southern Brazil. Anais Academia Brasileira de Ciéncias 70:221–229

Veronez P Jr, Bastos AC, Pizzin BF et al (2009) Sonar de varredura lateral e sísmica de alta resolução aplicados no estudo de ecofácies na Baía de Vitória – ES. Revista Brasileira de Geofísica 27(3):411–425

Villena HH, Pereira SD, Barros LC et al (2003) Baía de Sepetiba: considerações geológicas e oceanográficas com base em dados batimétricos e sedimentológicos. In 9th Congress of the Brazilian Association for Quaternary Studies, Recife, Pernanbuco, 4 p. http://www,abequa,org,br/trabalhos/quatmar_45,pdf. Accessed 05 Mar 2011

Woodworth PL, Teferle FN, Bingley RM et al (2009) Trends in UK mean sea level revisited. Geophys J Int 176(1):19–30

Zem RC (2005) Dinâmica sedimentar da Baia de Guaratuba – PR. BSc Monography, Oceanography, Universidade Federal do Paraná, 87 p. http://acervodigital,ufpr,br/bitstream/handle/1884/36684/ZEM%20R,C,%202005%20Dinamica%20Sedimentar%20da%20Baia%20de%20Guaratuba,pdf?sequence=1&isAllowed=y. Accessed 07 Mar 2012

Chapter 2
Benthic Estuarine Assemblages of the Brazilian North Coast (Amazonia Ecoregion)

José Souto Rosa Filho, Luci Cajueiro Carneiro Pereira, Daiane Aviz, César França Braga, Marcela Carneiro Monteiro, Rauquírio Albuquerque Marinho da Costa, Nils Edvin Asp, and Colin Robert Beasley

Abstract Despite its ecological and economical importance, the Brazilian North Coast (BNC) or Amazonia Ecoregion *sensu* Spalding et al. (Bioscience 57:573–583, 2007), was poorly studied until the 1980s, when major interdisciplinary coastal research programs began. The Amazon and other major rivers strongly influence the BNC, causing seasonally reduced surface salinity and significant sediment deposition. From February to March, monthly accumulated rainfall is 400 mm or more and estuarine salinity varies between 0 and 10, but reaches over 40 in the dry season. Mangrove vegetation, with narrow seaward fringes of salt marsh, dominates the BNC. Macrotides between 4 and 7 m expose large areas of muddy to sandy sediments at low tide. Tidal amplitude is twice as large during spring tides, inundating large areas of mangrove. Tidal export, as well as riverine discharge, determines BNC dissolved nutrient profiles. Despite high turbidity, BNC estuaries have high phytoplankton biomasses and washout of benthic microalgae may also contribute to high chlorophyll-*a* concentrations. Though benthic diversity is low, secondary pro-

J. S. Rosa Filho (✉)
Laboratório de Bentos, Departamento de Oceanografia, Centro de Tecnologia e Geociências – Universidade Federal de Pernambuco, Recife, PE, Brazil
e-mail: souto.rosa@ufpe.br

L. C. C. Pereira · M. C. Monteiro
Laboratório de Oceanografia Costeira e Estuarina – Instituto de Estudos Costeiros, Universidade Federal do Pará, Bragança, Pará, Brazil
e-mail: cajueiro@ufpa.br

D. Aviz
Laboratório de Oceanografia Biológica, Instituto de Geociências, Universidade Federal do Pará, Belém, PA, Brazil

C. F. Braga
Laboratório de Ecologia e Conservação Ambiental, Universidade Federal Rural da Amazônia, Capitão Poço, PA, Brazil
e-mail: cesar.braga@ufra.edu.br

P. C. Lana, A. F. Bernardino (eds.), *Brazilian Estuaries*, Brazilian Marine Biodiversity, https://doi.org/10.1007/978-3-319-77779-5_2

ductivity in sediments is high, and important for nutrient cycling, especially in mangrove forest. The uçá-crab (*Ucides cordatus*) is economically important in the region. Strong gradients in salinity, along both the coast and individual estuaries, determine the relative abundance of freshwater and marine benthic taxa but abundance and diversity are lower in the wet season. Although relatively conserved, there is increasing pressure on the BNC through urban expansion, organic pollution, mangrove logging, and over-harvesting of coastal resources. With only 17 protected and especially managed areas along the BNC, the implementation of conservation policies is, so far, unsatisfactory.

Keywords Northern Brazil Coast · Amazon · Benthos · Impacts

2.1 Introduction

The northern coast of Brazil, the so-called Brazilian North Coast (BNC), corresponding to the Amazonia Ecoregion of Spalding et al. (2007), extends from Cabo Orange in Amapá (AP, Amapá State) (05°N, 051°W) to Ponta do Tubarão in Maranhão (MA, Maranhão State) (04°S, 043°W) (Fig. 2.1) and encompasses dozens of estuaries and thousands of km^2 of mangrove forests (Souza Filho et al. 2009). This coastal region is strongly influenced by the Amazon river, which has an average discharge of approximately 1.73×10^5 m^3 s^{-1}, which represents about 30% of the freshwater entering the Atlantic Ocean and is responsible for reducing the surface salinity of the water in this area and for an annual input of approximately 754×10^6 tons of sediments (Martinez et al. 2009; Wisser et al. 2010).

The BNC is dominated by semidiurnal macrotides with tidal range of 4–11 m (DHN 2018). The climate is tropical humid with a mean annual temperature of 27.7 ± 1.1 °C and a mean annual precipitation (30-year record) from 2300 to 2800 mm (Moraes et al. 2005). Precipitation varies considerably over the year, with a well-marked rainy season from January to July, with mean total monthly rainfall of around 350 mm; and a dry season from August to December, with total rainfall of less than 50 mm (Moraes et al. 2005).

R. A. M. da Costa
Laboratório de Plâncton e Cultivo de Microalgas, Instituto de Estudos Costeiros, Universidade Federal do Pará, Bragança, Pará, Brazil
e-mail: raucosta@ufpa.br; rauquirio@pq.cnpq.br

N. E. Asp
Instituto de Estudos Costeiros, Universidade Federal do Pará, Bragança, Pará, Brazil
e-mail: nilsasp@ufpa.br

C. R. Beasley
Laboratório de Conservação da Biodiversidade e das Águas, Instituto de Estudos Costeiros – Universidade Federal do Pará, Bragança, PA, Brazil
e-mail: beasley@ufpa.br

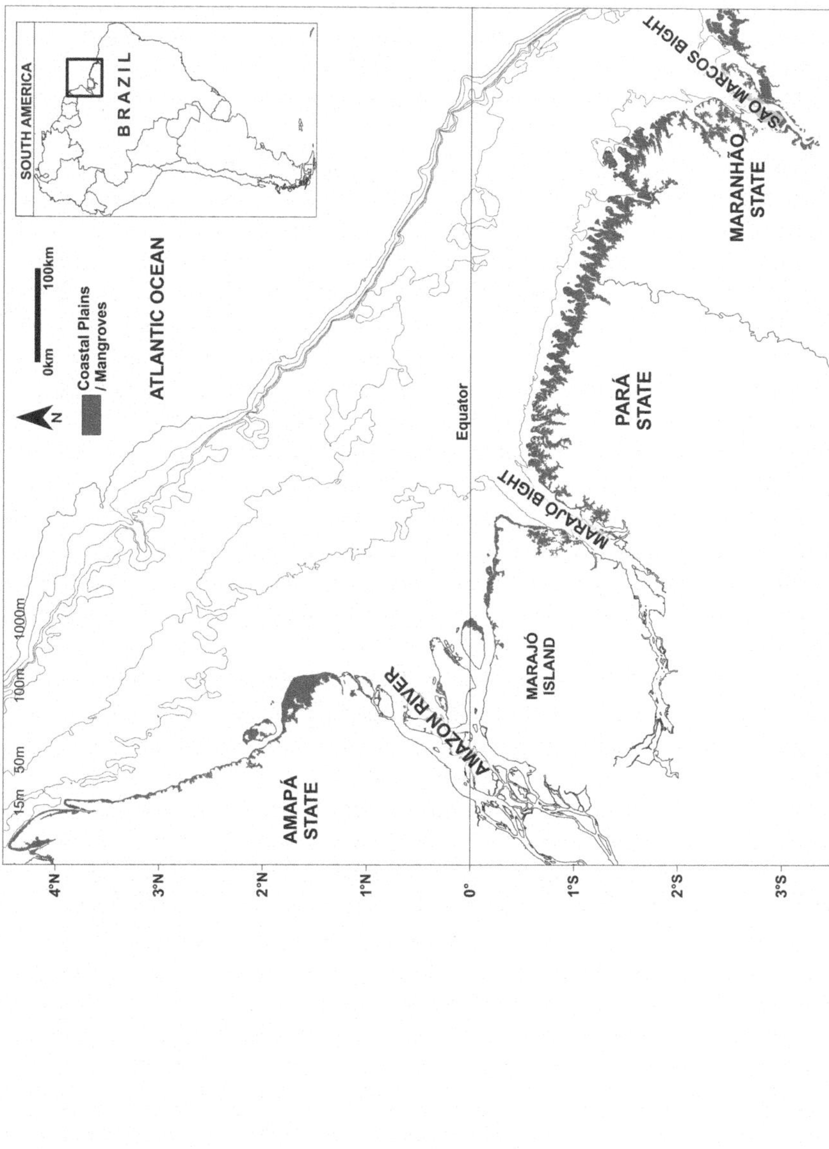

Fig. 2.1 Map of the Brazilian North Coast (BNC)

The estuaries of the Brazilian North Coast are composed of mangrove and muddy tidal flats with numerous narrow inlets forming estuaries and bays subject to strong coastal dynamics on the eastern side of the mouth of the Amazon (Souza Filho et al. 2009; Lara et al. 2010; Isaac and Ferrari 2017). Over 80% of Brazil's mangroves are found in the BNC (Kjerfve and Lacerda 1993). Mangroves can extend more than 40 km landward following the course of numerous small estuaries and bays (Souza Filho et al. 2009). Along the coast of the states of Pará (PA) and Maranhão, mangroves form a continuous belt of about 7423.6 km^2 (480 km long), representing the most extensive continuous belt of mangroves globally (Nascimento et al. 2013). The vegetation is dominated by mangrove, mainly *Rhizophora*, *Avicennia*, and *Laguncularia* (Menezes et al. 2008), with a seaward fringe of salt marsh (Braga et al. 2011) and large areas of muddy to sandy sediments exposed at low tide (Souza Filho et al. 2009).

The Brazilian North Coast has been visited by explorers and researchers since 1500, when Vicente Yañez-Pinzón explored the river Oiapoque and the mouth of the Amazon river, which he called "Mar dulce" (freshwater sea). Between 1541 and 1542 the expedition led by Francisco Orellana discovered the lands between the actual Amazon and Pará states, describing the potential riches and the Amazons (tribe of women warriors) who lived there. Closing the cycle of expeditions and aiming to take possession of the newly discovered territories, the Pedro Teixeira expedition (1637–1638) crossed the Amazon river from Cametá in Pará to Quito in Peru, beginning the colonization of the Brazilian Amazon (Corrêa 1997).

In the seventeenth century, several expeditions with a geographic and scientific interest visited the coastal zone of the states of Maranhão, Pará, and Amapá aiming at the demarcation of the territories conquered there by the Portuguese Crown as well as the observation and cataloging of the fauna and flora. Among these, one deserves special mention, the "Comissões Demarcadoras de Limites" (1754, 1780) that crossed the Amazon region producing maps with a high level of accuracy and Alexandre Rodrigues Ferreira's "Philosophical Journey" (1783–1792), which complemented the latter and had a more scientific focus. In both expeditions, the representation of landscapes reflects political and military objectives of Portugal. Some descriptions are predominantly naturalistic (zoological, botanical, or anthropological) and others essentially geographic (Sanjad and Pataca 2007).

Several artistic and scientific expeditions visited the BNC making records on both the physical environment and biota. Between 1819 and 1820, the naturalists Johann Spix (a botanist) and Karl Martius (a zoologist), members of the Scientific Mission of Natural History, known as the Austrian Mission, crossed the Amazon River from Belém to the border with Peru and Colombia. This expedition resulted in the works "Reise in Brasilien," published in 3 volumes (1823, 1828, and 1831), and "Flora brasiliensis," composed of 15 volumes published between 1840 and 1906. The German Eduard Poeppig crossed the Amazon River, from Peru to its mouth and published the botanical and zoological results of his trip in the books "Reise in Chile, Peru und auf dem Amazonenstrome: während der Jahre 1827–1832" and "Nova genera ac species plantarum, quas in regno Chilensi Peruviano et in terra Amazonica: annis MDCCCXXVII ad MDCCCXXXII." In the nineteenth century, the French Francis de Castelnau traveled 4800 km in the Amazon river from Peru to its mouth in the Atlantic Ocean making botanical, zoological, geographic, and geological records. The results of Castelnau's expedition were published in 15 volumes under the title "Expédition dans les parties centrales de l'Amérique du Sud: de Rio de Janeiro à Lima, et de Lima au Pará" (Sanjad and Pataca 2007).

From the early twentieth century until the 1980s, research in the Amazon was mostly directed towards rainforests (Eastern Amazonia) and there were very few coastal zone studies. Although in 1966 there was information on the geology and formation of the northern coast of Brazil (Ackermann 1966), and in the 1970s the RADAMBRASIL Project (Amazonas Radar Project after 1975, Radambrasil Project) presented data on geology, geomorphology, soil, vegetation, and land use in the BNC (Brasil 1973, 1974a, b), Mendes (2005) points out that until 1980 the estuaries of the Brazilian North Coast remained almost unstudied from a geological and sedimentological point of view. Similarly, Damazio et al. (1989) and Lana et al. (1996) pointed out that knowledge of biological communities from the Amazonian coastal zone was incipient (although increasing in number) and fragmented until the 1980s.

In general, systematic and multidisciplinary studies on Amazonian estuaries began (or grew substantially) from the 1980s onwards when national and international research projects were started. These studies brought together researchers from several Amazonian as well as other Brazilian and foreign institutions. Probably the first research initiative focusing on the Amazonian estuaries (although only considering geology and geomorphology) was the PROMAR project (Research Program in Marine Sciences—1982/1990) of the Center of Geosciences of the Federal University of Pará. Subsequently, several research projects were carried out, as: Atlantic Coast (1983/1989); GLOBESAR 2—Application of RADARSAT in the NE of Pará (coastal plain of Bragança) and Golfão Maranhense (1998/2000); MADAM (Mangrove Dynamics and Management—1995/2005); RECOS Millennium Institutes (Use and Appropriation of Coastal Resources—2000/2004); PIATAM mar (Potential Environmental Impacts of Transport of Petroleum and By-products in the Amazon Coastal Zone—2004/2008).

Following the general trend, studies of benthic communities from the BNC were also intensified in the 1980s. Between the 1960s and 2000s, research programs focused on the estuaries of the island of São Luis (MA) and the Ajuruteua Peninsula (PA, Pará State) and aiming at producing lists of species of macroinvertebrates, particularly molluscs and crustaceans (Coelho 1967/69; Matthews et al. 1977; Coelho and Ramos-Porto 1980; Costa and Silva-Mello 1983; Ferreira 1989; Reis 1995). From the 2000s, the number of studies grew exponentially (in both number of study areas and approaches), including meiofauna, with research being carried out in several estuaries, and covering subjects as diverse as spatial and temporal variability (Rosa Filho et al. 2011; Silva et al. 2011; Braga et al. 2013; Ataíde et al. 2014; Rodrigues et al. 2016), effects of anthropogenic activities (Paula et al. 2006; Aviz et al. 2012; Venekey and Melo 2016), genetics (Santos et al. 2005; Melo et al. 2010, 2013); population dynamics (Koch et al. 2005; Lucena-Frédou et al. 2012; Rosa

Fig. 2.2 Tidal range (m) recorded during spring tides by Centro de Hidrografia e Navegação (Brazilian Navy): Barra Norte do Rio Amazonas (BNRA), Porto de Santana (PS), Atracadouro de Breves (AB), Porto Vila do Conde (PVC), Porto de Belém (PB), Ilha do Mosqueiro (IM), Ilhas dos Guarás (IG), Fundeadouro de Salinópolis (FS), Porto de Itaqui (PI), Terminal do Alumar (TA), Terminal da Ponta da Madeira (TPM), São Luís (SL), Porto de Tutóia (PT), and by field campaigns: Paracauari (Par), Murubira (Mur), Colares (Col), Marudá (Mar), Princesa (Pri), Taperaçu (Tap), Ajuruteua (Aju), Caeté (Cae), and Gurupi (Gur). Micro: microtides, Meso: mesotides, Macro: macrotides, Hyper: Hypertides

Filho et al. 2013; Aviz et al. 2016), and trophic ecology of some coastal macroinvertebrates (Koch and Wolff 2002; Nordhaus et al. 2006; Koch and Nordhaus 2010).

2.2 General Physical and Chemical Characteristics

The main hydrodynamic driver along the Brazilian North Coast is the semidiurnal macrotidal regime (Fig. 2.2). The distance between the African and South American continents results in a condition of near-resonance, which enhances the tidal amplitude in the western Atlantic (Cartwright et al. 1991). Usually the tidal range is twice as large during spring tides. In addition, there is substantial tidal deformation as it propagates along the estuaries. A strong asymmetry is observed, progressively extending the ebb and shortening the flood phase upstream. A slight amplification usually occurs at the estuarine funnels, but substantial attenuation is observed by water levels between the lower- and upper-estuary and the major reduction in the tidal range is related primarily to the extensive outflow into the marginal mangroves and the friction produced by the partial dissipation of the tidal energy. Figure 2.3 shows the tides propagating during spring tides in Taperaçu estuary (PA).

Despite the general concept of stable and high air temperature in equatorial regions, rainfall and, consequently, riverine discharge vary substantially along the year; the Intertropical Convergence Zone (ITCZ) is the primary factor determining the seasonal climatic pattern on the Amazonian coast (Marengo 1995). Based on

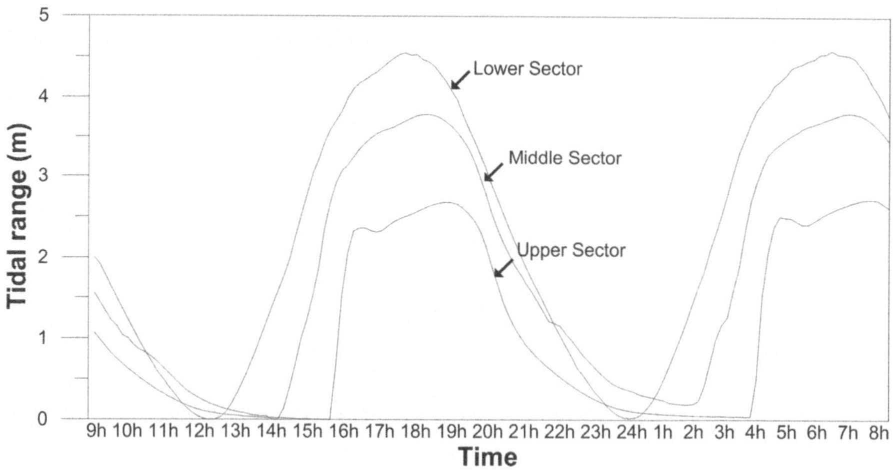

Fig. 2.3 Tides propagating during spring tides into Taperaçu (PA) estuary from lower- to upper-estuary. Water level oscillations were recorded every 10 min using a tide data logger (TWR 2050)

local historical records, there is a typically distinct rainy season between December and June when the ITCZ shifts to the southern hemisphere (near 10°S), causing an increase in rainfall along the Amazon coastal zone. Drier conditions predominate after July when the ITCZ shifts to the northern hemisphere. During these months, accumulated monthly rainfall (<50 mm) decreases and may often be close to zero, especially in September, October, and November along the eastern sector of the Amazonian coast. Even close to the Amazon river mouth and at the northwest sector, rainfall drops down to about one quarter of that observed during the rainy season (Fig. 2.4, INMET 2017).

The local tidal range also plays an important role in the determination of the dissolved nutrient profile in Amazonian estuaries, due to the flooding of extensive areas of mangrove during each tidal cycle. Nevertheless, a remarkable difference in conditions during spring and neap tides comes from the fact that mangroves are only effectively flooded during spring tides, which changes the entire hydrodynamic and biogeochemical scenario on a weekly basis (Dittmar and Lara 2001). In fact, in mangrove areas, the high tide may also reach certain nutrient-rich environments, such as those between the Caeté and Taperaçu rivers (Fig. 2.5), which are connected by tidal flooding of the Taici Creek (Asp et al. 2012; Araújo and Asp 2013), and the flooding of adjacent mangroves and wetland areas leads to an additional input of water richer in dissolved nutrients and chlorophyll-*a*, mainly in the upper-estuary.

Furthermore, seasonal riverine discharge pattern is other major force controlling oscillations in dissolved nutrient and chlorophyll-*a* concentrations (Santos et al. 2008; Pamplona et al. 2013). With the overall reduction in rainfall and consequently, riverine discharge during the dry season, the waters of the entire Amazon coastal zone are characterized by a tendency towards reduced concentrations of dissolved nutrients (DeMaster and Pope 1996; Geyer et al. 1996).

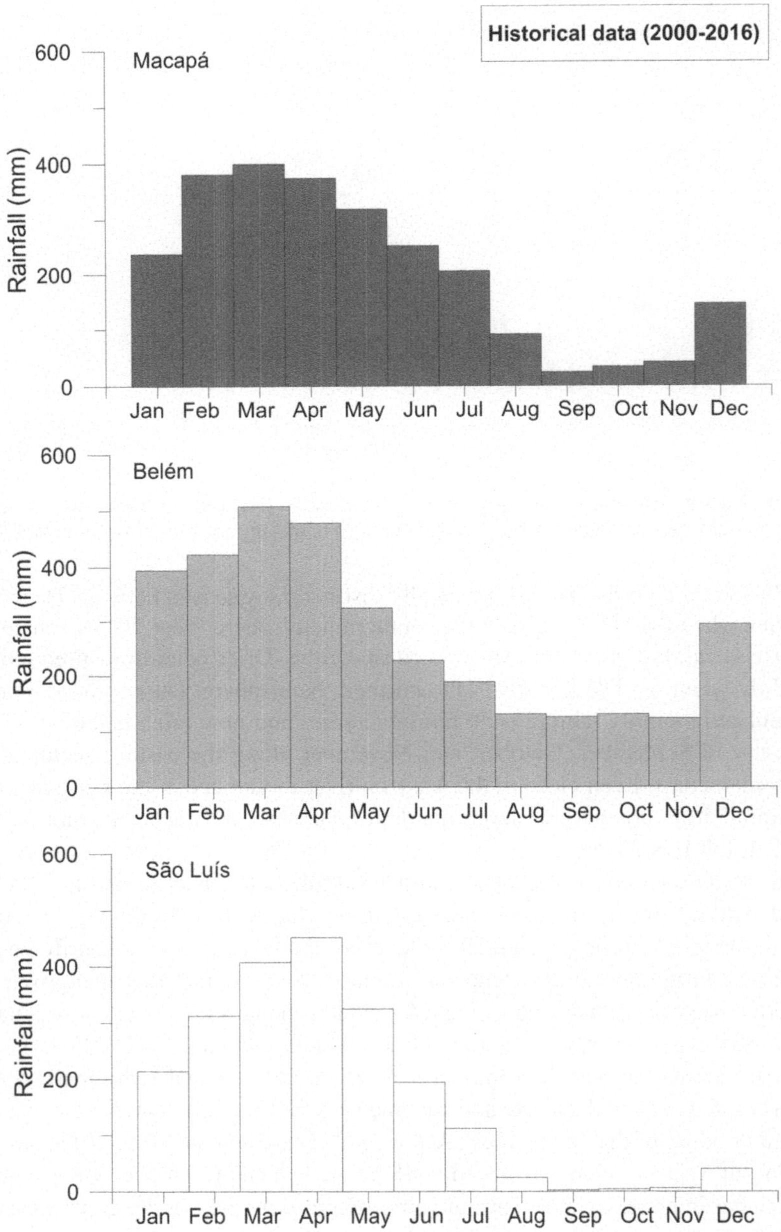

Fig. 2.4 Historical data (2000–2016) of mean rainfall per month (mm) in Macapá (AP, Amapá State), Belém (PA, Pará State), and São Luís (MA, Maranhão State) from the Brazilian Meteorology Institute (INMET)

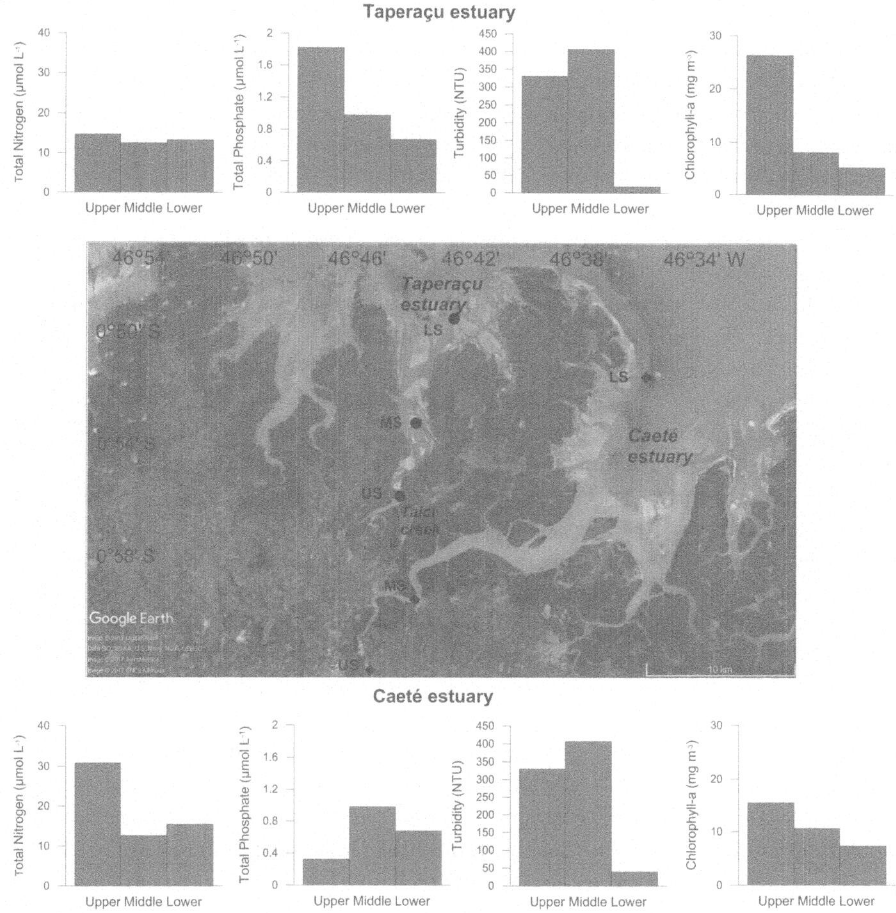

Fig. 2.5 Hydrological data (mean values) of Total Nitrogen, Total Phosphate, Turbidity, and Chlorophyll-a along Caeté and Taperaçu estuaries (PA). US: Upper-estuary, MS: Middle-estuary, LS: Lower-estuary

Annually, water temperature is high and stable (normally between 25 and 29 °C) around BNC. During the dry season, riverine discharge is reduced, and a larger volume of seawater enters the estuaries, with salinity reaching values around 40 (Pereira et al. 2013) in Atlantic coastal waters and decrease to 5–10 (Costa et al. 2012) in Marajó bay (Fig. 2.5). During the rainy season, salinity values can fall to zero, depending on both precipitation and riverine discharge. High turbidity is recorded during the rainy season (high riverine discharge) but also in the dry season, when combined higher tidal cycles and strong tidal currents re-suspend the fine material from the bottom into the water column. During the rainy season the suspended sediment concentrations (SSC) often reaches 1 g L^{-3} inside estuaries, with values up to 3 g L^{-3} at the maximum turbidity, whereas SSC around 0.25 g L^{-3} are typical of the dry season (Asp et al. 2016). Despite their high turbidity, which is considered a major factor limiting primary productivity, estuarine Amazon waters typically present high phytoplankton biomass. Nevertheless, it is likely that the presence of microphytobenthos, washed out of adjacent mangroves and wetland areas, contributes to the increase of chlorophyll-*a* concentrations along the entire Amazonian coast (Pamplona et al. 2013; Pereira et al. 2013).

BNC estuarine waters are normally well-oxygenated as consequence of both high levels of photosynthetic activity and intense water-atmosphere interactions generated by strong hydrodynamics and winds, primarily during the dry season. As might be expected, these hydrodynamic interactions are most intense in the lower-estuaries, which is more exposed to the wind and breaking waves (common in some local estuaries). Conversely, dissolved oxygen (DO) concentrations mostly presented daily oscillations in the upper-estuary, with high values recorded during the daylight period, when the sunlight stimulates the production of phytoplankton, and low ones during the night, when phytoplankton production is minimal (Dittmar and Lara 2001).

In the Taperaçu estuary (PA), for example, the presence of orthophosphate and total phosphorous are clearly associated with the influence of marine waters and the re-suspension of sediments during the dry season, when salinity is higher, coinciding with the strongest winds and currents. Water turnover takes longer where energy is low, as in the upper-estuary, which benefits phosphorous cycling and raises its concentration in the water column. An opposite pattern is normally recorded for nitrogen compounds. In contrast with the nitrogen and phosphorus cycles, which involve biological remineralization, the production of silicates in aquatic environments is basically a physical-chemical process. The Amazon River discharges approximately 10% of the total amount of silicates of fluvial origin deposited into the world's oceans (DeMaster et al. 1986), being other local rivers on the Amazon coast are also important sources of this material (Pereira et al. 2010; Pamplona et al. 2013). In general, a clear relationship is also found between high silicate concentrations and the increase in chlorophyll-*a* (Pereira et al. 2012). Thus, siliceous phytoplankton (diatoms) constitute one of the most abundant phytoplankton groups in estuarine and marine Amazon waters (Matos et al. 2011; Pereira et al. 2013).

2.3 Benthic Fauna

2.3.1 Assemblage Composition

The BNC benthos is one of the least known faunas in Brazil, both in terms of taxonomy and ecological processes (Lana et al. 1996; MMA 2002). Most studies carried out in BNC estuaries are recent, starting in the 2000s, and concentrated in the state of Pará. The taxonomy of many faunal groups is still undetermined, with many organisms identified only at the generic or supra-generic levels, therefore requiring specialist review. Thus, the different levels of taxonomic resolution, as well as sampling effort adopted in studies, must be considered when discussing patterns of benthic faunal biodiversity and distribution in the region. Mangroves are the best-known BNC coastal environment and intertidal areas are better sampled than sub-tidal zones. Most benthic studies generally comprise faunal inventories and present abundance and biomass data that are both spatially and temporally limited in scope, with little information about population dynamics, biomass, and secondary production, except for species of economic interest. The macrofauna is the most studied group and is dominated by crustaceans, annelids, and mollusks.

The mangrove habitat mosaic is responsible for high benthic production both within the mangrove and along the coast (Wolff et al. 2000; Isaac and Ferrari 2017). This is caused by a tightly coupled nutrient cycling (Wolff et al. 2000), along with a macrotidal regime (Souza Filho et al. 2006) and strong seasonal variation in freshwater input (Araújo and Asp 2013) and consequently temporally variable coastal nutrient input via mangroves (Dittmar and Lara 2001). However, the species richness of the BNC benthic faunal assemblages is relatively poor (Bernardino et al. 2015). General seasonal and spatial variation in benthic abundance and composition along the North Coast accompanies the environmental variability described above (Rosa Filho et al. 2009; Silva et al. 2011; Aviz et al. 2012; Melo et al. 2013; Andrade et al. 2014).

2.3.1.1 Mudflats

Although apparently homogeneous, large areas of muddy and/or sandy sediment may have significant variation in topography, with temporary pools of tidal water (Silva Júnior and El-Robrini 2001; Souza Filho et al. 2009; Krause 2010). Muddy flats are crossed by tidal channels of varying width and depth, which may have large areas of sand and/or gravel just beneath the muddy surface. A large number of man-made fishing structures, called "currais" (fixed tidal trap) (Barletta et al. 1998; Krause 2010), is set up along mudflats, thus creating a variety of habitats and micro-habitats for the benthic fauna. The epibenthic fauna in mudflats is dominated by gastropods and crabs. Large gastropods such as *Pugilina morio* and *Thais* sp. are detritivores and appear to require structurally solid refuges, being common around "currais," where they feed on the remains of fish (Beasley et al. 2005), and/or near fallen mangrove trees. Other smaller gastropods such as *Nassarius vibex* occur on the muddy surface and are especially common where tidal water is draining away or in the tidal creek itself. Various types of fiddler crabs (Koch et al. 2005) occur in open areas and exposed creeks, whereas swimming crabs such as *Callinectes danae* and *C. bocourti* are common in lower tidal areas and tidal creeks, where they are caught by artisanal fishermen (Bentes et al. 2013). In low-lying muddy flats, large areas may be covered with mats of the mussel *Mytella charruana*, which is heavily exploited in the region (Carranza et al. 2009; Santos et al. 2010).

The estuarine mudflat infauna is mainly composed of bivalves, crustaceans, and annelids. The clam *Anomalocardia flexuosa* is common in muddy areas with underlying gravel and is also harvested for food (Silva et al. 2014). Crustaceans such as the thalassinidean shrimps, which inhabit deeper parts of sandy/muddy sediment, are important bioturbators due to the deep and extensive burrows they create (Felder 2001). The muddy flats infauna is dominated by polychaetes with overall diversity lower than in the sediment of the adjacent mangrove vegetation (Rosa Filho et al. 2006).

2.3.1.2 Salt Marshes

Along the BNC, salt marshes composed of *Spartina alterniflora* colonize bare sediment on the seaward edge of the mangrove, forming a narrow strip of patches of salt marsh, which vary in area between 22 and 3700 m^2 (Braga et al. 2013). *Spartina* salt marshes harbor a wide variety of taxa, with dominance of polychaetes, such as *Laeonereis* sp., *Isolda* sp., *Namalycastis* sp., *Capitella* sp., *Heteromastus* sp., isopods and the gastropod *Neritina virginea* (Braga et al. 2009). Although no vertical zonation in macrofaunal abundance and composition was found (Braga et al. 2011), the number of individuals is positively correlated along the Pará coast with *Spartina* density, sediment water and organic matter contents, salinity and *Spartina* height (Braga et al. 2011). Higher macrofaunal abundance and diversity occur in the dry season, when the marine fauna is typically composed of malacostracid crustaceans (mostly tanaids and ostracods) and polychaetes (mostly Capitellidae), and higher water salinity and temperature, and sediment organic matter content occur (Braga et al. 2013). Along the Pará coast, abundance and numbers of macrofaunal taxa were associated with increasing pore water salinity in *Spartina* salt marshes, from west to east, coinciding with patterns of rainfall in the region (Braga et al. 2011).

2.3.1.3 Mangroves

Rhizophora is the most widely distributed mangrove tree genus. *R. mangle dominates* in estuaries more exposed to the ocean. The other two *Rhizophora* species, *R. racemosa* and *R. harrisonii,* occur in less saline estuaries. *Avicennia germinans* is more frequent in elevated and less inundated areas and *A. schaueriana* is recorded mainly near sandy beaches. *Laguncularia racemosa* occurs along the entire BNC in saline as well as in brackish water and mostly occupies forest edges, large gaps, riparian sites, and other open areas (Menezes et al. 2008).

The semi-terrestrial mangrove crab (*Ucides cordatus*), also called uçá-crab, is the most important component of the assemblage, with almost three times the biomass of other aquatic components in the mangrove (Wolff et al. 2000). This species is exploited for sale in markets both in Pará and along northeastern Brazil (Furtado-Junior et al. 2016). The most conspicuous members of the epifauna, besides *U. cordatus*, include predatory crabs such as *Callinectes* spp., *Eurytium limosum* and *Panopeus lacustris*, and fiddler crabs, *Uca* spp. (Sousa et al. 2015), which show distinct preferences for humidity, temperature, and organic matter contents among the mangrove forest, small creek, and large exposed creek habitats (Koch et al.

2005). *Ucides* is also ecologically important since it processes most of the mangrove detritus (Nordhaus et al. 2006) resulting in tightly coupled nutrient cycling among crabs, bacteria, and mangrove trees (Wolff et al. 2000). Other relevant components of the fauna include the boring bivalves (shipworms and pholadids), which occur in logs, the large gastropod *Thais* sp., which occur in creeks and around logs, infaunal and log-dwelling annelid worms, mainly polychaetes and oligochaetes, and shrimps and juvenile crustaceans in the tidal creeks (Wolff et al. 2000). The most common species of shipworm in the mangrove in this region is *Neoteredo reynei* (Santos Filho et al. 2008) and is harvested along the BNC and sold in markets and restaurants, as well as consumed locally. The shipworms *N. reynei* and *Bankia fimbriatula* occur in saltwater, whereas *Psiloteredo healdi* may be found in habitats with a strong freshwater influence (Santos et al. 2005). Mangrove oysters are found encrusting rocks in low intertidal areas and, although exploited extensively (Beasley et al. 2010), are increasingly cultivated for local and remote markets (Sampaio and Boulhosa 2007; Sampaio et al. 2017).

Mangrove trees are habitat for many small gastropods including *Littoraria angulifera* and *Melampus coffeus*, among others (Beasley et al. 2005). Decaying mangrove logs provide microhabitats for a diversity of macroinvertebrates. Aviz et al. (2009) recorded 31 taxa including Nemertea, Gastropods, Polychaetes, decapod crabs, amphipods, isopods, and a smaller number of insects. Andrade et al. (2014) found 85 macrofauna taxa in decaying mangrove logs in Pará, with a more marine fauna at an island location and a more terrestrial fauna at a peninsular site, as well as greater dominance and lower diversity in the wet season, like the pattern described for the sediment macroinfauna.

2.4 Spatial and Temporal Patterns of Faunal Diversity, Abundance, and Biomass

The benthic fauna of Amazonian estuaries is relatively poor in species, particularly in intertidal areas. Spatial and temporal variation of the benthos are mainly determined by freshwater discharge from dozens of rivers that flow to the coast, associated with marked seasonal variation in rainfall, as well as the macrotidal regime. In estuaries near the mouth of the Amazon River, which have muddy bottoms and low salinity, the benthic fauna is composed of euryhaline species, as well as freshwater or terrestrial animals (Fig. 2.6). Far from the influence of the large rivers, other types of substrate become more frequent and the fauna is of typically marine origin (Almeida 2008; Gomes 2008; Aviz et al. 2012; Rosa Filho and Aviz 2013). Strong salinity gradients are observed along estuaries. Communities typical of freshwater (e.g., dominated by *Nephtys fluviatilis* and tubificid oligochaetes) occur in the upper estuarine regions, a brackish-water fauna in the middle-estuary, and a more diverse and abundant marine fauna in the lower-estuary (Silva et al. 2011; Aviz et al. 2012).

Temporally, in most estuaries, there is wide variation in macrofaunal abundance and composition between the peak wet (February–April) and dry (September–

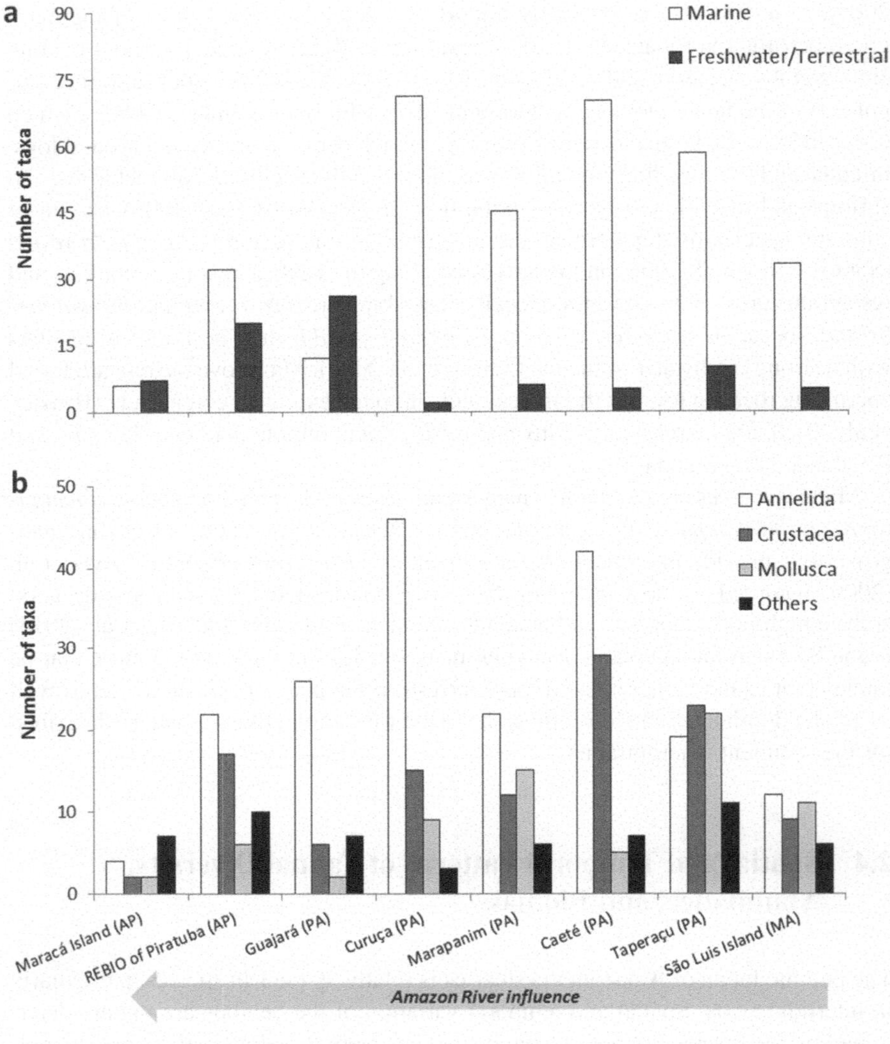

Fig. 2.6 Relative abundance and number of taxa of macrobenthic fauna in Amazonian estuaries by (**a**) habitat and (**b**) Phylum. Data source: Maracá Island (AP, Amapá state) (Fernandes 2003), REBIO Piratuba (AP) (Almeida 2008), Guajará (PA, Pará state) (Rosa Filho and Aviz 2013), Curuçá (PA) (Vasconcelos 2006), Caeté (PA) (Silva et al. 2011), Taperuçu (PA) (Braga et al. 2011, 2013), São Luis Island (MA, Maranhão state) (Rebelo-Mochel 1997; Oliveira and Mochel 1999)

November) periods of the year (Silva et al. 2011; Melo et al. 2013; Pamplona et al. 2013). In more saline estuaries, highest richness and abundance occur in the dry season (Amazonian summer—monthly accumulated rainfall <50 mm), since the significant decrease in salinity in the rainy season causes physiological stress to marine organisms. Estuarine areas permanently meso- to oligohaline, where freshwater species or those tolerant to low salinity are dominant, may present greater richness and abundance in the rainy season (Almeida 2008; Rosa Filho and Aviz 2013).

The most economically important benthic species are the uçá-crab (*Ucides cordatus*) and the mussels of the genus *Mytella*. Estimates of the density (based on number of burrows) and biomass of *U. cordatus* vary widely along the coast. In Amapá state, values vary from 1.08 ind m^{-2} (Macapá, Calçoene, and Oiapoque) (Amaral et al. 2014) to 4.95 ind m^{-2} (23.2 t ha^{-1}; Maracá Island) and 6.51 ind m^{-2} (22.3 t ha^{-1}; Vila de Sucuriju) (Fernandes and Carvalho 2007). In Pará, values are in a lower range and vary between 1.02 ind m^{-2} (1.5 t ha^{-1}; Vizeu; Cunha 2016), 1.21–1.36 ind m^{-2} (1.2 t ha^{-1}; Quatipú; Silva 2008), 1.48 ind m^{-2} (10.5 t ha^{-1}; Curuçá; Perote 2010), and 1.65 ind m^{-2} (1.4 t ha^{-1}; Furo Grandeç Bragança; Diele 2000). In Maranhão, values vary between 2.9 ind m^{-2} (3.5 t ha^{-1}; Rio dos Cachorros) and 3.0 ind m^{-2} (4.5–5.1 t ha^{-1}; Reentrâncias Maranhenses) (Castro 1986; Paiva 1997). Variation in densities has been associated with the topography and flooding frequency by tides, vegetation characteristics, and harvesting pressure. In general, the highest values of *U. cordatus* density and biomass occur in well-preserved mangrove areas and where harvesting pressure is lower.

Abundance and production data for *Mytella* are very scarce. In the Urumajó estuary (PA), density and biomass of *Mytella guyanensis* are highly variable throughout the year (Santos et al. 2010). Densities of *M. guyanensis* may peak to 2500 ind m^{-2} at the end of the dry season, and decrease to just 175 ind m^{-2} when the rainy season begins. Likewise, biomass can range from 1 (rainy season) to 650 g m^{-2} (dry season). Densities of *M. falcata* at two sites of the Curuca estuary (MA) varied between 300 and 9000 ind m^{-2} (Fernandes et al. 1983). Greater mussel density may be related to higher salinity in Maranhão state, due to lower rainfall and riverine discharges, in comparison with Pará state.

2.4.1 Amapá Estuaries

The Amapá coast is considered the best conserved and least populated area of the Brazilian coast, and it is also the one with the least known benthic fauna. The macrofauna is not very diverse, due to the physiological stress resulting from the permanently low water salinity caused by the high freshwater discharge from the Amazon and several other rivers to the coast. In addition, the high relief of the coast prevents the inundation of the mangrove forests by the tides (resulting in a low flooding frequency) and the action of waves and currents cause strong erosion of the coastline.

In the estuarine portion of Maracá Island, 15 macrofaunal taxa were identified which have distinct distributions in the mangrove forest, according to the flooding frequency and the dominant tree species (Fernandes 2003). The highest macrofaunal diversity is recorded in *Avicennia germinans* and *Laguncularia* forests that are only flooded during spring tides, the tanaid *Halmyrapseudes spaansi* being the dominant species (~96% of total individuals, 286 ind m^{-2}). In *Rhizophora harrisonii* forests, under drier conditions, densities are very low and only insects (Carabidae, Diptera e Collembola) occur (Fernandes 2003).

Strong salinity gradients occur along the coastal region of the Piratuba Lake Biological Reserve (REBIO Piratuba—AP), which encompasses floodplain lakes (in coastal floodplain areas) and mangroves. These water salinity gradients are accompanied by changes in the benthic fauna, which varies from freshwater to estuarine-marine species (Almeida 2008). A total of 54 taxa were identified at REBIO Piratuba. In the coastal lakes, 32 macrobenthic taxa were recorded (mean density of 668 ind m^{-2}), with dominance of the gastropod *Heleobia* sp. (36.6% of total individuals), Chironominae (15.6%), and Polymitarcidae (11.4%). In mangroves, the mean density was 1353 ind m^{-2}, and *Mediomastus californiensis* (46.8% of total individuals), Ostracoda (20%), and *Nephtys fluviatilis* (11.9%) were numerically dominant. In both coastal lakes and mangroves, species richness and abundance are higher in the dry season (Almeida 2008).

2.4.2 Pará Estuaries

Although along the Pará coast there are dozens of estuaries, only a few estuarine areas have been studied. Most studies were carried out in the Bragança Coastal Plain. In the Guajara estuary, where oligohaline waters occur throughout the year, the benthic fauna is abundant (>1000 ind m^{-2}) but not highly diverse (22–39 taxa), being composed mainly of annelids, which represent 99% of total abundance and include *Namalycastis abiuma*, *Nephtys fluviatilis*, *Mediomastus* sp., *Dero* (*Aulophorus*) sp., *Limnodrilus hoffmeisteri*, and *Paranadrilus desenlei* (Rosa Filho and Aviz 2013). Polychaetes are dominant in areas closer to the estuary mouth, whereas oligochaetes are more abundant in the middle- and upper-estuary, expanding their distribution to the lower-estuary in the rainy season (Aviz et al. 2012; Rosa Filho and Aviz 2013). In the Curuçá River estuary, the macrofauna is composed of 73 taxa, with annelids representing 95% of the fauna in the rainy season and 65% in the dry season. The tanaid *Halmyrapseudes spaansi* is the most abundant species accounting for 32% of the total individuals. Temporal variation in benthic macroinfaunal communities is not significant, although in the dry season, higher species richness and abundance occur (Vasconcelos 2006).

The benthic macroinfauna of intertidal mangroves, mudflats, and salt marshes of the Ajuruteua Peninsula is dominated by annelids, which account for more than 90% of the total abundance (Acheampong 2001; Rosa Filho et al. 2006; Braga et al. 2013). The epifauna is composed mainly of barnacles (*Fistulobalanus citerosum*), bivalves (*Crassostrea gasar*, *Mytella charruana* (syn. *falcata*)), and decapods (Koch and Wolff 2002; Beasley et al. 2010). Abundance and species richness vary enormously among areas. In unvegetated intertidal areas of the Caeté estuary, Rosa Filho et al. (2006) recorded values ranging from 2625 ind m^{-2} and 3 taxa, to 96,625 ind m^{-2} and 16 taxa (Rosa Filho et al. 2006). In vegetated intertidal areas, 17–27 taxa were found in salt marshes (Braga et al. 2011, 2013) and 23 taxa in mangroves (Braga et al. 2013). In the subtidal areas of the Caeté River, 83 macroinfaunal taxa were distributed along a salinity gradient (Silva et al. 2011). In the upper

Caeté estuary (salinity 0–9) richness and density are low and euryhaline annelids (*Namalycastis terrestris*, *Nephtys fluviatilis* and Tubificidae) dominate. In the middle (salinity 10–20) and lower (salinity 17–28) Caeté estuary, richness and density increase and, although polychaetes are still dominant, there is a shift towards dominance of species of marine origin (*Mediomastus* sp., *Nephtys simoni*, *Sigambra grubii* and *Armandia* sp.). Seasonally, 15 taxa were restricted to the rainy season, 21 to the dry season, and 47 were common to both seasons. Densities are also higher in the dry season (1308 ± 288 SE ind m^{-2}) than in the rainy season (679 ± 145 SE ind m^{-2}) (Silva et al. 2011).

A total of 29 crab species (aquatic and semi-terrestrial) are known from the Caeté estuary, which have marked vertical zonation (Diele et al. 2010). *Cardisoma guanhumi* is the only supratidal species and occurs in salt marsh vegetation. *Aratus pisonii*, *Armases rubripes*, *Sesarma rectum*, *Uca rapax*, *U. vocator*, and *Ucides cordatus* dominate the high intertidal zone of mangroves. For instance, *Eurytium limosum*, *Panopeus lacustris*, *Pachygrapsus gracilis*, *S. curacaoense*, *U. thayeri*, *U. cumulanta*, and *U. maracoani* are found mostly in the mid- and low-intertidal. In the subtidal zone, *Arenaeus cribrarius*, *Callinectes bocourti*, and *C. danae* are commonly found. A total of 30 species of bivalve and 19 species of gastropods were identified from the Ajuruteua Peninsula (Beasley et al. 2005). Of these, 14 occur only in mangroves, 16 in estuarine sandy beaches, and 16 in other muddy-sandy substrates; only three species of *Thaisella* are common to the three environments (Beasley et al. 2005). *Littoraria angulifera*, *Melampus coffeus*, and *Thais trinitatensis* (Beasley et al. 2005) are frequently encountered on leaves, trunk, and branches of *Rhizophora mangle*. The highest densities of molluscs are generally recorded in the dry season (e.g., *Anomalocardia brasiliana* and *Mytella guyanensis*), although some have highest population densities in the rainy season (e.g., *Tagelus plebeius*) (Beasley et al. 2005).

The salt marsh macrofauna from nine estuaries along the Pará coast was composed of 110 taxa, including 32 annelids, 29 molluscs, 24 crustaceans, and 19 insects (Braga et al. 2009, 2013). This fauna presents a lower diversity, when compared to other salt marshes in Brazil and around the world, although the taxonomic composition is similar (Braga et al. 2011). The fauna of salt marshes is more abundant than in other unvegetated estuarine habitats, such as tidal channels and beaches, with richness and density being positively related to the density and height of the vegetation (Braga et al. 2013). Taxon richness and density varied from six to 36 species and from 190 ind m^{-2} to 105,443 ind m^{-2} in Marapanim and Salinópolis, respectively. The higher densities in Salinópolis are due to the abundance of the tanaid *Halmyrapseudes spaansi*, which comprised 78% of the total abundance. A significant seasonal variation in salt marsh macrofaunal structure occurs, with a less diverse and abundant community in the rainy season.

Along the Pará coast, the macrofauna associated with galleries excavated by shipworms (Teredinidae) in mangrove trunks is generally well known. Dominated by crustaceans, this fauna is richer and more abundant than that found in mangroves and mudflats (Ferreira 1989; Aviz et al. 2009; Andrade et al. 2014). Ferreira (1989) recorded 40 taxa in galleries of teredinids (6 Mollusca, 7 Annelida, 27 Arthropoda) in nine estuaries, where the number of taxa tended to increase with salinity. There was a clear pattern of species zonation inside the galleries with Brachyura generally found near the gallery exit, Amphipoda occupying the innermost areas, and Gastropoda in intermediate positions. In estuaries with a strong freshwater influence, species richness is higher in the dry season than in the wet season, whereas in the more saline estuaries the opposite occurs. Aviz et al. (2009) found 31 macrofaunal species inhabiting Teredinidae galleries in São Caetano de Odivelas. Crustaceans were dominant (53.4% of total; particularly *Armases angustipes* and *Cirolana* sp.), followed by molluscs (27.5% of total) and polychaetes (10.8% of total—especially *Namalycastis abiuma*). These authors also recorded an increase in the richness and density greater salinity. Andrade et al. (2014) identified 83 taxa at two mangrove sites with different salinities and, like Ferreira (1989), macrofaunal abundance in galleries was higher during the rainy season, especially at the site with higher salinity.

In northeastern Pará, relatively small areas with rocky outcrops interspersed with mud or sand are occasionally found in estuaries. These rocky outcrops are areas of high species density and diversity, being one of the few types of consolidated substrates in the BNC. In rocky intertidal areas in Areuá Island, 85 macrofaunal taxa, both mobile and sedentary, were dominated by 32 polychaetes, 24 crustaceans, and 21 molluscs (Morais and Lee 2014). The crustaceans *Petrolisthes armatus*, *Parhyale* sp. and *Dynamenella tropica* are the most frequent and abundant species. Mean macrofaunal density and biomass in this area are 2690 ± 334 ind m^{-2} and 55.870 ± 8.139 g m^{-2}. Richness and biomass increase towards the subtidal zone; richness and density are higher during the rainy season, with no significant variation in biomass (Morais and Lee 2014).

Benthic communities in *Sabellaria wilsoni* (Polychaeta) reefs in Algodoal Island are probably the most diverse macrofauna along the Amazonian coast (Aviz 2015; Aviz et al. 2016). A total of 168 taxa were recorded, including 53 annelids, 46 molluscs, and 47 crustaceans, as well as other benthic groups such as Porifera, Sipuncula, Bryozoa, Pycnogonida, and Ophiuroidea, which are relatively rare along the BNC (Aviz 2015). Macrofauna densities in the reefs varied between 11,013 and 159,494 ind m^{-2}, whereas in the adjacent sediment, densities ranged from 127 to 1519 ind m^{-2} (Aviz 2015). In the rainy season, the density of the sabellarids is highest, whereas the associated macrofauna presents a lower density and richness. On the other hand, in the dry season, when the reefs are eroded and almost completely destroyed due to the increase in wave and wind action, the density of sabellarids decreases, and there is an increase in the abundance of associated macrofauna (Aviz 2015).

2.4.3 Maranhão Estuaries

In Maranhão state, almost all the information on benthic communities deals with faunal inventories carried out in the estuarine areas of the Island of São Luis. Mollusca is the best-known phylum, followed by Polychaeta. Matthews et al. (1977) present the largest list of mollusks from São Luis Island with 61 species identified (32 Bivalvia, 29 Gastropoda, and 1 Cephalopoda). In the infralittoral areas of the

São Marcos and São José bays, 38 species were registered (26 gastropods and 12 bivalves) (Silva-Mello and Costa 1993). Costa and Silva-Mello (1983/1984) identified 11 gastropod and 8 bivalve species from the Mosquitos Strait and Côco and Cachorros rivers in the intertidal zone of São Luis Island.

In the estuarine regions of São Luis Island, Ribeiro and Almeida (2014) reported the occurrence of 38 species (with dominance of the Nereididae polychaetes), citing, however, that the richness should be much greater, since most of the studies did not identify Polychaeta to species level. In Maranhão estuaries, 21 species of Crustacea occur, with dominance of Ocypodidae and Penaeidae (Coelho-Costa 2007; Sousa et al. 2015).

In Parnauaçú (São Luis Island) mangroves, 26 taxa of macrofauna were recorded (13 polychaetes, 5 molluscs, 5 insects, 1 polychaete, and unidentified species of nemertea and oligochaetes) (Oliveira and Mochel 1999; Mochel et al. 2001). The most abundant species were *Nereis oligohalina* (612 ind m^{-2}), *Isolda pulchella* (232 ind m^{-2}), *Notomastus lobatus* (120 ind m^{-2}), and *Perinereis vancaurica* (104 ind m^{-2}) (Mochel et al. 2001). Monteles et al. (2009) studied the molluscs and crustaceans exploited by artisanal fisheries in the Raposa river estuary and identified *Anomalocardia brasiliana*, *Phacoides* (= *Iphigenia*) *brasiliensis*, *Mytella falcata* [= *charruana*], *Crassostrea rhizophorae*, *Callinectes* sp., and *Ucides cordatus* as being those of greatest economic importance.

2.5 Assemblage Succession

Along the Amazon coast, one of the main successions involving benthic organisms occurs where bare sediment is colonized by salt marsh *Spartina* sp. followed by mangrove forest, similar to successions in other Brazilian mangrove areas (Fig. 2.7; Cunha-Lignon et al. 2009). Such changes may be relatively rapid. At the northern edge of Ajuruteua beach (Pará), this process occurred between 2009 and 2013 (Beasley pers. obs.). However, in 2014, deposition of sand completely covered the developing mangrove. Thus, substantial associated changes in drainage, nutrient dynamics, as well as in sediment type, may also occur. Changes in the benthic macrofauna may be subtle and may require use of several indicator variables (for example, diversity and dominance indices or abundances of specific indicator taxa) to detect changes (Tang et al. 2014). This is particularly important in the Brazilian North Coast, where although benthic composition and abundance change along a gradient from the mangrove down through *Spartina* salt marsh to the muddy beach and tidal flats, many taxa are found in more than one of these habitats.

Marques-Silva et al. (2006) and Beasley et al. (2010) presented the only two studies on settlement and succession of benthic invertebrates on wooden panels in Amazon estuaries. The authors followed the settlement of three encrusting epibenthic macrofaunal species (*Fistulobalanus citerosum*, *Crassostrea gasar*, and *Mytella falcata* (= *charruana*)) in two creeks (Furo do Meio and Furo do Café)

Fig. 2.7 Succession from sand to salt marsh and young mangrove at a northern Brazilian estuary (Chavascal—0°48′29.4″S, 46°37′53.3″W). (**a**) Location of Chavascal and position and direction view of photos taken (arrows). (**b**) Bare sandy sediment in 2008 with Chavascal tidal channel immediately before tree-line in background and left, and dune system on the right. (**c**) The same area in 2013 with muddy sediment in the foreground, predominance of salt marsh vegetation and young mangrove saplings. (**d**) Areas of dense young mangrove and salt marsh at Chavascal inundated by the tide, as seen from the top of the dune system. (**e**) The succession was reset in 2014 with a heavy deposition of sand over the developing vegetation. Image and photo credits: (**a**) © Openstreetmap contributors, 2017, (**b–d**) Colin Robert Beasley, (**e**) Daniela de Nazaré Torres de Barros

and one adjacent mangrove of the Ajuruteua Peninsula (PA). In both creeks, peak settlement of *F. citerosum* occurred during the wet season, whereas *C. gasar* settled mainly during the dry season. By contrast, settlement of *M. falcata* was generally low during the entire year. Overall, settler density was usually greater on the underside of panels close to the creek bottom. When comparing settlement between creeks and mangrove, there were significant differences in number of settlers of the three species between habitats with very low levels of settlement, or none, in the mangrove forest, due to low inundation frequencies in the latter (Beasley et al. 2010).

2.6 Processes Mediated by Benthic Organisms

Although the importance of benthic animals and plants as structuring forces in marine and estuarine environments has been known for a long time (Wild 1976; Mann 1982; Bouma et al. 2009), information on the role of benthic organisms in mediating biological, ecological, and geochemical processes in the estuaries of the Brazilian North Coast are still scarce. The few studies in Amazonian estuaries investigated the trophic ecology of epibenthic macrocrustaceans in Caeté mangroves (PA) (Koch and Wolff 2002; Nordhaus et al. 2006; Koch and Nordhaus 2010) and the role of *Sabellaria wilsoni* reefs in associated macrofaunal communities in Algodoal Island (PA; Fig. 2.8) (Ataíde et al. 2014; Aviz 2015).

In the Caeté mangroves (Pará), a total of 29 species of aquatic and terrestrial crabs were recorded and Wolff et al. (2000) stated that much of the energy and nutrient fluxes are either directly channeled through or dependent on these species. In this area, *Ucides cordatus* is the most important benthic leaf-consuming crab (Nordhaus et al. 2006), occurring at average densities of 1.7 individuals m^{-2} (Koch and Wolff 2002). *U. cordatus* has a small foraging radius (max. 1 m) and mainly collects mangrove litter near its burrow entrance, feeding on mangrove leaves (61.2 ± 17.5%), unidentified plant material and detritus (28.0 ± 17.0%), roots (4.9 ± 6.3%), sediment (3.3 ± 3.4%), bark (2.5 ± 4.1%), and animal remains (0.1 ± 0.4%) (Nordhaus et al. 2006). Most mangrove litter and propagules (81.15%) are eaten by *U. cordatus* (Nordhaus et al. 2006). A large part of the litter and propagule production (43.51% in a *R. mangle*-dominated forest stand) is returned to the sediment as finely shredded detritus through feces (7.13 t dm ha^{-1} $year^{-1}$) (Nordhaus and Wolff 2007). In turn, crabs' feces have a low C/N ratio and provide an excellent growth medium for microorganisms (Koch and Nordhaus 2010).

Koch and Wolff (2002) hypothesized that interactions among mangrove trees, crabs, and bacteria form a positive feedback, where increased activity of any participant also benefits either directly or indirectly the activity of all others. They also postulated a positive feedback effect on primary production, since (1) nutrients and energy are retained, (2) nutrient remineralization is enhanced, and (3) the soil is aerated by the burrowing activity of the crabs, decreasing anoxic conditions and production of phytotoxins in the sediment. Although seemingly well established, the

Fig. 2.8 *Sabellaria wilsoni* reefs in Algodoal Island (Pará State). (**a**) Platform shape and (**b**) ball shape

feedback proposed by Koch and Wolf (2007) was not confirmed by Püllmans et al. (2016) who, in the same area, found that none of the measured abiotic sediment variables (salinity, organic matter content, reduction potential, CO_2 efflux rates from the sediment) and biotic variables (leaf production estimated by the stipule fall rate) differed between treatments (crab removal, experimental control and control). It is important to point out that the experiment conducted by Püllmans et al. (2016) only lasted 1 year and used "crab removal" rather than "crab exclusion" since no fences or other artificial borders were placed around the experimental plots.

In Algodoal Island (PA), extensive reefs of *Sabellaria wilsoni* occur in the lower intertidal and in the shallow subtidal zones. These structures may have ball or platform-like shapes and can extend for more than 500 m with widths of more than 50 m. In this area, the reefs cause a clear change in the bottom topography, providing a three-dimensional consolidated substrate above the surface that is colonized by invertebrates and vertebrates (Aviz 2015). The presence of reefs caused significant changes in the structure of the meiofaunal and macrofaunal communities associated with them.

The reef meiofauna is richer and more diverse than in the sandy bottoms without reefs (Ataíde et al. 2014). Aviz (2015) identified eight times more macrofauna species inside than outside reefs (reef, 76 taxa; sandy sediment, 10 taxa) and recorded that density varied from 11,013 to 159,494 ind m^{-2} in reefs in comparison to 127–1519 ind m^{-2} outside the reef. These significant changes in benthic communities structure between reefs and adjacent sandy sediments are due to substratum structural complexity (which increases the availability of space for colonization and decreases competition and predation); food supply (represented by deposited organic matter, dead organisms and feces, the sabellariids themselves, and chlorophyll-*a*); hydrodynamic changes as the polychaete tubes act as barriers reducing the effects of waves and tidal currents (Aviz 2015); and desiccation reduction since large amount of water is retained inside the tubes (Ataíde et al. 2014).

2.7 Impacts and Conservation Issues

The Brazilian North Coast is characterized by a large number of municipal territories with a dense coastal human population, which along with increasing pressure on coastal resources and the effect of global climate changes, must be taken into account in coastal management plans (Szlafsztein 2012). Some estuaries are surrounded by the largest metropolitan urban centers of the BNC, such as Macapá-Santana (Amapá), Belém (Pará), and São Luís (Maranhão), with a total population over three million inhabitants (IBGE 2016). There are also estuaries in areas that are moderately occupied or isolated with few or no inhabitants. The economic activities of the urban centers include a range of manufacturing and food processing industries, shipping, fisheries, tourism, commercial trading, real estate, and cattle ranching. In the smaller coastal communities, the local economy is based primarily on agriculture and/or fisheries (Pereira et al. 2009).

Between 2006 and 2016, the population of the Amazonian coastal zone has grown about 20% (IBGE 2010). However local infrastructure and services have not kept up with population growth, and public sanitation systems and water supplies are grossly inadequate in most cases. This generates additional anthropogenic stresses on natural environments, with negative impacts on local ecosystems including estuarine environments (Pereira et al. 2010; Gomes et al. 2011; Trindade et al. 2011).

Inadequate use and unplanned occupation are among the main anthropogenic problems of Amazon estuaries. Major problems are related to the discharge of untreated sewage, since only 4% of the homes in Amazon coastal states (Amapá, Pará, and Maranhão) have sewage treatment system (SNIS 2013). Although forbidden by resolution number 357/2005 (article 24) of the Brazilian National Environment Council (CONAMA), untreated sewage is commonly discharged into aquatic environments, with negative consequences for water quality (Pereira et al. 2010; Silva et al. 2013; Monteiro et al. 2016).

Increasing organic pollution from expanding urban areas along the BNC may have a similar effect since, in the absence of industrial pollution along most of the Amazon coast, intense hydrodynamics and high freshwater runoff volumes tend to dilute organic pollution (Aviz et al. 2012), resulting in a beneficial effect on benthic macrofaunal assemblages. This has already been reported for Guarajá Bay, in front of Belém (PA), where greater macrofaunal diversity and abundance has been associated with higher freshwater runoff and nutrient input in the wet season (Aviz et al. 2012).

Mangrove logging is common in the region, generally on a local, subsistence scale (Glaser et al. 2010), the timber being used for fuel, construction, and fishing artifacts (Menezes et al. 2008). The local disturbance caused by the removal of mangrove trees results in immediate effects at logged sites, such as increased exposure to the sun, higher temperatures, and loss of humidity from the sediment, as well as nutrient loss due to decay of leaves, branches, and twigs that remain after logging. Logging is expected to cause an overall decrease in benthic diversity and function unless management practices are implemented to maintain mangrove ecosystem services (Ellison 2008; Ferreira and Lacerda 2016).

Harvesting of mangrove crabs may be one of the greatest potential impacts on the mangrove and coastal ecosystems of the Brazilian North Coast, supplying 97% of the production to regional markets and 3% to national markets (Diele et al. 2010). Although currently managed and classified as sustainable (Furtado-Junior et al. 2016), a rise in demand may cause an increase in the crab harvest, beyond the bounds of sustainability (Diele et al. 2010), disrupting the mangrove nutrient cycle described above with consequences for the benthic assemblage and primary and secondary productivity in the mangrove and coastal area.

Aquaculture in the BNC is still incipient, with records of small shrimp and oyster farms in the state of Maranhão and Pará (Sampaio and Boulhosa 2007). However, production of cultivated oysters is growing rapidly in Pará state (Sampaio et al. 2017). Although carried out on a small scale, aquaculture in the Amazonian estuaries can have local effects on the nutrient status of coastal habitats and can change benthic communities as was shown by Paula et al. (2006) and Venekey and Melo (2016).

Sea-level rise, due to the effects of global warming, is predicted to increase the inundation frequency of the mangrove forest, increasing the export of organic matter to estuarine and coastal areas (Wolff et al. 2000), resulting in an increase in diversity, abundance, and production of the benthic infauna in lower estuarine areas. An observed long-term trend in rising temperatures may increase thermal stress, reducing benthic faunal diversity and production in the North Coast but predicted reduction in rainfall may increase salinity, resulting in greater diversity (Bernardino et al. 2015, 2016).

To protect and preserve the valuable natural resources of the Amazon, as well as optimizing the use and conservation of coastal environments, while also reducing anthropogenic impacts, a number of conservation initiatives have been implemented for the protection of coastal ecosystems and the traditional populations that inhabit them. One of the main results of this initiative has been the creation of protected areas involving either no-take or sustainable use. At present 17 protected areas have been established along the Amazon coast, including no-take Reserves and National Parks, Habitat/Species Management Areas, Protected Landscape/Seascapes, and

Protected Areas with sustainable use of natural resources (the most common type of protected area in the BNC). The Brazilian National Coastal Management Plan (GERCO) also represents an important practical tool for the conservation of the Amazon coastal zone. Unfortunately, recent research (Szlafsztein 2012; Gomes et al. 2014) has shown that the performance of the public sectors in the states of Amapá, Pará, and Maranhão responsible for the implementation of the policies and measures proposed by the plan has been unsatisfactory.

2.7.1 *The Case Study of the Hydrographic Basin of the Caeté River (Pará)*

To better understand the Amazon impacts, the coastal hydrographic basin of the Rio Caeté (PA), located in the northeast of the Brazilian state of Pará, was used as case study. The hydrographic basin of the Caeté River has a total area of 2195 km^2 of which 6% (190 km^2) is covered by mangrove (Dittmar and Lara 2001), and the main river is 149 km long from its source (in the municipality of Bonito) to its mouth (in Bragança and Augusto Corrêa) (Fig. 2.9a). The Caeté basin drains seven municipalities (Bonito, Tracuateua, Ourém, Capanema, Santa Luzia do Pará, Bragança, and Augusto Corrêa) with a total population of about 300,000 inhabitants (IBGE 2010). Around 40% of this population is found in 18 riverside communities (known as the Caeteuara communities) located on the margins of the Caeté (Gorayeb 2008).

Some of the economic activities and environmental impacts in the hydrographic basin of the Caeté River are shown in Fig. 2.9. The principal economic activities of the Caeteuara communities of the middle and upper course of the Caeté are smallholder farming (subsistence agriculture, and orchards, Fig. 2.9d), quarrying, cattle ranching, extraction of plant resources, and artisanal fishing. On the lower Caeté, the principal activities are semi-industrial fisheries and the artisanal harvesting of fish, crustaceans, and mollusks (Guimarães et al. 2011; Krause and Glaser 2003; Glaser and Diele 2004).

The degradation of aquatic environments is one of the main anthropogenic impacts observed in the Caeté basin. Daily, tons of crabs (*Ucides cordatus*) are extracted from the mangrove area to supply both local and regional demand (Fig. 2.9b). The inadequate disposal of solid waste and the discharge of untreated wastewater directly into local water bodies are other common problems. Public sanitation is absent from the whole of the hydrographic basin, and solid waste can be observed in open spaces almost everywhere (Gorayeb et al. 2011; Guimarães et al. 2011). An estimated 28.6 t of garbage is produced per day in the town of Bragança, of which, approximately 70% ends up on the municipal dump (Fig. 2.9c, Gorayeb 2008), which is located less than 3 km from the Chumucuí River, which discharges into the Caeté.

In addition, ornamental fish are illegally caught from the Caeté river at the middle and upper sectors (Fig. 2.9e). Deforestation in mangrove and secondary forest is

Fig. 2.9 Economic activities and environmental impacts in the hydrographic basin of the Caeté River. The seven municipalities draining the Caeté basin. (**a**) Crabs extracted from the mangrove, (**b**) Solid waste on the municipal dump, (**c**) Subsistence agriculture (manioc), (**d**) Capture of aquarium fish, (**e**) Deforestation in secondary forest to supply wood to ovens, (**f**) Recreational area, and (**g**) Asphalt factory

carried out to supply wood ovens in local bakeries, pizzerias, potteries, etc. (Fig. 2.9f) and bathing areas (e.g., wood pools, Fig. 2.9g) used for recreation are present through the estuary. An asphalt factory is situated in a pristine forest in the upper sector of the basin (Fig. 2.9h) and strategies to avoid or mitigate future impacts are necessary. These issues have been registered, because environmental laws and codes of conduct are usually not enforced in these areas.

Nine Bragantinian communities (Fig. 2.10a) in the municipalities of Bragança and Augusto Corrêa (Maranhãozinho, Fazendinha, Camutá, Vila que Era, Sítio Grande, Bacurituea, Caratateua, Ponta do Urumajó, and Vila dos Pescadores) are found at the margins of the Caeté estuary, with an approximate total of 50,000 inhabitants (Guimarães et al. 2011). Of this total, around 85% are found in the upper, more urbanized sector, which includes the city of Bragança, the region's most important urban and commercial center (Monteiro et al. 2016).

Fig. 2.10 Location of rural communities and Bragança downtown along Caeté estuary, Pará state. (**a–d**) Precarious infrastructure in Bragança downtown, and subsistence activities (**e**, **g**, **h**) fishing and (**f**) manioc flour production

Fishing is considered the most important economic activity in the region (Fig. 2.10f, h, i). Currently, more than 6000 t of fish are landed annually in the Caeté estuary, which ranks as the third most important fishing center in Pará and the largest artisanal fishery in Brazil (Isaac et al. 2008; IBAMA 2005). Although industrial development in the region is limited, installations such as ice factories, fish processing plants, fish markets, and dry docks for boat repairs can be found in many parts of the estuary, mainly in the city of Bragança.

The middle and lower sectors of the estuary have a low population density totaling about 6000 inhabitants, which is dependent on the public services provided mainly by the city of Bragança in the upper sector of the estuary. In the middle sector, subsistence agriculture is the main activity. Cassava, rice, and beans are the main local produces (Fig. 2.10e, g). In the lower sector, fishing is the main subsistence activity (Krause and Glaser 2003; Glaser and Diele 2004; Gorayeb 2008).

On May 20th 2005, the Caeté-Taperaçu Extractive Reserve was created to protect and preserve the region's natural resources and human traditional activities.

This reserve has a total area of 42,068 ha and about 3000 inhabitants, encompassing the whole of the Caeté estuary and the communities found along its margins, which exploit its natural resources through cooperative management.

Krause and Glaser (2003) and Glaser and Diele (2004) showed that approximately 75% of the households in the villages surrounding the Caeté estuary are economically dependent on the rich natural resources of the mangrove, mainly by harvesting fish, crabs, and mollusks. The most important mangrove resource is the *uçá*-crab *Ucides cordatus*. These crabs are captured for sale live in local or regional markets or to produce processed crab meat, for sale to regional and national consumers (Krause et al. 2001; Magalhães et al. 2007). Other important resources are fish, shrimp, and other invertebrates, which are also harvested for consumption by the local population.

Despite the extraordinary environmental value of these natural resources and their enormous potential for economic development, the unplanned growth of urban settlements along the estuary and the lack of adequate infrastructure have led to increasing anthropogenic stress on the estuary (Monteiro et al. 2016). There is no public sewage treatment system in the city of Bragança, for example, or any of the surrounding settlements, leading to the wastewater from 26,221 residences (IBGE 2010) being released directly into the local aquatic environments (including the Caeté estuary) or into the ground in public areas or private properties (Fig. 2.10b, c).

Domestic effluents, waste from fishing boats, fish processing plants, filling stations, and hospitals are discharged directly into the estuary or into open public spaces, leading to the degradation of the estuary's waters, making them inappropriate for most types of use. In addition to these sources of pollution, the Cereja River which flows through most of the urban area of Bragança discharges into the upper Caeté estuary. This tributary is 5 km long and has a depth of 1.40 m and a width of 6 m, crossing the southern margin of Bragança to reach the eastern margin of the Caeté. These environmental impacts are exacerbated by the deforestation of the local mangroves and the over-exploitation of local natural resources (Guimarães et al. 2009). Several measures to improve the current situation are needed, such as: (i) regulation of land use, to reduce the environmental impact of the economic sector; (ii) implementation of management programs for the sustainable exploitation of natural resources (fish, crabs, clay, vegetation, etc.), (iii) installation of public services, particularly water supplies and sanitation; (iv) better monitoring and penalization of the illegal exploitation of local natural resources.

2.8 Concluding Remarks

The northern coast of Brazil, the so-called Brazilian North Coast (BNC) extends from Cabo Orange in Amapá (05°N, 051°W) to Ponta do Tubarão in Maranhão (04°S, 043°W), encompasses dozens of estuaries and thousands of km^2 of mangrove forests, and harbors over 80% of Brazil's mangroves. Since its discovery in 1500 the BNC has been visited by explorers and researchers, who recognize its great zoological, botanical, and anthropological richness. Despite the recognition of the high ecological and social importance of the Amazon coast, from the early twentieth century until the 1980s these estuaries remained almost unstudied, and the knowledge of biological communities in these areas was incipient and fragmented. At present, even after almost 40 years of the first studies on the benthic communities of Amazonian estuaries, the benthic fauna of this region remains the least studied of the Brazilian coast.

In general, the benthic fauna of Amazonian estuaries is poor in species. Spatial and temporal variation of the benthos are mainly determined by freshwater discharge from dozens of rivers that flow to the coast, associated with marked seasonal variation in rainfall, as well as the macrotidal regime. Strong salinity gradients are observed along most estuaries and communities typical of freshwater occur in the upper estuarine regions, a brackish-water fauna in the middle-estuary, and a more diverse and abundant marine fauna in the lower-estuary. Temporally, in more saline estuaries highest richness and abundance occur in the dry season (July to December) and estuarine areas permanently meso- to oligohaline may present greater richness and abundance in the rainy season (January to June).

Although the Amazonian estuaries are the best preserved on the Brazilian coast, the impacts of increasing pressure on coastal resources and the effect of global climate changes have been causing concern. The main threats to estuaries in the NBC are inadequate land use and unplanned occupation, mangrove logging, harvesting of mangrove crabs, and sea-level rise (still poorly studied in the area). Aquaculture is still incipient, although production of cultivated oysters is growing rapidly in Pará state. To protect and preserve the valuable natural resources of the Amazon, one of the main government initiatives has been the creation of protected areas. At present 17 protected areas have been established along the Amazon coast, including no-take Reserves and National Parks, Habitat/Species Management Areas, Protected Landscape/Seascapes, and Protected Areas with Sustainable Use of Natural Resources.

For the future, the main challenges for researchers in the benthic communities of the Amazonian estuaries are the expansion of the number of zoological groups considered, the increase of the scales (spatial and temporal) of the studies, and the ability to predict the effects of the increasing environmental impacts of natural and anthropic origin in the structure and functioning of communities.

Acknowledgements Thanks to the Conselho Nacional para Desenvolvimento Científico e Tecnológico of Brazil for the grants to Nils E. Asp (CNPq—309845/2015-7), Luci C.C. Pereira (310909/2014-7 and 200629/2014-0), and Rauquírio A.M. Costa (309527/2014-7 and 200622/2014-5).

References

Acheampong E (2001) Distribution of macrozoobenthos abundance and biomass in intertidal soft sediments of North-East Brazil. Dissertation, Universidade Federal do Pará

Ackermann FL (1966) A depredação dos solos da região Bragantina e na Amazônia. Dissertation, Universidade Federal do Pará

Almeida MF (2008) Comunidades macrobentônicas da reserva biológica do Lago Piratuba (Amapá – Brasil). Dissertation, Universidade Federal do Pará

Amaral KDS, Vieira IM, Osório FM, Rocha JM, Lima JF (2014) Bioecology of the crab *Ucides cordatus* (Crustacea, Decapoda) in mangroves influenced by the Amazon River, Brazil. Acta Amazon 44(2):213–222. https://doi.org/10.1590/S0044-59672014000200007

Andrade JTM, Palhano NB, Tagliaro CH, Beasley CR (2014) Spatial and temporal variation in the abundance and taxonomic composition of estuarine and terrestrial macrofauna associated with mangrove logs. J Mar Biol Assoc UK 94(1):35–42. https://doi.org/10.1017/S002531541300121

Araújo WP, Asp NE (2013) Hydrodynamic connectivity between two macrotidal Amazonian estuaries. J Coast Res SI 65:1086–1091

Asp NE, Schettini CAF, Siegle E, Silva MS, Brito RNR (2012) The dynamics of a frictionally-dominated Amazonian estuary. Braz J Oceanogr 60(3):391–403

Asp NE, Gomes VJC, Ogston A, Borges JCC, Nittrouer CA (2016) Sediment source, turbidity maximum, and implications for mud exchange between channel and mangroves in an Amazonian estuary. Ocean Dyn 66:285–297

Ataíde MB, Venekey V, Rosa Filho JS, Santos PJP (2014) Sandy reefs of Sabellaria wilsoni (Polychaeta: Sabellariidae) as ecosystem engineers for meiofauna in the Amazon coastal region, Brazil. Mar Biodivers 44(3):403–413

Aviz DE (2015) Os recifes arenosos de *Sabellaria wilsoni* (Polychaeta: Sabellariidae) na ilha de Algodoal-Maiandeua, Pará: dinâmica dos construtores e da macrofauna bentônica associada. Thesis, Universidade Federal do Pará

Aviz D, Mello CF, Silva PF (2009) Macrofauna associada às galerias de *Neoteredo reynei* (Bartsch, 1920) (Mollusca: Bivalvia) em troncos de *Rhizophora mangle* Linnaeus durante o período menos chuvoso, em manguezal de São Caetano de Odivelas, Pará (costa norte do Brasil). Bol Mus Para Emílio Goeldi, sér Ciências Naturais 4:47–55

Aviz D, Carvalho ILR, Rosa Filho JS (2012) Spatial and temporal changes in macrobenthic communities in the Amazon coastal zone (Guajará Estuary, Brazil) caused by discharge of urban effluents. Sci Mar 76:381–390

Aviz D, Pinto AJA, Ferreira MAP, Rocha RM, Rosa Filho JS (2016) Reproductive biology of *Sabellaria wilsoni* (Sabellariidae: Polychaeta), an important ecosystem engineer on the Amazon coast. J Mar Biol Assoc UK. doi: https://doi.org/10.1017/S0025315416001776

Barletta M, Barletta-Bergan A, Saint-Paul U (1998) Description of the fisheries structure in the mangrove-dominated region of Braganc¸a (State of Para–North Brazil). Ecotropica 4:41–54

Beasley CR, Fernandes CM, Gomes CP, Brito BA, Santos SML, Tagliaro CH (2005) Molluscan diversity and abundance among coastal habitats of northern Brazil. Ecotropica 11:9–20

Beasley CR, Fernandes MEB, Figueira EAG, Sampaio DS, Melo KR, Barros RS (2010) Mangrove infauna and sessile epifauna. In: Saint-Paul U, Schneider H (eds) Mangrove dynamics and management in North Brazil, Ecological Studies. Springer, Berlin, Heidelberg, pp 109–123

Bentes AB, Pereira L, Pereira J, Cruz P, Marques C, Bentes B (2013) Estimativa da Abundância e Estrutura da População de Portunidae (Crustacea: Decapoda) em um estuário da costa norte do Brasil. Biota Amazônia 3:138–148

Bernardino AF, Netto SA, Pagliosa PR, Barros F, Christofoletti RA, Rosa Filho JS, Colling A, Lana PC (2015) Predicting ecological changes on benthic estuarine assemblages through decadal climate trends along Brazilian Marine Ecoregions. Estuar Coast Shelf Sci 166:74–82. https://doi.org/10.1590/S1679-875920160849064sp2

Bernardino AF, Pagliosa PR, Christofoletti RA, Barros F, Netto SA, Muniz P, Lana PC (2016) Benthic estuarine communities in Brazil: moving forward to long term studies to assess climate change impacts. Braz J Oceanogr 64(2):81–96

Bouma TJ, Olenin S, Reise K, Ysebaert T (2009) Ecosystem engineering and biodiversity in coastal sediments: posing hypotheses. Helgol Mar Res 63:95–106

Braga CF, Beasley CR, Isaac VJ (2009) Effects of plant cover on the macrofauna of Spartina marshes in northern Brazil. Braz Arch Biol Technol 52:1409–1420. https://doi.org/10.1590/S1516-89132009000600013

Braga CF, Monteiro VF, Rosa Filho JS, Beasley CR (2011) Benthic macroinfaunal assemblages associated with Amazonian saltmarshes. Wetl Ecol Manag 19:257–272

Braga CF, Silva RF, Rosa Filho JS, Beasley CR (2013) Spatio-temporal changes in macrofaunal assemblages of tropical saltmarshes, northern Brazil. PanAmJAS 8:282–298

Carranza A, Defeo O, Beck M, Castilla JC (2009) Linking fisheries management and conservation in bioengineering species: the case of South American mussels (Mytilidae). Rev Biol Fish 19:349–366

Cartwright DE, Ray RD, Sanchez BV (1991). Oceanic tide maps and spherical marmonic coefficients from Geosat altimetry. NASA Technical Memorandum 104544. Goddard Space Flight Center, Greenbelt, MD, p. 75

Castro ACLA (1986) Aspectos bioecológicos do caranguejo-uçá, *Ucides cordatus* (Linnaeus, 1763) no estuário do rio dos Cachorros e estreito do Coqueiro, São Luís do Maranhão. Bol Lab Hidrobiol 7(1):7–27

Coelho PA (1967/9) A distribuição dos crustáceos decápodos reptantes do norte do Brasil. Trab Oceanogr UFPE 9/11:223–238

Coelho-Costa, CM (2007) Distribuição Espacial e Temporal dos macrozoobentos de Habitats Entre-Marés do Canal da Raposa, Baía de São Marcos, Maranhão, Brasil. Dissertation, Universidade Federal do Maranhão

Coelho PA, Ramos-Porto M (1980) Crustáceos decápodos da costa do Maranhão, Brasil. Bol Inst Oceanogr 29(2):135–138

Corrêa MMS (1997) Da construção do olhar europeu sobre o novo mundo ao (re) descobrimento do reino tropical. Thesis, Universidade Federal de Goiás

Costa ML, Silva-Mello RLS (1983) Inventário dos moluscos estuarinos do sudoeste da Ilha de São Luís, Estado do Maranhão, Brasil. Bol Lab Hidrobiol 5(6):30–33

Cunha PRF (2016) Associação do caranguejo-uçá Ucides cordatus (Linnaeus, 1763) com bosques de mangue e análise do seu potencial extrativo no município de Viseu—Pará. Dissertation, Universidade Federal do Pará

Cunha-Lignon M, Mahiques MM, Schaeffer-Novelli Y, Rodrigues M, Klein DA, Goya SC, Menghini RP, Tolentino CC, Cintrón-Molero G, Dahdouh-Guebas F (2009) Analysis of mangrove forest succession, using sediment cores: a case study in the Cananéia-Iguape coastal system, São Paulo Brazil. Braz J Oceanogr 57:161–174

Damazio E, Ferreira-Correa MM, Medeiros TCC (1989) Golfo do Maranhão: levantamento bibliográfico sobre oceanografia, biologia marinha e ecologia marinha. Caderno de Pesquisas (São Luís) 5(2):4–8

DeMaster DJ, Pope RH (1996) Nutrient dynamics in Amazon shelf waters: results from amassed. Cont Shelf Res 16(3):263–289

DeMaster DJ, Kuehl SA, Nittrouer CA (1986) Effects of suspended sediments on geochemical processes near the mouth of the Amazon River: examination of biological silica uptake and the fate of particle-reactive elements. Cont Shelf Res 6:107–125

Diele K (2000) Life history and population structure of the exploited mangrove crab *Ucides cordatus* (Linnaeus, 1763) (Decapoda: Brachyura) in the Caeté Estuary, North Brazil. Thesis, University of Bremen

Diele K, Araújo ARR, Glaser M, Salzmann U (2010) Artisanal fishery of the mangrove crab *Ucides cordatus* (Ucididae) and first steps toward a successful co-management in Bragança, North Brazil. In: Saint-Paul U, Schneider H (eds) Mangrove dynamics and management in North Brazil, Ecological Studies. Springer, Berlin, Heidelberg, pp 287–297

Dittmar T, Lara RJ (2001) Driving forces behind nutrient and organic matter dynamics in a mangrove tidal creek in North Brazil. Estuar Coast Shelf Sci 52:249–259

Ellison AM (2008) Managing mangroves with benthic biodiversity in mind: moving beyond roving banditry. J Sea Res 59:2–15

Felder DL (2001) Diversity and ecological significance of deep-burrowing macrocrustaceans in coastal tropical waters of the Americas (Decapoda: Thalassinidea). Interciencia 26:440–449

Fernandes MEB (2003) Macroendofauna bêntica de substrato móvel. In: Fernandes MEB (ed) Os manguezais da costa norte brasileira, vol 1. Fundação Rio Bacanga, São Luis, pp 87–103
Fernandes MEB, Carvalho ML (2007) Bioecologia de *Ucides cordatus* Linnaeus, 1763 (Decapoda: Brachyura) na costa do estado do Amapá. B Lab Hidro 20(1):15–22
Fernandes LMB, Castro ACL, Fernandes GL, Mendes NG, Juras IAG (1983). Parte II – Prospecção Pesqueira (Moluscos), Sururu (*Mytella falcata*). In: SUDAM/UFMA. Caracterização ambiental e prospecção pesqueira do estuário do Rio Cururuca. SUDAM, Maranhão, pp 30–44
Ferreira CP (1989) Manguezais do estado do Pará: fauna de galerias perfuradas por teredo em toras de *Rhizophora*. Thesis, Universidade de Campinas
Ferreira AC, Lacerda LD (2016) Degradation and conservation of Brazilian mangroves, status and perspectives. Ocean Coast Manag 125:38–46
Furtado-Junior I, Abrunhosa FA, Holanda FCAF, Tavares MCS (2016) Selectivity curves of the capture of mangrove crab (*Ucides cordatus*) on the northern coast of Brazil using bayesian inference. Braz J Biol 76:412–419
Geyer WR, Beardsley RC, Lentz SJ, Candela J, Limeburner R, Johns WE, Castro BM, Soares ID (1996) Physical oceanography of the Amazon shelf. Cont Shelf Res 16(5/6):575–616
Glaser M, Diele K (2004) Asymmetric outcomes: assessing central aspects of the biological, economic and social sustainability of a mangrove crab fishery,*Ucides cordatus* (Ocypodidae), in North Brazil. Ecol Econ 49:361–373
Glaser M, Krause G, Oliveira RS, Fontalvo-Herazo M (2010) Mangroves and people: a social-ecological system. In: Saint-Paul U, Schneider H (eds) Mangrove dynamics and management in North Brazil, Ecological studies. Springer, Berlin, Heidelberg, pp 307–351
Gomes TP (2008) Meiofauna e nematofauna da zona costeira da Reserva Biológica do Lago Piratuba, Amapá, Brasil. Dissertation, UFPA
Gomes RKS, Takiyama LR, Pereira LCC, Silva URL, Ferreira RCM (2011) Socialdiagnosis and guidelines for coastal management in environmental protection areas off the Amazon Littoral (Amapa, Brazil). J Coast Res SI 64:1331–1334
Gomes RKS, Sousa FBB, Amanajás VV, Santos GC, Pereira LCC (2014) Guidelines for the management of the coastal zone of the Amazon estuary in the Brazilian state of Amapá. J Coast Res SI 70:401–406
Gorayeb A (2008) Análise Integrada da paisagem na Bacia Hidrográfica do Rio Caeté-Amazônia Oriental. Thesis, Universidade Estadual de São Palo
Gorayeb A, Lombardo AM, Pereira LCC (2011) Natural conditions and environmental impacts in a coastal hydrographic basin in the Brazilian Amazon. J Coast Res SI 64:1340–1344
Guimarães DO, Pereira LCC, Monteiro MC, Gorayeb A, Costa RM (2009) Effects of urban development on the Cereja river and Caeté estuary (Amazon coast, Brazil). J Coast Res SI 56:1219–1223
Guimarães DO, Pereira LCC, Gorayeb A, Costa RM (2011) Exploitation and management of natural resources by rural communities in the Caeté River Basin in Northern Brazil. J Coast Res SI 64:1228–1232
IBAMA – Instituto Brasileiro do Meio Ambiente e dos Recursos Naturais Renováveis (2005) Estatística da Pesca 2005. Brasil: Grandes regiões e unidades da federação. MMA, Brasília
IBGE (2010) Instituto Brasileiro de Geografia e Estatística. 2010. Censo 2010. http://www.ibge.gov.br/home/estatistica/populacao/censo2010/default.shtm. Assessed 10 Jan 2017
IBGE (2016) Instituto Brasileiro de Geografia e Estatística, 2016. Estimativas da população de2015. http://www.ibge.gov.br/home/estatistica/populacao/estimativa2015. Assessed 06 Jan 2017
INMET (2017) Instituto Nacional de Meteorologia. Dados históricos (2006–2016) de precipitação pluviométrica mensal. http://www.inmet.gov.br/portal/index.php?r=avisoMeteorologico/listarAvisos&offset=0. Assessed 10 Jan 2017
Isaac VJ, Ferrari SF (2017) Assessment and management of the North Brazil Shelf Large Marine Ecosystem. Env Dev 22:97–110. https://doi.org/10.1016/j.envdev.2016.11.004
Isaac VJ, Espírito-Santo RV, Nunes JLG (2008) A estatística pesqueira no litoral do Pará: resultados divergentes. PanAmJAS 3(3):205–213

Kjerfve B, Lacerda LD (1993) Mangroves of Brazil. In: Lacerda LD (ed) Conservation and sustainable utilization of mangrove forests in Latin America and Africa regions. Part I. Latin America, vol 2. ISME, Okinawa, pp 245–272

Koch V, Nordhaus I (2010) Feeding ecology and ecological role of North Brazilian Mangrove Crabs. In: Saint-Paul U, Schneider H (eds) Mangrove dynamics and management in North Brazil, Ecological studies. Springer, Berlin, Heidelberg, pp 265–273

Koch V, Wolff M (2002) Energy budget and ecological role of mangrove epibenthos in the Caeté estuary, North Brazil. Mar Ecol Prog Ser 228:119–130. https://doi.org/10.3354/meps228119

Koch V, Wolff M, Diele K (2005) Comparative population dynamics of four fiddler crabs (Ocypodidae, genus *Uca)* from a North Brazilian mangrove ecosystem. Mar Ecol Prog Ser 291:177–188

Krause G (2010) The geography of the Bragança coastal region. In: Saint-Paul U, Schneider H (eds) Mangrove dynamics and management in North Brazil, Ecological studies. Springer, Berlin, Heidelberg, pp 19–34

Krause G, Glaser M (2003) Co-evolving geomorphical and socio-economic dynamics in a coastal fishing village of the Bragança region (Pará, North Brazil). Ocean Coast Manag 46:859–874

Krause G, Schories D, Glaser M, Diele K (2001) Spatial patterns of mangrove ecosystems: the Bragantinian mangroves of northern Brazil (Bragança, Pará). Ecotropica 7:93–107

Lana PC, Camargo MG, Brogim RA, Isaac VJ (1996) O bentos da costa brasileira. Avaliação crítica e levantamento bibliográfico (1958–1996). FEMAR, Rio de Janeiro

Lara RJ, Cohen M, Szlafsztein C (2010) Drivers of temporal changes in mangrove vegetation boundaries and consequences for land use. In: Saint-Paul U, Schneider H (eds) Mangrove dynamics and management in North Brazil, Ecological studies. Springer, Berlin, Heidelberg, pp 127–141

Lucena-Frédou F, Rosa Filho JS, Silva MCN, Azevedo EF (2012) Population dynamics of the river prawn *Macrobrachium amazonicum* (Heller, 1862) (Decapoda, Palaemonidae) on Combu Island (Amazon Estuary). Crustaceana 83(3):277–290

Magalhães A, Costa RM, Silva R, Pereira LCC (2007) The role of women in the mangrove crab (*Ucides cordatus*, Ocypodidae) production process in North Brazil (Amazon region, Pará). Ecol Econ 61:559–565

Mann KH (1982) Ecology of coastal waters: a systems approach. University of California Press, Berkeley

Marengo JA (1995) Interannual variability of deep convection over the tropical South American sector as deduced from ISCCP C2 data. Int J Climatol 15(9):995–1010

Marques-Silva NS, Beasley CR, Gomes CP, Gardunho DCL, Tagliaro CH, Schories D, Mehlig U (2006) Settlement dynamics of the encrusting epibenthic macrofauna in two creeks of the Caeté mangrove estuary (North Brazil). Wetl Ecol Manag 14:67–78

Martinez JM, Guyot JL, Filizola N, Sondag F (2009) Increase in suspended sediment discharge of the Amazon River assessed by monitoring network and satellite data. Catena 79:257–264

Matos JB, Sodré DKL, Costa KG, Pereira LCC, Costa RM (2011) Spatial and temporal variation in the composition and biomass of phytoplankton in an Amazonian estuary. J Coast Res SI 64:1525–1529

Matthews HR, Ferreira-Correia MM, Sousa NR (1977) Levantamento da fauna aquática da Ilha de São Luís (Estado do Maranhão, Brasil) I: Mollusca. Bol Lab Hidrobiol 1(1):9–22

Melo AGC, Varela ES, Beasley CR, Schneider H, Sampaio I, Gaffney PM, Reece KS, Tagliaro CH (2010) Molecular identification, phylogeny and geographic distribution of Brazilian mangrove oysters (*Crassostrea*). Genet Mol Biol 33(3):564–572

Melo MAD, Silva ARB, Beasley CR, Tagliaro CH (2013) Multiplex species-specific PCR identification of native and non-native oysters (Crassostrea) in Brazil: a useful tool for application in oyster culture and stock management. Aquac Int 21:1325–1332

Mendes AC (2005) Geomorfologia e sedimentologia. In: Fernandes MEB (ed) Os manguezais da costa norte brasileira, vol 2. Fundação Rio Bacanga, Belém, pp 13–31

Menezes MPM, Berger U, Mehlig U (2008) Mangrove vegetation in Amazonia: a review of studies from the coast of Pará and Maranhão States, north Brazil. Acta Amaz 38:403–420

Ministério do Meio Ambiente (2002) Avaliação e identificação de áreas e ações prioritárias para conservação, utilização sustentável e repartição dos benefícios da biodiversidade nos biomas brasileiros. MMA/SBF, Brasília

Mochel F, Cutrim MVJ, Ferreira-Correia MM (2001) Degradação dos manguezais na Ilha de São Luís (MA): processos naturais e impactos antrópicos. In: Prost MT, Mendes AC (eds) Ecossistemas costeiros: impactos e gestão ambiental. Museu Paraense Emílio Goeldi, Belém, pp 27–42

Monteiro MC, Jimenez JA, Pereira LCC (2016) Natural and human controls of water quality of an Amazon estuary (Caeté-PA, Brazil). Ocean Coast Manag 124:42–52

Monteles JS, Castro TCS, Viana DCP, Conceição FS, França VL, Funo ICSA (2009) Percepção socioambiental das marisqueiras no município da Raposa-MA. REPESCA 4(2):34–45

Moraes BC, Costa JMN, Costa ACL, Costa MH (2005) Variação espacial e temporal da precipitação no Estado do Pará. Acta Amaz 35:207–214

Morais GC, Lee JT (2014) Intertidal benthic macrofauna of rare rocky fragments in the Amazon region. Rev Biol Trop 62(1):69–86. https://doi.org/10.15517/rbt.v62i1.8425

Nascimento WR, Souza Filho PWM, Proisy C, Lucas RM, Rosenqvist A (2013) Mapping changes in the largest continuous Amazonian mangrove belt using object-based classification of multisensor satellite imagery. Estuar Coast Shelf Sci 117:83–93. https://doi.org/10.1016/j.ecss.2012.10.005

Nordhaus I, Wolff M, Diele K (2006) Litter processing and population food intake of the mangrove crab Ucides cordatus in a high intertidal forest in northern Brazil. Estuar Coast Shelf Sci 67:239–250

Nordhaus I, Wolff M (2007) Feeding ecology of the mangrove crab Ucides cordatus (Ocypodidae): food choice, food quality and assimilation efficiency. Mar Biol 151(5):1665–1681

Oliveira VM, Mochel FR (1999) Macroendofauna bêntica de substratos móveis de um manguezal sob impacto das atividades humanas no sudoeste as ilha de São Luís, Maranhão, Brasil. Bol Lab Hidrobiol 12(1):75–93

Paiva MP (1997) Recursos pesqueiros estuarinos e marinhos do Brasil. UFC Edições, Fortaleza

Pamplona FC, Paes ET, Nepomuceno A (2013) Nutrient fluctuations in the Quatipuru river: a macrotidal estuarine mangrove system in the Brazilian Amazonian basin. Estuar Coast Shelf Sci 133:273–284

Paula JHC, Rosa Filho JS, Souza ALB, Aviz D (2006) A meiofauna como indicadora de impactos da carcinicultura no estuário de Curuçá (PA). Bol Lab Hidrobiol 19:61–72

Pereira LCC, Ribeiro CMM, Monteiro MC, Asp N (2009) Morphological and sedimentological changes in a macrotidal sand beach in the Amazon littoral (Vila dos Pescadores, Pará, Brazil). J Coast Res 1(SI56):113–117

Pereira LCC, Monteiro MC, Guimarães DO, Costa RM (2010) Seasonal effects of wastewater to the water quality of the Caeté river estuary. An Acad Bras Cienc 82(2):467–478

Pereira LCC, da Silva NIS, da Costa RM, Asp NE, da Costa KG, Vila-Concejo A (2012) Seasonal changes in oceanographic processes at an equatorial macrotidal beach in northern Brazil. Cont Shelf Res 43:95–106

Pereira LCC, Oliveira SMO, Costa RM, Costa KG, Vila-Concejo A (2013) What happens on an equatorial beach on the Amazon coast when La Niña occurs during the rainy season? Estuar Coast Shelf Sci:135116–135127

Perote SMO (2010) Estrutura Populacional da Floresta de Mangue e do Caranguejo-uçá, Ucides cordatus (Linnaeus, 1763), na Reserva Extrativista Marinha "Mãe Grande" de Curuçá, Curuçá—Pará. Dissertation, Universidade Federal do Pará

Püllmans N, Mehlig U, Nordhaus I, Saint-Paul U, Diele K (2016) Mangrove crab Ucides cordatus removal does not affect sediment parameters and stipule production in a one year experiment in northern Brazil. PLoS One 11(12):e0167375

Rebelo-Mochel F (1997) Mangrove ecosystems in São Luis Island, Maranhão, Brazil. In: Kjerfve B, Lacerda LD, Diop EHS (eds) Mangrove ecosystem studies in Latin America and Africa. UNESCO, Paris, pp 145–154

Reis REML (1995) Moluscos bivalves perfuradores de madeira do estado do Pará, Brasil: caracterização taxonômica, distribuição e resistência das madeiras. Bol Mus Para Emílio Goeldi, sér. Fortschr Zool 11(2):125–203

Ribeiro RP, Almeida ZS (2014) Anelídeos Poliquetas do estado do Maranhão, Brasil: síntese do conhecimento. Bioikos 28(1):45–55

Rodrigues CAL, Ribeiro RP, Santos NB, Almeida ZS (2016) Patterns of mollusc distribution in mangroves from the São Marcos Bay, coast of Maranhão State, Brazil. Bol Lab Hidrobiol 46(4):391–400

Rosa Filho JS, Aviz D (2013) Macrobenthic communities of an Amazonian estuary (Guajará Bay, Brazil): temporal and spatial changes. J Coast Res 65(SI):123–128. https://doi.org/10.2112/SI65-022.1

Rosa Filho JS, Busman DV, Viana AP, Gregório AM, Oliveira DM (2006) Macrofauna bentônica de zonas entre-marés não vegetadas do estuário do rio Caeté (Bragança-PA). Bol Mus Para Emílio Goeldi, sér Ciências Naturais 1:85–96

Rosa Filho JS, Almeida MF, Aviz DE (2009) Spatial and temporal changes in the benthic fauna of a macrotidal Amazon sandy beach, Ajuruteua, Brazil. J Coast Res SI56:1796–1780

Rosa Filho JS, Gomes TP, Almeida MF, Silva RF (2011) Benthic fauna of macrotidal sandy beaches along a small-scale morphodynamic gradient on the Amazon coast (Algodoal Island, Brazil). J Coast Res SI 64:435–439

Rosa Filho JS, Girard TC, Lucena-Frédou F (2013) Population dynamics of the burrowing shrimp *Lepidophthalmus siriboia* Felder and Rodrigues, 1993 (Reptantia: Axiidea: Callianassidae) on the Amazonian coast. J Crustac Biol 33(4):503–511

Sampaio D, Boulhosa RLM (2007) Energia que vem da ostra: do extrativismo para o cultivo. In: Araújo Duarte RB (ed) Histórias de Sucesso: agronegócios: aqüicultura e pesca. SEBRAE, Brasília

Sampaio DS, Schneider H, Tagliaro CH, Beasley CR (2017) Oyster culture on the Amazon mangrove coast: asymmetries and advances in an emerging sector. Rev Aquac. https://doi.org/10.1111/raq.12227

Sanjad N, Pataca EM (2007) As fronteiras do ultramar: engenheiros, matemáticos, naturalistas e artistas na Amazônia, 1750–1820. In: Proceeding of the VII Colóquio Luso-brasileiro de história da arte, UFRS, 2007

Santos Filho C, Tagliaro CH, Beasley CR (2008) Seasonal abundance of the shipworm *Neoteredo reynei* (Bivalvia, Teredinidae) in mangrove driftwood from a northern Brazilian beach. Iheringia Sér Zoologia 98:17–23

Santos SML, Tagliaro CH, Beasley CR, Schneider H, Sampaio I, Santos C, Müller ACD (2005) Taxonomic implications of molecular studies on northern Brazilian Teredinidae (Mollusca, Bivalvia) specimens. Genet Mol Biol 28:175–179

Santos MLS, Medeiros C, Muniz M, Feitosa MLS, Schwamborn R, Macedo SJ (2008) Influence of the Amazon and Pará Rivers on water composition and phytoplankton biomass on the adjacent shelf. J Coast Res 24(3):585–593

Santos HSS, Beasley CR, Tagliaro CH (2010) Changes in population characteristics of *Mytella falcata* (d'Orbigny, 1846) beds, an exploited tropical estuarine mussel. B Inst Pesca SP 36:85–97

Silva MMT (2008) Bioecologia e produção comercial do caranguejo-uçá (*Ucides cordatus Linnaeus*, 1763) em Quatipurú—Pará. Dissertation, Universidade Federal do Pará

Silva Júnior OG, El-Robrini M (2001) Morfoestratigrafia do noroeste da planície costeira de São João de Pirabas (Nordeste do Pará). Braz J Oceanogr 49:98–114

Silva-Mello RL, Costa ML (1993) Contribuição preliminar ao estudo dos moluscos da região sul-sudoeste da Ilha de São Luís, Estado do Maranhão, Brasil. Bol Mus Malacolog 1:51–64

Silva RF, Rosa Filho JS, Souza SR, Souza Filho PW (2011) Spatial and temporal changes in the structure of soft-bottom benthic communities in an Amazon estuary (Caeté estuary, Brazil). J Coast Res SI64:440–444

Silva IR, Pereira LCC, Trindade WN, Magalhães A, Costa RM (2013) Natural and anthropogenic processes on the recreational activities in urban Amazon beaches. Ocean Coast Manag 76:75–84

Silva LS, Tagliaro CH, Beasley CR (2014) How precise are estimates of intertidal macroinfaunal density and spatial dispersion when converted to larger areas? PanAmJAS 9:112–125

Sousa DB, Santos NB, Oliveira VM, Carvalho-Neta RNF, Almeida ZS (2015) Benthic estuarine carcinofauna of two mangroves of the Maranhão Amazon coast, Brazil. Iheringia, Sér Zoologia 105:339–347

Souza Filho PWM, Cohen MCL, Lara RJ et al (2006) Holocene coastal evolution and facies model of the Bragança macrotidal flat on the Amazon Mangrove Coast, Northern Brazil. J Coast Res SI39:306–310

Souza Filho PWM, Lessa GC, Cohen MCL, Costa FR, Lara RJ (2009) The subsiding macrotidal barrier estuarine system of the eastern Amazon coast, northern Brazil. In: Dillenburg SF, Hesp PA (eds) Geology and geomorphology of Holocene coastal barriers of Brazil, Lecture notes in earth sciences. Springer, Berlin, Heidelberg, pp 347–375

Spalding MD, Fox HE, Allen GR et al (2007) Marine ecoregions of the world: a bioregionalization of coastal and shelf areas. Bioscience 57:573–583. https://doi.org/10.1641/B570707

Szlafsztein CF (2012) The Brazilian Amazon coastal zone management: implementation and development obstacles. J Coast Conserv 16:335–343

Tang Y, Fang Z, Zhang Z, Zhong Y, Chen K, Liu B, Fan J, Xiong T, Lu X (2014) Ecological indicators showing the succession of macrofauna communities in *Sonneratia apetala* artificial mangrove wetlands on Qi'ao Island at Zhuhai, South China. Acta Oceanol Sin 33:62–72

Trindade WN, Pereira LCC, Guimarães DO, Costa RM (2011) The effects of sewage discharge on the water quality of the beaches of São Luís (Maranhão, Brazil). J Coast Res SI 64:1425–1429

Vasconcelos TR (2006) Respostas às variações espaço temporais das comunidades macrobêntonicas de fundos inconsolidados do estuário de Curuçá (Curuçá-PA). Dissertation, Universidade Federal do Pará

Venekey V, Melo TPG (2016) Nematodes as indicators of shrimp farm impact on an amazonian estuary (Curuçá, Pará, Brazil). Braz J Oceanogr 64(1):75–88

Wild PAWJ (1976) The benthic boundary layer from the point of view of a biologist. In: McCave IN (ed) The benthic boundary layer. Plenum Press, New York, pp 81–94

Wisser D, Fekete BM, Vörösmarty CJ, Schumann AH (2010) Reconstructing 20th century global hydrography: a contribution to the Global Terrestrial Network-Hydrology (GTN-H). Hydrol Earth Syst Sci 14:1–2

Wolff M, Koch V, Isaac V (2000) A trophic flow model of the Caeté mangrove estuary (North Brazil) with considerations for the sustainable use of its resources. Estuar Coast Shelf Sci 50:789–780

Chapter 3
Benthic Estuarine Assemblages of the Northeastern Brazil Marine Ecoregion

Rafaela Camargo Maia, José Souto Rosa Filho, Cristina de Almeida Rocha-Barreira, Helena Matthews-Cascon, Emanuel Soares dos Santos, Heleny Noronha David, and Alisson Sousa Matos

Abstract The estuaries of the Northeastern Brazil Ecoregion are dominated by mangrove ecosystems and tend to present high salinity values due to elevated temperatures and low rainfall. Nevertheless, they present a high photosynthetic productivity as a result of continental nutrient input and intense insolation. Those singular environmental conditions (shallow estuaries) are reflected in the composition and structure of marine invertebrate assemblages and consequently in the way in which these organisms interact with the environment. Although there is reasonable knowledge about benthic biodiversity from estuaries of the Northeastern Marine ecoregion, ecological knowledge on benthic assemblages, including their spatial and temporal dynamics, is scarce and mostly restricted to a few estuaries. However, general estuarine gradients such as an increase in benthic species richness towards euhaline sectors are observed. Species richness and macrofaunal abundance are higher during dry seasons. Benthic assemblage zonation is observed in unconsolidated bottoms along the intertidal zone, with mangrove trees being colonized by their own set of organisms. Estuarine areas in this ecoregion are being greatly

R. C. Maia (✉) · E. S. dos Santos
Instituto Federal de Educação, Ciência e Tecnologia do Ceará, Acaraú, CE, Brazil
e-mail: rafaelamaia@ifce.edu.br

J. S. Rosa Filho
Universidade Federal de Pernambuco, Centro de Tecnologia, Departamento de Oceanografia, Recife, PE, Brazil
e-mail: souto.rosa@ufpe.br

C. d. A. Rocha-Barreira · H. N. David
Universidade Federal do Ceará, Instituto de Ciências do Mar, Fortaleza, CE, Brazil

H. Matthews-Cascon · A. S. Matos
Universidade Federal do Ceará, Centro de Ciências, Departamento de Biologia, Fortaleza, CE, Brazil
e-mail: hmc@ufc.br

P. C. Lana, A. F. Bernardino (eds.), *Brazilian Estuaries*, Brazilian Marine Biodiversity, https://doi.org/10.1007/978-3-319-77779-5_3

impacted by economic activities, with shrimp farming advancing fast in some areas. Thus, environmental education will be of great relevance towards a sustainable use of ecosystem goods in the Northeastern States in Brazil.

Keywords Animal–sediment relationships · Benthic fauna · Mangroves · Impacts

3.1 Introduction

The estuaries of the Northeastern Brazil Marine Ecoregion (NEME; Spalding et al. 2007) are located between Piauí (PI) and Sergipe (SE) (Fig. 3.1). As in other tropical ecoregions, these estuaries support extense and well-developed mangrove forests with a strong influence of hydrodynamic processes. The NEME has a typical semiarid dry tropical climate with low levels of precipitation (Alvares et al. 2014). Two well-defined seasons predominate during the year: a rainy season extending from January to June when high levels of rainfall and low occurrence of winds are observed; and a dry season extending from July to December, with less rain and winds. Many of these estuaries fit into the definition of a negative or inverse behavior, with their salinity gradients increasing towards the upper estuary. Salt water enters the estuary through the surface water layers and despite mixing with fresh water, it becomes hypersaline due to evaporation. High density surface hypersaline water in turn sinks, and mixes with bottom fresh water. As a result, these estuaries tend to be predominantly euhaline (up to 47). Some estuaries may exhibit high primary productivity as the result of continental nutrient input, low turbidity, and high insolation (Santiago et al. 2005), although low in comparison to the North Brazil.

Estuarine benthic assemblages are well adapted to high variability in water salinity, temperature, and tidal cycles. Estuarine benthic assemblages are highly productive; they mediate biogeophysical and biogeochemical interactions that serve as a support to ecosystem balance through organic degradation of carbon, and the trophic transfer and remineralization of nutrients (McLusky and Elliott 2004). Estuarine benthic animals have a great ecological importance since they act as decomposers of organic matter and strongly influence nutrient recycling. In addition to their contribution to ecosystem dynamics, benthic organisms are a source of food for several other animals with commercial value (Vannucci 2001). The benthic estuarine fauna has also great economic importance because they guarantee the livelihood of local populations providing important fishing resources and exerting a strong influence on the economy. Consequently, the strong anthropogenic pressure, including pollution and mangrove clearing, has led to serious environmental problems and threaten a sustainable exploitation of its natural resources. Currently, there is also a great concern about global climate change, which will likely affect benthic estuarine assemblages. The impacts of climate change can be worrisome when the physiological tolerance of estuaries is exceeded such as during extreme temperature events (Sunday et al. 2014; Bernardino et al. 2015).

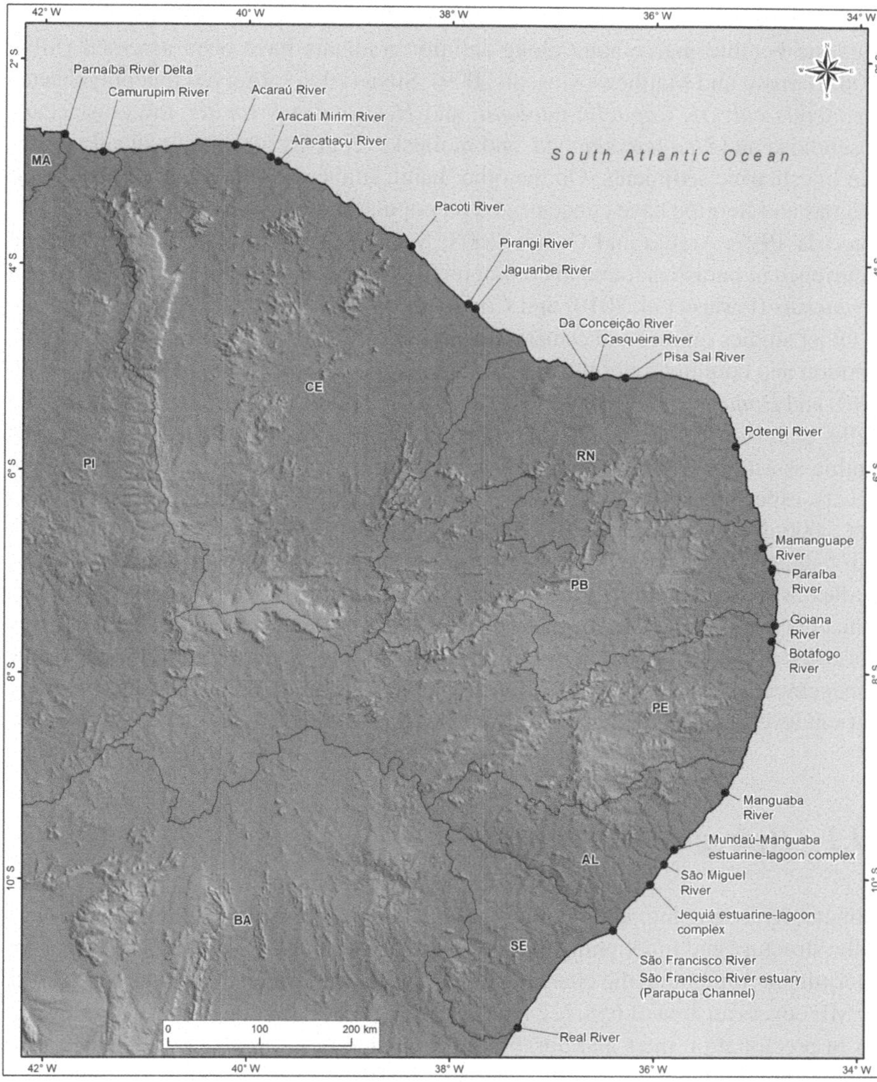

Fig. 3.1 Map indicating estuaries from the Northeastern Ecoregion (NE) in the states of Maranhão (MA), Ceará (CE), Rio Grande do Norte (RN), Paraíba (PB), Pernambuco (PE), Alagoas (AL), and Sergipe (SE)

Estuaries from the NEME are largely understudied when compared to southern estuarine ecosystems (Bernardino et al. 2016). Past and current research mostly includes descriptive approaches to local biodiversity efforts that were nonetheless essential to describe and discriminate varying estuarine patterns among sites. For example, studies reporting the occurrence of benthic megafauna including crabs *Calappa nitida* (Brachyura, Calappidae) (Góes and Fernandes-Góes 2007) and dominant vegetation within estuaries from Parnaíba River Delta are common (Silva et al. 2016). In other regions of the NEME, the biodiversity and distribution of estuarine benthic macrofauna along salinity gradients have been assessed (Silva 2006; Barroso and Matthews-Cascon 2009). Silva (2006) observed that polychaetes *Laeonereis culveri*, *Capitella capitata*, and *Heteromastus similis*, the crustaceans Apseudidae and *Uca leptodactyla*, and mollusks (*Tagelus plebeius*) typically dominate in estuarine sediments. On the other hand, studies on the estuarine benthos of Alagoas and Sergipe have concentrated on population dynamics of *Ucides cordatus* (Lacerda 1999; Araújo and Calado 2008; Santa Fe and Araújo 2013) and on the occurrence of parasites found in commercially exploited bivalves, including *Mytella guyanensis* (Farias et al. 2010) and *Crassostrea* sp. (Silva et al. 2016).

Other studies on regional estuarine benthos covered ecological aspects of the distribution and community of the bivalves *Anomalocardia brasiliana* (Rodrigues et al. 2010) and *Donax striatus* (Medeiros et al. 2015) in saline estuaries, and species of the fiddler crab *Uca* (Silva et al. 2016) in a semiarid estuary. The knowledge of estuarine benthic assemblages in the Paraíba State is mostly concentrated on ethnobiological aspects, especially on the species *Ucides cordatus* (Alves and Nishida 2004; Nishida et al. 2006; Nascimento et al. 2011). According to these authors, people involved with crab fishing activity can provide important information to support scientific studies. Studies assessing the quality of estuaries using benthic assemblages as ecosystem indicators are common in the literature (Valença and Santos 2012). The recovery processes of macrozoobenthic assemblages in tropical estuaries are relevant for the management and protection of tropical estuarine habitats threatened by excessive nutrient levels caused by pollution (Botter-Carvalho et al. 2011).

3.1.1 Hydrographic Patterns

Climate, hydrographic, and geomorphological characteristics are the main drivers of the structure and functioning of a given estuary. Rainfall regimes in NE Brazil are determinant to regulate the characteristics of estuarine waters. Considering that the NEME covers an area of 658,722 km^2, different rainfall regimes are observed, varying in precipitation rates and duration of rainy and dry periods. As a result, a pronounced environmental variability among NEME estuaries is observed (Table 3.1).

The minimum salinity value was 0.03 in the São Francisco River estuary (AL/SE) and the maximum was 47 in the Camurim River (PI). The lowest and highest temperature values occurred in the Botafogo River estuary in Pernambuco. The pH values presented in Table 3.1 are also greatly variable. The Paraíba River estuary (PB) showed pH of 6.7 (the lowest reported among the evaluated studies) and the Mamanguape River (PB) showed pH of 9.5 (the highest reported). These results suggest the influence of the saline environment in the evaluated estuaries, where there is a predominance of the bicarbonate-borate buffer system, which tends to maintain an alkaline environment around pH 8.0 and 8.5. Differently, in freshwater environments the predominant buffer system is the carbonate-bicarbonate, which is weaker and allows greater variation of pH values around the neutrality values (5.0–9.0).

Table 3.1 Physical–chemical characteristics (salinity, temperature, pH, transparency, DO, chlorophyll *a*) of estuaries in Northeastern Brazil Marine Ecoregion

Estuary	Municipality-State	Salinity	Temperature (°C)	pH	Transparency (m)	DO (mg L^{-1})	Chlorophyll *a* (mg m^{-3})	References
Camurim River	Luis Correia/Cajueiro da Praia-PI	25.0–47.0	≈27.0–≈29.0*	7.1–9.0	0.0–1.1	#	#	Rolemberg et al. (2008) Mai and Rosa (2009)
Catu River	Aquiraz-CE	≈3.0–≈19.0	25.5–33.5	7.7–9.0	0.4–1.7	4.6–8.5	#	Pereira (2007)
Pisa Sal River	Galinhos-RN	34–41	26–29.8	7.0–8.58	0.75–3.5	3.73–12.4	0.19–106.6	Santiago et al. (2005)
Mamanguape River	Rio Tinto/Marcação-PB	≈1.0–≈37.0	≈26.9–≈29.3	≈7.2–≈9.5	≈0.5–≈1.2	#	≈2.0	Alves et al. (2016)
Paraíba River	Cabedelo-PB	≈2.0–≈35.0	≈27.2–≈29.3	≈6.7–≈9.2	≈0.4–≈1.4	#	≈2.0–≈65.0	Alves et al. (2016)
Botafogo River	Itamaracá-PE	4.73–37.94	21.5–39.8	6.92–8.68	0.3–2.4	1.99–6.5	2.15–59.75	Otsuka et al. (2014)
São Miguel River	Barra de São Miguel-AL	7.5–26	25–30.5	7.2–7.6	0.15–0.25	#	2.28–10.16	Vilar et al. (2012)
São Francisco River	Piaçabuçu-AL/Brejo Grande-SE	0.03–28.47	25.78–29.38	7.3–8.3	0.3–2.5	4.24–7.43	nd–28.04	Barbosa (2011)

pH hydrogen potential, *DO* dissolved oxygen, *nd* not detectable; # no data

The values of water transparency measured with a Secchi disk ranged from zero (0.0 m) in the Camurim River (PI) to 3.5 m in the Pisa Sal River (RN) (Table 3.1). The lowest reported dissolved oxygen (DO) concentration was 1.99 mg of O_2 L^{-1} in the Botafogo River (PE) and the highest was 12.4 mg of O_2 L^{-1} in the Pisa Sal River (RN) (Table 3.1). The concentration values of chlorophyll *a* were below the detection limits in the São Francisco River estuary (AL/SE) and the highest (106.6 mg m^{-3}) in the Pisa Sal River (RN) (Table 3.1).

These three indicators show a close relationship with each other, i.e., low water transparency results from the presence of suspended particles or phytoplankton; (1) if it is suspended mineral material, the concentrations of chlorophyll *a* and consequently of DO will be low because without penetration of light there is no photosynthesis; (2) if it is excess of phytoplankton, the concentrations of chlorophyll *a* and, consequently, DO will be high. Conversely to these references, the Pisa Sal River estuary showed the greatest water transparency and the highest concentrations of DO and chlorophyll *a*. However, it is worth noting that this pattern possibly occurred due to the peculiar morphological characteristics of this estuary and the moment of measurement of these indicators in relation to the tidal prism (Santiago et al. 2005).

3.2 Benthic Fauna

3.2.1 Spatial and Temporal Patterns of Biodiversity and Faunal Abundance and Biomass

Most of the available information on benthic assemblages from the NEME is restricted to taxonomic lists from a few estuaries. Quantitative data at species level is mostly restricted to some estuaries in Pernambuco and Ceará, which have a stronger tradition of faunal surveys and ecological studies. The estuaries from the western sector of the NEME are bounded by extensive mangrove forests extending to approximately 22,936 ha distributed among 12 large rivers (Miranda et al. 1990). Till the end of the last century, benthic data was only available for the Cocó River (Menezes and Menezes 1968; Matthews-Cascon and Martins 1999), Ceará River (Alcântara-Filho 1978; Miranda et al. 1988; Vilanova and Chaves 1988), and Pacoti River (Oliveira et al. 1988). More recently, Rocha-Barreira et al. (2005) surveyed 21 estuarine regions in Ceará state and recorded diverse macrobenthic assemblages, numerically dominated mainly by polychaetes, mollusks, and crustaceans. Silva (2006) recorded 31 benthic macrofauna taxa, comprising 15 polychaete, 8 mollusk, and 8 crustacean species, along tidal flats of the Pacoti and Pirangi estuaries. The most abundant species were the polychaete *Laeonereis culveri* with 10,422 ind m^{-2} in the Pacoti River and the bivalve *Tagelus plebeius* with 910 ind m^{-2} in the Pirangi River. Beyond the latter species, *Tagelus plebeius*, *Lucina pectinata*, *Tellina versicolor*, *Mytella charruana*, *Anomalocardia brasiliana*, *Macoma constricta*, *Corbula cubaniana*, and *Crassostrea mangle* were the most common molluskan species.

Some of these species form extensive banks in soft bottoms with high densities and are artisanally exploited by local fisherfolk (Barreira and Araújo 2005; Farias 2008; Santana 2010). The gastropod *Neritina virginea* is found on tidal flats while *Melampus coffeus* colonizes roots of *Rhizophora mangle* in all estuaries of Ceará, reaching average densities of approximately 1123 ind m^{-2} (Silva 2015) and 80–600 ind m^{-2} (Maia et al. 2011), respectively. Bezerra et al. (2006) described the distribution of *Uca* crabs and found that *U. leptodactyla* and *U. rapax* inhabit sectors of the fluvial-marine plains composed of medium sand while *U. thayeri* and *U. maracoani* are found predominantly in muddy areas. The benthic macrofauna distribution in Ceará estuaries is structured by salinity gradients; assemblages are more diversified in areas where salinity is high and more constant (Barroso and Matthews-Cascon 2009). However, temporal variations associated to dry and wet seasons also influence the richness and abundance of benthic animals in local estuaries.

Studies have been carried out since the 1960s along the estuaries of Pernambuco, with most studies being carried on the Santa Cruz Channel until the 1990s. These studies were mostly faunistic surveys of crustaceans and mollusks. Since the 1990s, there was an increasing interest in assessing benthic assemblages as a whole. About 170 species of crustaceans (100 Decapoda, 27 Brachyura, 23 Amphipoda, 10 Cirripedia, and 10 species of other groups), 160 species of mollusks (90 Gastropoda, 50 Bivalvia, and 20 species of other classes), 82 species of Annelida (67 Polychaeta, 15 Oligochaeta, and 1 Hirudinea), and 33 species belonging to other phyla (Cnidaria, Platyhelminthes, Sipuncula, Nemertea, Phoronida, Echinodermata, Bryozoa, and Chordata) were identified among the most abundant and diverse benthic groups (Coêlho 1965/1966, 2000; Soares 1979, 1980; Silva-Mello and Tenório 2000; Farrapeira et al. 2009; Araújo et al. 2014).

Richness, diversity, and density were higher in the sublitoral, with 6 species of mollusks being sampled in the supralittoral, 58 species in the mid-littoral, and 79 species in the sublitoral (Silva-Mello and Tenório 2000). Temporal variations in richness, diversity, and density were lower in the sublitoral (Carvalho 2004), revealing that assemblages are more stable in this zone. Species richness increases towards the river-sea gradient as observed by Araújo et al. (2014) for Brachyura, and Farrapeira (2006, 2008) for Cirripedia. Richness and abundance increase in dry periods (between July and December). In unconsolidated (sandy and mostly muddy bottoms) and consolidated substrates (mainly trunks and mangrove roots), clear patterns of vertical fauna zoning are observed. In unconsolidated substrates in the Santa Cruz Canal (PE), Silva-Mello and Tenório (2000) and Coêlho (2000) reported the typical species of each zone as: Supralittoral—*Cardisoma guanhumi*, *Goniopsis cruentata*, *Armases angustipes*, *Sesarma crassipes*, *S. rectum*, *Pachygrapsus gracilis*, and *P. transversus*; Mesolittoral—*Lepidophthalmus syriboia*, *Uca* spp., *Ucides cordatus*, *G. cruentata*, *Neritina virginea*, *Tagelus plebeius*, and *Anomalocardia brasiliana*; Sublittoral—*Alpheus* spp., *Salmoneus ortmanni*, *Upogebia omissa*, *U. noronhensis*, *Cyrtoplax spinidentata*, *Callinectes* spp., *Clibanarius vittatus*, *Pugilina morio*, *Mytella guyanensis*, *M. charruana*, *Lucina pectinata*, and *Iphigenia brasiliensis*. Rosa-Filho and Farrapeira-Assunção (1998) and Farrapeira et al. (2000, 2009) identified four zones with distinct fauna on mangrove tree trunks: 1.

Supralittoral—*Littoraria angulifera, L. flava,* and *Aratus pisonii*; 2. Upper Mesolittoral—(*Chthamaletum*) dominated by *Microeuraphia rhizophorae* and *Chthamalus proteus*; 3. Medium Mesolittoral (Balanoid Zone) dominated by *Amphibalanus amphitrite, A. improvisus, A. reticulatus*, and *Crassostrea rhizophorae*; and 4. Lower Mesolittoral—*Bostrichietum* (macroalgae *Bostrichia, Catenella,* and *Calloglossa*).

Many mollusk and crustacean species are of great importance for artisanal fisheries in the estuarine areas of Pernambuco (Quinamo 2006). Among the species of commercial interest there are the crabs *Goniopsis cruentata, Cardisoma guanhumi, Ucides cordatus*, and *Callinectes* spp.; and the bivalves *Anomalocardia brasiliana, Crassostrea mangle, Tagelus plebeius, Mytella falcata, Lucina pectinata, Leukoma pectorina, Iphigenia brasiliana*, and *Mytilopsis leucophaeta*. Despite their socioeconomic relevance, very little is known about their stocks and levels of exploitation to which they are submitted. The densities of some species are known in certain localities, such as *U. cordatus* (1.28 and 1.37 burrows m^{-2} in the Mambucaba and Ariquindá Rivers, respectively), *Callinectes danae* (64.6 ind h^{-1} effort from one fisherman), *Mytilopsis leucophaeta* (176,800 ind m^{-2}; in the area adjacent to the Port of Recife), *Uca thayeri* (685 ind m^{-2}) (Araújo et al. 2014), and *Anomalocardia brasiliana* (3779 ind m^{-2}; in Itamaracá) (El-Deir 2009).

Although there are six estuaries along the Paraíba coast, the benthic fauna of only the Mamanguape and Goiana Rivers was studied. Franklin-Júnior (2000) identified 35 macrofaunal species with a high dominance of annelids (>90% of the total number of organisms) in Mamanguape. A clear vertical zonation was observed within the sediment; 85% of all individuals and 90.3% of the faunal richness were recorded in the top 5-cm sediment layer. There were marked variations in the fauna from different substrates. The enchytraeids oligochaetes, the polychaete *Aricidea fragilis*, and the gastropod *Caecum estuatum* dominated sandy bottoms. The polychaetes *Lumbrinerides gesae, Aricidea ramosa*, and *Mediomastus californiensis* and the edible bivalve *Anomalocardia brasiliana* were dominant on muddy bottoms. The distribution of shipworms was greatly affected by the salinity gradient; *Neoteredo reynei* occurring in low salinity areas, while *Nausitora fusticula* and *Bankia fimbriatula* being more abundant in areas with intermediate salinity, and *Teredo bartschi* and *Martesia striata* dominating near the sea (Leonel et al. 2002). *Anomalocardia brasiliana* occurs at densities of 5798 ind m^{-2} in the Mamanguape River estuary while the density of *U. cordatus* reached up to 1.7 ind m^{-2} in the Goiana River (Alves and Nishida 2004; Silva-Cavalcanti 2011).

The first study on the estuarine benthos from Alagoas was done in the second half of the nineteenth century, when Smith (1869) recorded the occurrence of the shrimp *Bithinus acanthurus* (= *Macrobrachium acanthurus*) and the mangrove tree crab *Aratus pisonii* in the Mundaú Lagoon. About 150 years after Smith's records, knowledge on the estuarine benthic fauna remains restricted to the Mundaú-Manguaba estuarine-lagoon complex. The other local estuaries, including the Manguaba estuary and Jequiá estuarine-lagoon complex, have been only sporadically surveyed. A total of 70 species of crustaceans (55 Decapoda, 5 Cirripedia, 7 Isopoda, and 3 other groups), 21 species of mollusks (11 Bivalvia and 10 Gastropoda), and

16 species of polychaetes are known in the Mundaú-Manguaba estuary-lagoon complex (Smith 1869; Silva and Prereira-Barros 1987; Pereira-Barros 1987; Sovierzoski 1994; Calado and Sousa 2003). Throughout the complex, species richness and abundance are lower in muddy areas than in sandy bottoms.

There is a clear faunal zonation along salinity gradients. In general, high salinity areas are more diversified than low salinity areas. Silva (1994) identified four zones with distinct salinities and benthic assemblages throughout the complex: Pre-marine (average salinity of 19.1)—*Tagelus plebeius*, *Anomalocardia brasiliana*, *Macoma constricta*, *Balanus* sp., *Callinectes danae*, and *Capitella* sp.; Estuarine (average salinity of 9.3)—*Mytella charruana*, *Neritina virginea*, *Palaemon pandaliformis* (in the rainy season), and *Laeonereis culveri*; Preliminal (average salinity of 3)—*Mytella charruana*, *Neritina virginea*, *Palaemon pandaliformis*, *Macrobrachium acanthurus*, and *Laeonereis culveri*; and Limnetic (average salinity of 0.8)—*Pomacea lineata*, *Palaemon pandaliformis*, *Callinectes bocourti*, and *Laeonereis culveri*. The occurrence and density of some species may change depending upon prevailing hydrographic condition. The appearance of *Pomacea lineata* in the Manguaba Lagoon and the disappearance of *Balanus* in all areas are observed during low salinity periods. *Macrobrachium acanthurus* migrates to river mouths and *Callinectes bocourti* is restricted to the core of the Manguaba Lagoon where salinity is lower in the drought.

The bivalves *Mytella charruana*, *Tagelus plebeius*, and *Anomalocardia brasiliana*, and the crabs *Callinectes danae* (Mundaú Lagoon) and *C. bocourti* (Manguaba Lagoon) are locally exploited. However, very little is known about their natural stocks, except for *M. charruana*, although the available data are mostly from studies in the 1980s and 1990s. Silva (1994) reported densities of exploited species in the order of 740.1 ind m^{-2} (*M. charruana*), 43.5 ind m^{-2} (*T. plebeius*), 21.7 ind m^{-2} (*Anomalocardia brasiliana*), 17.4 ind m^{-2} (*C. danae*), and 13.1 ind m^{-2} (*C. bocourti*). The high economic importance of *M. charruana* for populations in the Mundaú-Manguaba Complex area was highlighted by Pereira-Barros (1987) who reported extractions of around 3000 t year (which corresponds to approximately 10% of the stock). This bivalve is an important source of food and income for local populations in this estuary. According to this author, the species is present in about 15% of the total area of the complex, which is approximately 54 km^2, and reaches maximum densities of 1770 ind m^{-2} (Silva 1994).

Benthic composition and distribution patterns are still largely unknown along the states of Piauí, Rio Grande do Norte, and Sergipe. In the Camurupim River estuary in Piauí, 83 taxa were registered with the dominance of Polychaeta, Crustacea, and Mollusca represented in 37%, 38%, and 14% of all taxa and 47%, 33%, and 11% of all individuals, respectively (Rolemberg et al. 2008). Costa (2016) evaluated the environmental heterogeneity in the seagrass meadow in a tidal plain of the Timonha-Ubatuba Estuarine Complex and showed an environmental gradient in the macrofauna distribution with mollusks being the most frequent and accounting for 77% of all individuals distributed among 48 species; the 16 species of crustaceans represented 13% of total sampled individuals. Polychaetes accounted for 8.9% of the species recorded in the area, being represented by 25 species. The bivalves *Anomalocardia brasiliana*, *Tagelus plebeius*, *Mytella guyanensis*, and *Crassostrea rhizophorae*, and the crustaceans *Callinectes* spp. and *Ucides cordatus* are used by the population as food and/or source of income (Nascimento and Sassi 2007). Costa (2016) briefly described the artisanal exploitation of *Crassostrea mangle* and *Crassostrea brasiliana* in the Parnaiba River delta (PI).

Ferreira and Sankarankutty (2002) recorded 70 species of decapods in the Potengi, Casqueira and Conceição (Macau) River estuaries from Rio Grande do Norte and noticed that the number of species is higher in the sublitoral with sandy substrata with fragments of shells and coral skeletons than in other types of substrates. In Sergipe estuaries, polychaetes are the best-known group with 81 recognized taxa belonging to 32 families. Polychaete diversities are maximal near the mouth of the estuary and minimal in the limnetic zone. In the São Francisco River estuary (Parapuca Channel), Souza et al. (2015) reported the occurrence of the polychaete *Streblospio benedicti* at densities of approximately 14,000 ind m^{-2}. Rosa and Almeida (2012) recorded the occurrence of the crustacean *Axianassa australis* in the Real and São Francisco Rivers estuaries.

3.2.2 Processes Mediated by Benthic Organisms

Animal–sediment relationships in estuarine sedimentary environments are complex. According to Newell et al. (2001), they are not necessarily linear and tend to be much more variable than is commonly recognized. Behavioral activities of most benthic invertebrates may alter the physicochemical properties of sediments and may also modify the composition of the surrounding community. Franklin-Júnior (2000) observed that the occurrence of benthic animals on tidal flats of the Mamanguape River Estuary in Paraíba is limited to a small vertical zone, not exceeding 10–20 cm in depth. Bioturbation studies on some estuaries have also tried to evaluate sedimentary effects by crustaceans (Araújo-Júnior et al. 2012, 2016; Pülmanns et al. 2014), bivalves (Farias 2008; Silva 2006), and polychaetes (Rasmussen et al. 1998; Eça et al. 2013).

Silva (2006) identified two sediment conditions severely constrained by the presence of benthic invertebrates burrowing in sandy-muddy banks in the Pacoti River estuary in Ceará. The high density of the tube-building polychaete *Leoneries culveri* (10,422 ind m^{-2}) in areas with a high percentage of sand contributed to an increased movement of water through the sediment that resulted in high concentrations of oxygen in the sediment. Additionally, areas with more balanced percentages of silt and sand favored high densities of apseudid tanaidaceans (1641 ind m^{-2}), which act on the disarticulation of sediment grains by removing organic matter that aggregate sediment grains. Oliveira et al. (1988) also observed this close relationship between the occurrence of tanaidacean crustaceans and shellfish-eating mollusks in mixed or muddy sediments with high organic matter content in the estuarine plains of the Pacoti River in Ceará.

The bioturbation also plays an important role in the denitrification of marine sediments, recycling and transportation of nutrients from the sediment to the water column, and decomposition of organic matter, while also widening the distribution of contaminants (Cesário 2007). Such processes, therefore, change in the chemical

Fig. 3.2 *Tagelus plebeius* tube presenting the oxygenated halo (arrows) in the sediment around inhalant and exhalation channels in a sandy-muddy bank of the Ceará River estuary

composition of the sediment and promote microbial activity (Rasmussen et al. 1998). The crabs *Uca maracoani* and *Ucides cordatus* are considered important bioturbators in Brazilian semiarid mangrove soils; these species are capable of enhancing organic matter decomposition and shifting the dominant pathway of organic matter degradation in the mangrove environment (Araújo-Júnior et al. 2012, 2016; Bezerra 2005). The importance of the crab's burrowing activity has been well recognized as affecting the oxygen input in the sediment by overlying water with mixed reduced and oxidized sediment layers (Ferreira et al. 2007).

The vertical mixing particles and pore water exchange conducted by respiratory pump filtrating organisms are important bioturbation processes. Farias (2008) studied a population of *Tagelus plebeius* in the Ceará River estuary and found that this filter-feeding bivalve builds deep tubular galleries, which allow its vertical movement through the sediment. In addition, while buried at an intermediate depth, the species projects siphons forming inhalant and exhalant channels of water circulation that promotes an oxygenated halo in the sediment (Fig. 3.2). The oxidized burrow walls create microenvironments with steep gradients between reduced and oxidized compounds. These transition zones support increased microbial activities by providing ideal conditions for reoxidation processes (Bertics and Ziebis 2009).

The presence of dense concentrations of empty shells in the water–sediment interface of estuarine plains is another biogenic factor that can change the physical and chemical characteristics of the sediment and protect it from erosion. The amount of shells on the surface of the sediment is related to both sedimentation and shell production rates. Pinheiro (2003) and Silva (2006) found high levels of calcium carbonate ($CaCO_3$) and gravel in the sediment of sand-muddy banks in the Pirangi River estuary (Ceará) due to the presence of mollusk shell fragments. These accumulations of disarticulated valves can form dense packages that overlay the sediment interface and may contribute to the provision of $CaCO_3$ in addition to the decomposition of biogenic organic matter and other components such as silicon dioxide (SiO_2). These reactions influence the environment's pH and redox potential (Eh) and cause depletion or formation of solid or fluid compounds in estuarine sediments.

Studies with the gastropod *Melampus coffeus*, a common gastropod species in Atlantic mangroves that feeds on mangrove leaves, indicated that they accelerate litter decomposition processes and facilitate the consumption to plant debris by other organisms and consequently contribute to recycling organic matter in an increased rate (Tavares et al. 2011, 2015).

Carvalho et al. (2010) demonstrated that the microbiota associated with the digestive tract of *Neoteredo reynei* shows cellulolytic activity, thus contributing to cellulose digestion of fragments of wood. Therefore, similarly to the gastropod *Melampus coffeus*, teredinids contribute to the degradation process of plant material accumulated in local mangroves. Additionally, teredinids contribute to increase habitat heterogeneity because their galleries are colonized by different invertebrates, providing them with shelter and food, as observed by Santos (2008) and Carmo (2015) in the Pacoti and Acaraú Rivers mangroves, respectively.

Although the processes mediated by benthic organisms, which are determined by already well-established patterns, are well known in many parts of the world and in Brazil, more studies on the subject are needed in the NE Marine Ecoregion of Brazil to address their local relative importance. This region is mostly located in the semi-arid zone and shows peculiar environmental conditions (shallow estuaries, low freshwater inflow, and high salinization), which are reflected in the composition and structure of marine invertebrate assemblages and, consequently, in the way in which these organisms interact with the environment.

3.3 Impacts and Conservation Issues

Estuaries from the NEME host habitats of great economic potential, which explains why human interventions are the main threats to these ecosystems. Anthropogenic impacts such as urbanization, clearing of mangrove forests to shrimp farming, and pollution cause either direct or indirect ecosystem losses (Queiroz et al. 2013; Suárez-Abelenda et al. 2014; Kauffman et al. 2018; Bernardino et al. 2018).

The 3300 km Northeastern coast in Brazil is responsible for 94% of all Country's shrimp production, and 70% of cultivation facilities use estuarine and mangrove habitats encompassing a total of 19,610 hectares (ABCC/MPA 2013). The edafoclimatic conditions in the region are favorable and attractive for the cultivation of the exotic shrimp species *Litopenaeus vannamei* (ABCC/MPA 2013). Shrimp farming activities in the estuaries in northeastern Brazil can be implemented as long as the requirements of the new Forest Code (Law 12.651/12) are observed. Nevertheless, this activity has caused numerous damages to local estuarine systems (Meireles et al. 2007; IBAMA 2005; Araújo et al. 2009; Tancredo et al. 2011), among them: I—sociocultural impacts that led to changes in the standards of living and livelihood activities of riverbank assemblages; II—plant cover removal and degradation of mangroves and restinga vegetation; III—biodiversity loss in birds, fish, and invertebrates associated to mangroves, due to their dependence for refuge, mating, breeding, recruitment, nutrition, and fattening; IV—soil impoverishment as a consequence of salinization; V—eutrophication, increase of biological oxygen demand, total solids in suspension, and toxicity, and possibility of bioaccumulation or resistance to chemical substances; VI—changes in the hydrodynamic equilibrium of estuaries and contribution of terrigenous sediments and suspended and dissolved materials.

Benthic macroinvertebrates reflect these changes in the environment with an increase in the presence of infaunal species that are related to increased food availability and oxygen content as observed in the Jaguaribe estuary in Ceará (Rolemberg 2009). Lima (2015) studied the Acaraú, Aracatimirim, and Aracatiaçu Rivers estuaries in Ceará and found that areas impacted by shrimp farming presented higher richness of macrofauna species while conserved areas presented lower species richness and abundance due to the increase of nutrients from the effluents of shrimp farming, which in the long term may lead to the compromise of the local fauna. Nevertheless, some chemicals used in shrimp farming can cause negative effects in the early stages of the development of several species leading to tissue damage as observed in *Anomalocardia brasiliana* in the Potengi River estuaries (Rio Grande do Norte) (Emerenciano et al. 2008).

Another activity that causes a strong impact in Brazilian northeastern estuaries is the deforestation of mangroves. The extraction of wood by traditional communities to use as firewood and charcoal, residential constructions, building of fishing artifacts, and the use of the area for diversified developments are the main drivers. The building of ports and industrial infrastructure, as in the Suape estuary in Recife, Pernambuco, can also lead to deforestation resulting from landfills, dredging, and damming, which alter the local hydrology and drastically change the landscape. Studies carried out in Ceará indicate that a high density of dead individuals and dominance of *Laguncularia racemosa* (Paula 2015) are observed in deforested forests under regeneration. Thus, deforestation affects the structure of mangrove forests and results in loss of the inherent characteristics of these environments such as the muddy texture of their sediments. These changes promoted, for example, changes in the distribution and abundance of mollusk species that are typical of mangrove forests, such as *Littoraria angulifera* (Maia and Tanaka 2007) and *Melampus coffeus* (Maia and Coutinho 2013), as the result of changes in the availability and quality of resources, and thus compromising the mangrove structure and functioning in the Acaraú River estuary in Ceará.

The disposal of wastewater due to factors such as population increase and increasing urban development, tourism, and industrialization can expose the environment with pollutants according to the tidal regime, flow and the frequency of discharge, and the estuary's productivity. Berlamino et al. (2012) evaluated the contamination of solid waste in mangrove areas associated with the Potengi River estuary in the city of Natal (Rio Grande do Norte) and established a relationship between this pollutant and the location, levels of land use and occupation, and type of use of these areas demonstrating the need for planning and infrastructure. Another study was carried out in the municipality of Acaraú, located on the western coast of Ceará State in Brazil, to evaluate impacts caused by solid waste on the mangrove ecosystem by emphasizing losses in the vegetation structure (Souza 2016). The results indicated a great predominance of residues from vessels such as wood and styrofoam, which reflected in a high mortality of plants due to the imprisonment of these residues in pneumatophores preventing respiration and leading to the death of species in the most impacted areas. Lima (2015) observed that areas impacted by solid residues showed a dominance of polychaetes from the Nereididae and Capitellidae families when compared to conserved areas.

Therefore, the estuaries of the Northeastern Ecoregion are among the most affected by the human occupation process in Brazil and need actions that guide an integrated management to maintain an ecosystem sustainability. One of the strategies used for environmental conservation and management of these environments is the creation of protected areas. For example, the Environmental Protection Area (APA) of the Parnaíba Delta (PI), APA of Barra do Mamanguape (PB), and the Cocó River State Park (CE). However, although environmental legislation has evolved significantly in these areas, it has been shown to be of low effectiveness and with gaps in many states. Thus, environmental education will be of great relevance to inform local communities towards the need to preserve mangroves and estuarine habitats towards a sustainable use of ecosystem goods in the Northeastern States in Brazil.

3.4 Final Considerations

The amount and extent of biological and ecological knowledge, or even basic taxonomic resolution, are heavily unbalanced in estuaries from the Brazilian northeastern ecoregion. The available information is often restricted to taxonomic information (in some cases only species lists) for most estuaries. Few studies have generated consistent data on abundance, biomass or assemblage structure and function. Therefore, even the most basic current background information is still lacking for most regional estuaries. Future surveys need to include data on assemblage structure and function, both in protected and unprotected areas, to serve as a reference for effective management planning and a tool for decision-makers. Environmental education will have to assume a greater regional relevance among researchers, government, and society to counterbalance the loss of estuarine goods and services in areas increasingly impacted by human activities.

References

ABCC/MPA (2013) Levantamento da Infraestrutura Produtiva e dos Aspectos Tecnológicos, Econômicos, Sociais e Ambientais da Carcinicultura Marinha no Brasil em 2011. Associação Brasileira de Criadores de Camarão. http://abccam.com.br/site/wp-content/uploads/2013/12/LEVANTAMENTO-DA-INFRAESTRUTURA-PRODUTIVA.pdf. Accessed 29 Jun 2017

Alcantara-Filho P (1978) Contribuicão ao estudo da biologia e ecologia do caranguejo-ucá, Ucides cordatus cordatus (Linnaeus, 1763) (Crustacea, Decapoda, Brachyura), no manguezal do rio Ceará (Brasil). Arquivos de Ciências do mar 18:1–42

Alvares CA, Stape J, Sentelhas PC, Gonçalves JM, Sparovek G (2014) Kop-pen's climate classification map for Brazil. Meteorol Z 22(p):711–728

Alves RRN, Nishida AK (2004) Population structure of the mangrove crab Ucides cordatus (Crustacea: Decapoda; Brachyura) in the estuary of the Mamanguape River, Northeast Brazil. Trop Oceanogr 32:23–37

Alves VEN, Patrício J, Dolbeth M, Pessanha A, Palma ART, Dantas EW, Vendel AL (2016) Do different degrees of human impact affect the diet of Atherinella brasiliensis (Brazilian silverside) in two Brazilian estuaries? J Fish Biol 89:1239–1257

Araújo MS, Calado T (2008) Bioecologia do caranguejo-uçá *Ucides cordatus* (Linnaeus) no complexo estuarino lagunar Mundáu/Manguaba (CELMM), Alagoas, Brasil. Revista de Gestão Costeira Integrada 8(2)

Araújo MV, Freire GSS, Costa SSL, Portela JP (2009) Análise geoambiental da área estuarina do rio Acaraú, município de Acaraú – Ceará – Brasil, usando técnicas de sensoriamento remoto. Paper presented at the Simpósio Brasileiro de Sensoriamento Remoto, Instituto Nacional de Pesquisas Espaciais, Natal, 25–30 April 2009

Araújo MSLC, Tenório DO, Castiglioni DS (2014) Diversity and distribution of the Crustacea Brachyura from the mangroves of Ariquindá and Mamucabas Rivers, South Coast of Pernambuco, Brazil. J Integr Coast Zone Manag 14(3):483–499

Araújo-Júnior JMC, Otero XL, Marques AGB, Nóbrega GN, Silva JRF, Ferreira TO (2012) Selective geochemistry of iron in mangrove soils in a semiarid tropical climate: effects of the burrowing activity of the crabs Ucides cordatus and Uca maracoani. Geo-Mar Lett 32(4):289–300

Araújo-Júnior JMDC, Ferreira TO, Suarez-Abelenda M, Nóbrega GN, Albuquerque AGBM, Carvalho-Bezerra A, Otero XL (2016) The role of bioturbation by Ucides cordatus crab in the fractionation and bioavailability of trace metals in tropical semiarid mangroves. Mar Pollut Bull 111:194–202

Barbosa WFA (2011) Estrutura da comunidade fitoplanctônica do estuário do rio São Francisco (Alagoas/Sergipe, Brasil). Dissertation, Universidade Federal de Pernambuco

Barreira CAR, Araújo MLR (2005) Ciclo reprodutivo de Anomalocardia brasiliana (Gmelin, 1791) (Mollusca, bivalvia, Veneridae) na praia do Canto da Barra, Fortim, Ceará, Brasil. Bol Inst Pesca 31:9–20

Barroso CX, Matthews-Cascon H (2009) Distribuição espacial e temporal da malacofauna no estuário do rio Ceará, Ceará, Brasil. Pan-Am J Aquat Sci 4:79–86

Berlamino PHP, Silva SM, Dantas IPB, Silva AS, Rufener MC, Araújo MCB (2012) Resíduos sólidos em áreas de manguezal no rio Potengi (Natal-RN): relação com a localização e uso das áreas. Paper presented at the Congresso Brasileiro de Oceanografia, Rio de Janeiro, 13–16 November 2012

Bernardino AF, Netto SA, Pagliosa PR et al (2015) Predicting ecological changes on benthic estuarine assemblages through decadal climate trends along Brazilian Marine Ecoregions. Estuar Coast Shelf Sci 166:74–82. https://doi.org/10.1016/j.ecss.2015.05.021

Bernardino AF, Pagliosa PR, Christofoletti RA, Barros F, Netto SA, Muniz P, Lana PC (2016) Benthic estuarine assemblages in Brazil: moving forward to long term studies to assess climate change impacts. Braz J Oceanogr 64:83–97

Bernardino AF, Gomes LEO, Hadlich HL et al (2018) Mangrove clearing impacts on macrofaunal assemblages and benthic food webs in a tropical estuary. Mar Pollut Bull 126:228–235. https://doi.org/10.1016/j.marpolbul.2017.11.008

Bertics VJ, Ziebis W (2009) Biodiversity of benthic microbial assemblages in bioturbated coastal sediments is controlled by geochemical microniches. ISME J 3:1269–1285

Bezerra LEA (2005) Distribuição espacial e Aspectos Populacionais de caranguejos do gênero Uca Leach, 1814 (Crustácea: decapoda) no Manguezal do Rio Pacoti (Aquiraz-CE). Dissertation, Universidade Federal do Ceará

Bezerra LEA, Dias CB, Santana GX, Matthews-Cascon H (2006) Spatial distribution of fiddler crabs (genus Uca) in a tropical mangrove of northeast Brazil. Sci Mar 70(4):759–766

Botter-Carvalho ML, Carvalho PVVC, Santos PJP (2011) Recovery of macrobenthos in defaunated tropical estuarine sediments. Mar Pollut Bull 62:1867–1876

Calado TCS, Sousa EC (2003) Crustáceos do Complexo Estuarino-Lagunar Mundaú/Manguaba – Alagoas. FAPEAL, Maceió

Carmo EL (2015) Efeitos da salinidade na distribuição e infestação de teredinidae e na composição da macrofauna associada às suas galerias em manguezais de Acaraú – CE. Dissertation, Instituto Federal de Educação, Ciência e Tecnologia do Ceará

Carvalho PVVDB (2004) O macrozoobentos na avaliação da qualidade ambiental de áreas estuarinas no litoral norte de Pernambuco. Dissertation, Universidade Federal de Pernambuco

Carvalho AFU, Farias DF, Rocha-Bezerra LCB, Fernandes DAO, Queiroz LR, Viana MG, Paes FA, Barroso HS, Mesquita PCM, Melo VMM, Rocha-Barreira CA (2010) Preliminary study on cellulolytic activity of Neoteredo reynei (Bartsch, 1920) (Mollusca: Bivalvia: Teredinidae). Arquivos de Ciências do Mar 43(2):40–44

Cesário RIT (2007) Efeito da bioturbação na distribuição de metais e de isótopos estáveis de Pb em sedimentos de sapal, no estuário do Tejo. Dissertation, Universidade de Lisboa

Coêlho PA (1965/1966) Os Crustáceos Decápodos de Alguns Manguezais de Pernambuco. Trabalho do Instituto Oceanográfico da Universidade Federal de Pernambuco 7–8:71–89

Coêlho PA (2000) Carcinofauna. In: Barros HM, Macedo SJ, Eskinazi-Leça E, Lima T (eds) Gerenciamento Participativo de Estuários e Manguezais, 1st edn. UFPE, Recife, pp 119–142

Costa FN (2016) A heterogeneidade ambiental em angiospermas marinhas e os efeitos sobre a variação espacial da macrofauna associada no complexo estuarino Timonha-Ubatuba, Piauí, Brasil. Dissertation, Universidade Federal do Ceará

Eça GF, Pedreira RM, Hatje V (2013) Trace and major elements distribution and transfer within a benthic system: polychaete Chaetopterus variopedatus, commensal crab Polyonyx gibbesi, worm tube, and sediments. Mar Pollut Bull 74:32–41

El-Deir SG (2009) Estudo da mariscagem de Anomalocardia brasiliana (Mollusca: Bivalvia) nos bancos de Coroa do Avião, Ramalho e Mangue Seco (Igarassu-Pernambuco, Brasil), Dissertation, Universidade Federal de Pernambuco

Emerenciano DP, Silva HFO, Carvalho GC, Cruz AMF, Moura MFV (2008) Análise da ocorrência de metais: bário, cádmio, chumbo, cobre, cromo, estanho, níquel e zinco, em mexilhão (Anomalocardia brasiliana) coletados no Estuário Potengi/Jundiaí-RN. Revista Publica Natal 4:1–9

Farias MF (2008) Ciclo reprodutivo, distribuição populacional e condições microbiológicas de Tagelus plebeius (Lightfoot, 1786) (Mollusca, Bivalvia, Solecurtidae) no estuário do Rio Ceará, Fortaleza – Ceará. Dissertation, Universidade Federal do Ceará

Farias KL, Trindade RDC, Alcântara AV (2010) Ocorrência de E. coli (Epec e Eiec) no sururu, Mytella guayanensis Lamarck, e na água do estuário do rio Vaza Barris (Sergipe, Brasil). Arquivos de Ciências do Mar 42(2):66–70

Farrapeira CMR (2006) Barnacles (Cirripedia Balanomorpha) of the estuarine region of Recife, Pernambuco, Brazil. Trop Oceanogr 34(2):100–119

Farrapeira CMR (2008) Cirripedia Balanomorpha del estuario del Río Paripe (Isla de Itamaracá, Pernambuco, Brasil). Biota Neotropica 8(3):31–39

Farrapeira CMR, Coelho-Filho PA, Santos MA (2000) Zonação vertical da macrofauna bêntica de substratos consolidados do estuário do Rio Paripe (Itamaracá-PE). Paper presented at the MANGROVE 2000: Sustentabilidade de estuários e manguezais: desafios e perspectivas, Universidade Federal Rural de Pernambuco, Recife, October 2000

Farrapeira CMR, Ramos CAC, Barbosa DF, Melo AVOM, Pinto SL, Verçosa MM, Oliveira DAS, Francisco JA (2009) Zonación vertical del macrobentos de sustratos sólidos del estuario del río Massangana, Bahía de Suape, Pernambuco, Brasil. Biota Neotropica 9(1):87–100

Ferreira AC, Sankarankutty C (2002) Estuarine Carcinofauna (Decapoda) of Rio Grande do Norte, Brazil. Revista Nauplius 10(2):121–129

Ferreira TO, Otero XL, Vidal-Torrado P, Macías F (2007) Effects of bioturbation by root and crab activity on iron and sulfur biogeochemistry in mangrove substrate. Geoderma 142(1):36–46

Franklin-Júnior W (2000) Macrofauna Bentônica da Região entremarés de bancos areno-lamosos em um estuário Tropical: Rio Mamanguape, Paraíba, Brasil. Dissertation, Universidade Federal da Paraíba

Góes JM, Fernandes-Góes LC (2007) First occurrence of the crab Calappa nitida Holthuis, 1958 (Brachyura, Calappidae, Calappinae) on the coast of the state of Piauí, Brazil. Biotemas 20(1):123–125

IBAMA (2005) Diagnóstico da atividade de carcinicultura no Estado do Ceará. Ministério do Meio Ambiente. http://www.mma.gov.br/port/conama/processos/0B19D3B1/DIAGDACARCINICULTURACEARA.pdf. Accessed 30 Jun 2017

Kauffman JB, Bernardino AF, Ferreira TO, Bolton NW, de Gomes LE, O, Nobrega GN (2018) Shrimp ponds lead to massive loss of soil carbon and greenhouse gas emissions in northeastern Brazilian mangroves. Ecol Evol. doi.org/10.1002/ece3.4079

Lacerda LD (1999) Os manguezais do Brasil. In: Vannucci M (ed) Os manguezais e nós: uma síntese de percepções, 1st edn. Editora da USP, São Paulo, pp 185–196

Leonel RMV, Lopeso SGBC, Aversari M (2002) Distribution of wood-boring bivalves in the Mamanguape River estuary, Paraíba, Brazil. J Mar Biol Assoc U K 82:1039–1040

Lima BKS (2015) Uso de macroinvertebrados bentônicos como bioindicadores de impactos ambientais em manguezais estuarinos. Dissertation, Instituto Federal de Educação, Ciência e Tecnologia do Ceará

Mai ACG, Rosa IML (2009) Ecological aspects of the seahorse Hippocampus reidi in the Camurupim/Cardoso estuary, Piauí State, Brazil, as subsidies for the implementation of an Environmental Protection Area. Biota Neotropical 9(3):85–91

Maia RC, Coutinho R (2013) The influence of mangrove structure on the spatial distribution of Melampus coffeus (Gastropoda: Ellobiidae) in Brazilian estuaries. Pan-Am J Aquat Sci 8(1):21–29

Maia RC, Tanaka MO (2007) Avaliação de efeitos de espécies de mangue na distribuição de Melampus coffeus (Gastropoda, Ellobiidae) no Ceará, Nordeste do Brasil. Iheringia Série Zoologia 97(4):379–382

Maia RC, Rodrigues AB, Rocha-Barreira CA (2011) Distribuição espaço-temporal de Melampus coffeus (Mollusca: Gastropoda) em manguezais estuarinos da região metropolitana de Fortaleza, Ceará, Brasil. Conexões – Ciência e Tecnologia 5(3):28–35

Matthews-Cascon H, Martins IX (1999) Notes on the reproduction of Neritina virginea (Linnaeus, 1758) in northeast Brazil (Mollusca: Gastropoda: Neritidae). Arquivos de Ciências do Mar 32:129–132

McLusky DS, Elliot M (2004) The estuarine ecosystem: ecology, threats and management. Oxford University Press, New York

Medeiros EL, Fernandes GV, Henry-Silva GG (2015) Distribution and density of the mollusk Donax striatus (Linnaeus, 1767) in a tropical estuarine region in the brazilian semi-arid. Braz J Biol 75(4):914–922

Meireles AJA, Cassola RS, Tupinambá SV, Queiroz LS (2007) Impactos Ambientais Decorrentes das Atividades da Carcinicultura ao Longo do Litoral Cearense, Nordeste do Brasil. Mercator – Revista de Geografia da UFC 6(12):83–106

Menezes RS, Menezes MF (1968) Estudo preliminar sobre a flora e fauna de águas estuarinas do Estado do Ceará. Arquivos dos Estudos de Biologia Marinha da Universidade Federal do Ceará 8(1):101–106
Miranda PTC, Gurgel FFG, Liberato MAF, Oliveira MT, Arruda TLB (1988) Comunidades bentônicas em raízes de Rhizophora mangle Linnaeus, no manguezal do Rio Ceará (Ceará-Brasil). Arquivos de Ciências do Mar 27:101–110
Miranda PTC, Maciel NC, Eskinazi-de-Oliveira AM (1990) Política estadual para preservação de manguezais e estuários do Ceará: proposta. SEMACE, Fortaleza
Nascimento MSV, Sassi R (2007) Análise da atividade pesqueira e das condições socioeconômicas dos pescadores artesanais de Cajueiro da Praia, Estado do Piauí, Brasil. Gaia Scientia 1(2):141–154
Nascimento DM, Mourão JS, Alves RRNA (2011) Substituição das técnicas tradicionais de captura do caranguejo-uçá (Ucides cordatus) pela técnica "redinha" no estuário do rio Mamanguape, Paraíba. Sitientibus série Ciências Biológicas 11(2):113–119
Newell RC, Seiderer LJ, Robinson JE (2001) Animal:sediment relationships in coastal deposits of the eastern English Channel. J Mar Biol Assoc U K 81:1–9
Nishida AK, Nordi N, Alves RRN (2006) The lunar-tide cycle viewed by crustacean and mollusc gatherers in the State of Paraíba, Northeast Brazil and their influence in collection attitudes. J Ethnobiol Ethnomed 2(1):1–12
Oliveira AME, Irving MA, Lima HH (1988) Aspectos bioecológicos do estuário do rio Pacoti, Ceará, Brasil. Arquivos de Ciências do Mar 27:91–100
Otsuka AY, Feitosa FAN, Flores-Montes MJ, Honorato-da-Silva M, Travassos RK (2014) Condições Ambientais do estuário do rio Botafogo (Itamaracá-Pernambuco-Brasil): clorofila a e algumas variáveis ambientais. Trop Oceanogr 42:112–128
Paula ALS (2015) Estrutura vegetal e grau de perturbação ambiental em bosques de mangue estuarinos. Dissertation, Instituto Federal de Educação, Ciência e Tecnologia do Ceará
Pereira ACA (2007) Bioindicadores fitoplanctônicos da qualidade da água no estuário do rio Catu, Aquiraz, Ceará. Dissertation, Universidade Federal do Ceará
Pereira-Barros JB (1987) Bibliografia comentada sobre o sururu Mytela falcata e a bioecologia do complexo estuarino Mundaú-Manguaba, AL. Parte I. Boletim de Estudos de Ciências do Mar 6:36–48
Pinheiro DR (2003) Análise do comportamento hidrodinâmico e sedimentológico do Estuário do Rio Piranji-CE (NE/BRASIL). Dissertation, Universidade Federal do Rio Grande do Sul
Pülmanns N, Diele K, Mehlig U, Nordhaus I (2014) Burrows of the semi-terrestrial crab Ucides cordatus enhance CO2 release in a North Brazilian mangrove forest. PLoS One 9(10):1–13
Queiroz L, Rossi S, Meireles J, Coelho C (2013) Shrimp aquaculture in the federal state of Ceará, 1970–2012: trends after mangrove forest privatization in Brazil. Ocean Coast Manag 73:54–62
Quinamo TS (2006) Artisanal fishing and environment in mangrove areas in the estuary-coast complex of Itamaracá, Pernambuco: the case of Itapissuma. Dissertation, Universidade Federal da Paraíba
Rasmussen AD, Banta GT, Andersen O (1998) Effects of bioturbation by the lungworm Arenicola marina on cadmium uptake and distribution in sandy sediments. Mar Ecol Prog Ser 164:179–188
Rocha-Barreira CA, Silva HL, Vidal JMA, Silva AF, Santos CLT, Bezerra DF (2005) Macroinfauna bentonica de ambientes inconsolidados estuarinos no estado no Ceará. Governo de Estado do Ceará. Programa Ecológico e Econômico da Zona Costeira do Estado do Ceará, Fortaleza
Rodrigues AM, Azevedo CB, Silva GHG (2010) Aspectos da biologia e ecologia do molusco bivalve Anomalocardia brasiliana (Gmelin, 1791) (Bivalvia, Veneridae). Revista Brasileira de Biociências 8(4):377–383
Rolemberg KF (2009) Estudo dos Macrozoobentos em viveiros de engorda do camarão marinho Litopenaeus vannamei (Boone, 1931) (Crustacea, Decapoda, Penaeidae) no entorno do estuário do Rio Jaguaribe, Ceará, Brasil. Dissertation, Universidade Federal do Ceará

Rolemberg KF, Rocha-Barreira CA, Arzabe C (2008) Caracterização do macrozoobentos nas áreas sob a influência da carcinicultura no entorno do estuário do Rio Camurupim, Piauí, Brasil. Arquivo de Ciências do Mar 41:36–47

Rosa LC, Almeida AO (2012) Axianassa australis Rodrigues and Shimizu, 1992 (Crustacea: Decapoda: Gebiidea: Axianassidae): first record from Sergipe, NE Brazil. Check List 8(6):1321–1322

Rosa-Filho JS, Farrapeira-Assunção CM (1998) Zonação vertical da macrofauna bentônica dos andares supra-litoral e médio-litoral dos manguezais da região da Ilha de Itamaracá (Pernambuco – Brasil). Caderno Ômega, Série Ciências Aquáticas 4:59–68

Santana LMBM (2010) Biologia reprodutiva e considerações sobre parasitismo em Lucina pectinata (Mollusca: Bivalvia) em um estuário tropical. Dissertation, Universidade Federal do Ceará

Santa Fé ÚMG, Araujo ARR (2013) Seletividade e eficiência das artes de pesca utilizadas na captura de *Ucides cordatus* (Linnaeus, 1763), Sergipe, Brasil. Acta Fish Aquat Res 1(1):29–44

Santiago MF, Passavante JZO, Silva-Cunha MGG (2005) Caracterization of physical, chemical and biological parameters in environment hyperhaline, rio Pisa Sal river estuary (Galinhos, Rio Grande do Norte, Brasil). Trop Oceanogr 33(1):39–55

Santos CLT (2008) Análise experimental da infestação dos Teredinidae Rafinesque, 1815 (Mollusca: Bivalvia) e os invertebrados associados às suas galerias no manguezal do estuário do Rio Pacoti, Ceará, Brasil. Dissertation, Universidade Federal do Ceará

Silva CS (1994) Distribuição e abundância da fauna macrobentônica do complexo estuarino Mundaú/Manguaba (Alagoas-Brasil). Boletim de Estudos de Ciências do Mar 8:45–64

Silva AF (2006) Caracterização da macrofauna bentônica de bancos areno-lamosos dos estuários dos rios Pacoti e Pirangi – Ceará, Brasil. Dissertation, Universidade Federal do Ceará

Silva PA (2015) Plasticidade fenotípica da concha de Neritina virginea (Linnaeus, 1758) (Gastropoda: Mollusca) em uma praia estuarina em Itarema, Ceará Dissertation, Instituto Federal de Educação, Ciência e Tecnologia do Ceará

Silva FMRO, Ribeiro FB, Bezerra LEA (2016) Population biology and morphometric sexual maturity of the fiddler crab *Uca* (Uca) *maracoani* (Latreille, 1802) (Crustacea: Decapoda: Ocypodidae) in a semi-arid tropical estuary of northeastern Brazil. Lat Am J Aquat Res 44(4):671

Silva CS, Prereira-Barros JB (1987) Inventário da malacofauna do Complexo Lagunar Mundaú/ Manguaba, Alagoas. Boletim de Estudos de Ciências do Mar 6:65–74

Silva FMRO, Ribeiro FB, Bezerra LEA (2016) Population biology and morphometric sexual maturity of the fiddler crab Uca (Uca) maracoani (Latreille, 1802) (Crustacea: Decapoda: Ocypodidae) in a semi-arid tropical estuary of northeastern Brazil. Lat Am J Aquat Res 44(4)

Silva-Cavalcanti JS (2011) Biologia e exploração de Anomalocardia brasiliana por populações de ribeirinhos no estuário do Rio Goiana (PE-PB), Nordeste do Brasil. Dissertation, Universidade Federal de Pernambuco

Silva-Mello RL, Tenório NO (2000) Malacofauna. In: Barros HM, Macedo SJ, Eskinazi-Leça E, Lima T (eds) Gerenciamento Participativo de Estuários e Manguezais. UFPE, Recife, pp 103–118

Smith SI (1869) Notice of the Crustacea collected by Prof. C.R. Hartt on the coast of Brazil, in 1867. Trans Connecticut Acad Arts Sci 2:1–14

Soares CME (1979) Estudo ecológico da região de Itamaracá, Pernambuco, Brasil. III. Anfípodos das famílias Talitridae e Ampithoidae. Trabalhos Oceanográficos da Universidade Federal de Pernambuco 14:93–104

Soares CMA (1980) Estudo Ecológico da Região de Itamaracá, Pernambuco, Brasil. VI. Anfípodos da Família Gammaridae. Trabalhos Oceanográficos da Universidade Federal de Pernambuco 15:263–276

Souza KNS (2016) Impactos ambientais da disposição inadequada dos resíduos sólidos em manguezais estuarinos no Ceará. Dissertation, Instituto Federal de Educação, Ciência e Tecnologia do Ceará

Souza JWS, Carvalho WVS, Souza JA, Menezes AR, Guimarães CRP (2015) Streblospio benedicti (Annelida: Polychaeta: Spionidae) da laguna principal do Canal do Parapuca como bioindicador da qualidade das águas do Rio São Francisco, Sergipe, Brasil. Paper presented at 2nd Congresso Internacional RESAG, Universidade Tiradentes, Aracaju, 9–11 December 2015

Sovierzoski HH (1994) Nota preliminar sobre a ocorrência de Polychaeta no litoral de Alagoas. Boletim de Estudos de Ciências do Mar 8:23–28

Spalding MD, Fox HE, Allen GR, Davidson N, Ferdana ZA, Finlayson M, Halpern BS, Jorge MA, Lombana A, Lourie SA, Martin KD, Mcmanus E, Molnar J, Recchia CA, Robertson J (2007) Marine ecoregionsof the world: a bioregionalization of Coastal and Shelf Areas. Bioscience 57:573–583

Suárez-Abelenda M, Ferreira TO, Camps-Arbestain M et al (2014) The effect of nutrient-rich effluents from shrimp farming on mangrove soil carbon storage and geochemistry under semiarid climate conditions in northern Brazil. Geoderma 213:551–559

Sunday JM, Batesc AE, Kearneye MR, Colwellf RK, Dulvyb NK, Longinoh JT, Hueyi RB (2014) Thermal-safety margins and the necessity of thermoregulatory behavior across latitude and elevation. PNAS 111(15):1–6

Tancredo KR, Nobrega RO, Dias T, Lapa KR (2011) Impactos Ambientais da Carcinicultura Brasileira. Paper presented at the 3rd International Workshop Advances in Cleaner Production: Cleaner Production Initiatives and Challenges for a Sustainable World. Universidade Paulista, São Paulo, 18–20 May 2011

Tavares DS, Maia RC, Rocha-Barreira CA (2011) Contribuição de Melampus coffeus (Gastropoda, Ellobiidae) na degradação da serapilheira do médio estuário do rio Pacoti, Ceará, Brasil. Iheringia, Série Zoologia 101(1–2):56–60

Tavares DS, Maia RC, Rocha-Barreira CA, Matthews-Cascon H (2015) Ecological relations between mangrove leaf litter and the spatial distribution of the gastropod Melampus coffeus in a fringe mangrove forest. Iheringia, Série Zoologia 105(1):35–40

Valença APMC, Santos PJP (2012) Macrobenthic community for assessment of estuarine health in tropical areas (Northeast, Brazil): review of macrofauna classification in ecological groups and application of AZTI Marine Biotic Index. Mar Pollut Bull 64:1809–1820

Vannucci M (2001) What is so special about mangroves? Braz J Biol 61:599–603

Vilanova MFV, Chaves EMB (1988) Contribuição para o conhecimento da viabilidade do cultivo de ostra-do-mangue, Crassotrea rhizophorae (Guilding, 1828) (Mollusca: Bivalvia), no estuário do rio Ceará, Ceará, Brasil. Arquivos de Ciências do Mar 27:111–125

Vilar TC, Tenório DO, Feitosa FAN (2012) Criação experimental da ostra-do-mangue Crassostrea rhizophorae (Guilding, 1828) em barra de São Miguel, Alagoas. Trop Oceanogr 40(2):296–308

Chapter 4
Benthic Estuarine Assemblages of the Eastern Marine Brazilian Ecoregion (EME)

Angelo Fraga Bernardino, Alice Reis, Antônio Carlos Dórea Pereira Filho, Luiz Eduardo de Oliveira Gomes, Lorena Bonno Bissoli, and Francisco Carlos Rocha de Barros Jr

Abstract The Eastern Brazil Marine Ecoregion includes over 50 estuaries along roughly 1200 km of coastline with latitudinal changes in mean rainfall and average yearly atmospheric temperatures. Estuarine ecosystems within this ecoregion have been relatively well studied with respect to the impacts from human pollution and the benthic biodiversity in mangrove forests, estuarine channels, and tidal flats. Benthic estuarine assemblages exhibit typical spatial changes with salinity gradients, with higher diversity towards euhaline sectors. Macrofaunal abundance and biomass are typically higher within mud and organic-rich sediments along tidal flats, although spatial patterns often differ within sectors (euhaline to oligohaline) and between estuaries in the ecoregion. The largest coastal bays and estuaries of the Eastern Marine Ecoregion are impacted by variable levels of sewage and industrial discharge and mangrove forest removal. Although the effects of these impacts likely result in changes in the amount or quality of water supply, decrease of fish stocks, and transformations of food webs, there is limited understanding of the potential loss of estuarine services. Climate change effects including higher mean atmospheric temperatures and lower rainfall are predicted to significantly impact estuarine benthic assemblages in the Eastern ecoregion and localized effects of higher salinity are already in place at some areas. Further studies need to understand accurately what are the most important estuarine functions and services in order to evaluate how different local (biological invasion, habitat destruction, pollution) and global (climate change) impacts will affect these systems.

A. F. Bernardino (✉) · L. E. de O. Gomes · L. B. Bissoli
Grupo de Ecologia Bêntica, Departamento de Oceanografia e Ecologia, CCHN, Universidade Federal do Espírito Santo, Espírito Santo, Brazil
e-mail: angelo.bernardino@ufes.br

A. Reis · A. C. D. Pereira Filho · F. C. R. de Barros Jr
Laboratório de Ecologia Bentônica, Programa de Pós-Graduação em Ecologia e Biomonitoramento, Instituto de Biologia, Centro Interdisciplinar de Energia e Ambiente, Universidade Federal da Bahia, Salvador, Bahia, Brazil
e-mail: francisco.barros@pq.cnpq.br

P. C. Lana, A. F. Bernardino (eds.), *Brazilian Estuaries*, Brazilian Marine Biodiversity, https://doi.org/10.1007/978-3-319-77779-5_4

Concomitantly, estuarine areas for conservation must be identified, implemented and managed in Eastern Brazil Marine Ecoregion.

Keywords Eastern Brazil · Benthic ecology · Impacts · Mangroves · Estuaries

4.1 Introduction

The Eastern Brazil Marine Ecoregion (EME) is located within the Tropical Southwestern Atlantic province, and includes over 50 estuaries from the states of Bahia and Espírito Santo along roughly 1200 km of coastline (11°28'S to 21°18'S; Fig. 4.1; Spalding et al. 2007). The EME has low seasonal oscillations in temperature (20–26 °C), but with higher mean annual values in the northern sector on the coast of Bahia state (24–26 °C), if compared to the southern sector along the coastal Espírito Santo state (20–24 °C; Alvares et al. 2014; Bernardino et al. 2015). Annual rainfall is also markedly variable within the EME, with higher volumes within the northern sector (1600–2200 mm/year) whereas southern Bahia and Espírito Santo are drier, with 1000–1600 mm/year (Alvares et al. 2014). Atmospheric historical data from the last 40 years evidences a mean temperature increase of 0.1 °C per decade along the EME, with frequent positive temperature anomalies (i.e., annual mean temperatures above historical means) in the last 10 years (Bernardino et al. 2015). Therefore, although the EME encompasses estuaries along the eastern coast of Brazil with a typical tropical climate, regional (and latitudinal) differences in annual rainfall and mean temperatures are evident and may result in distinct long-term ecological changes in benthic assemblages related to temperature and estuarine salinity.

Although there is a large number of estuaries (N > 50) within the EME, most scientific work has been focused on those near large urban areas including the Todos os Santos and Vitória Bays, within the metropolitan areas of Salvador and Vitória, respectively (Fig. 4.1). On the northern region of the EME, the Todos os Santos Bay (TSB) encompasses several smaller estuaries. Water salinity gradients from freshwater to marine conditions are only observed on rivers that drain in the TSB, and therefore true estuarine regions are restricted to these smaller rivers. The main estuaries in TSB are the Paraguaçu, Jaguaripe, Subaé, and minor systems including the Mataripe and São Paulo estuaries. Many other estuaries are situated in the coast of Bahia State but, with a few exceptions (e.g., Hatje et al. 2008), with virtually no published studies. On the southern region of the EME, Vitória Bay (VB) is within a large metropolitan area and exhibits a marked salinity gradient from the main river (Santa Maria de Vitória) towards the ocean, with a significant presence of mangrove forests in the inner portions of the estuarine system (Rigo 2004). Several smaller estuaries within the southern region of the EME are located either near smaller towns or near marine protected areas on the coast. Two of these smaller estuaries have been relatively well sampled, the Benevente (BE) and the Piraquê-Açu-Mirim (PAM) estuaries. Tidal flats and mangrove areas occur along all of those estuaries and the available data on their benthic biodiversity will be reviewed in this chapter (Fig. 4.2).

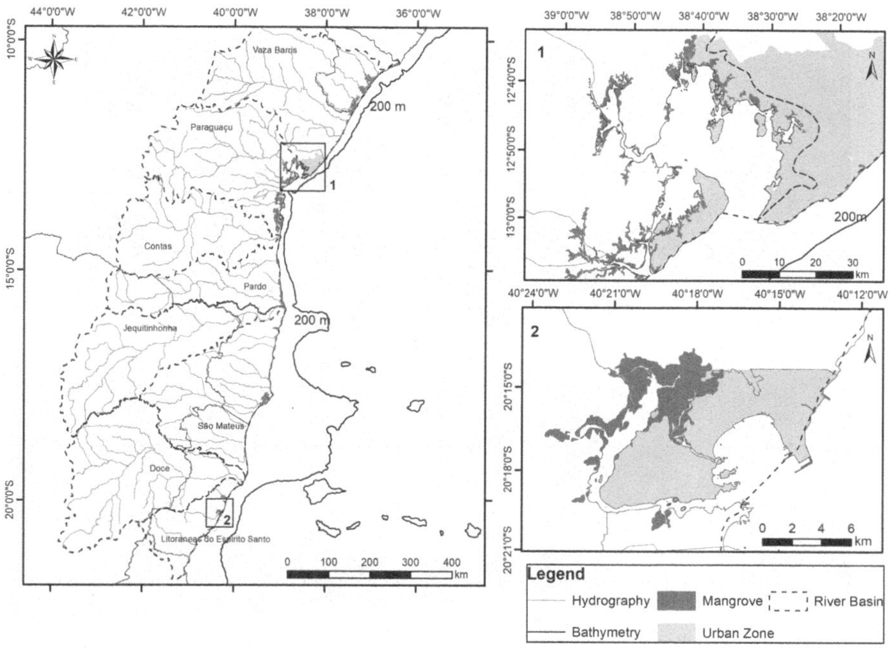

Fig. 4.1 Map (left) indicating river basins within the Eastern Brazil Marine Ecoregion (EME), with detail of the Todos os Santos Bay (1) and Vitoria Bay (2) near the metropolitan areas of Salvador and Vitória, respectively

There is considerable work on the structure of estuarine benthic assemblages from a few estuaries along the EME. Most studies evaluated short-term patterns of benthic assemblages across salinity gradients and between vegetated and unvegetated habitats (Bernardino et al. 2016). Some effort towards impact assessments from the heavily urbanized estuaries have been made and will be reviewed here (e.g., Hatje et al. 2006; Barros et al. 2008; Hatje and Barros 2012; Krull et al. 2014). Ecological and functional approaches with variable spatial and temporal scales have also been carried out in recent years, and significantly advanced the knowledge of fairly important tropical estuarine ecosystems within the EME.

As observed in other coastal regions, estuaries and their associated habitats within the EME may provide significant ecological services to human populations through water quality, biodiversity and food provision, coastal protection and climate regulation (McLeod et al. 2011; Pendleton et al. 2012). Mangrove forests are an important habitat of estuaries along the EME, but their specific ecological services to the population within this region is poorly quantified (but see Servino et al. 2018). For example, even heavily urbanized estuaries near VB and TSB have extensive mangrove forests (Costa et al. 2015) that provide food for traditional communities and habitat for estuarine and marine organisms (Rondinelli and Barros 2010), and they likely have fundamental role on water quality. In a similar way, many traditional communities depend on services provided by smaller estuaries along the EME, and many conflicts on the management of those resources exist, including the lack of proper

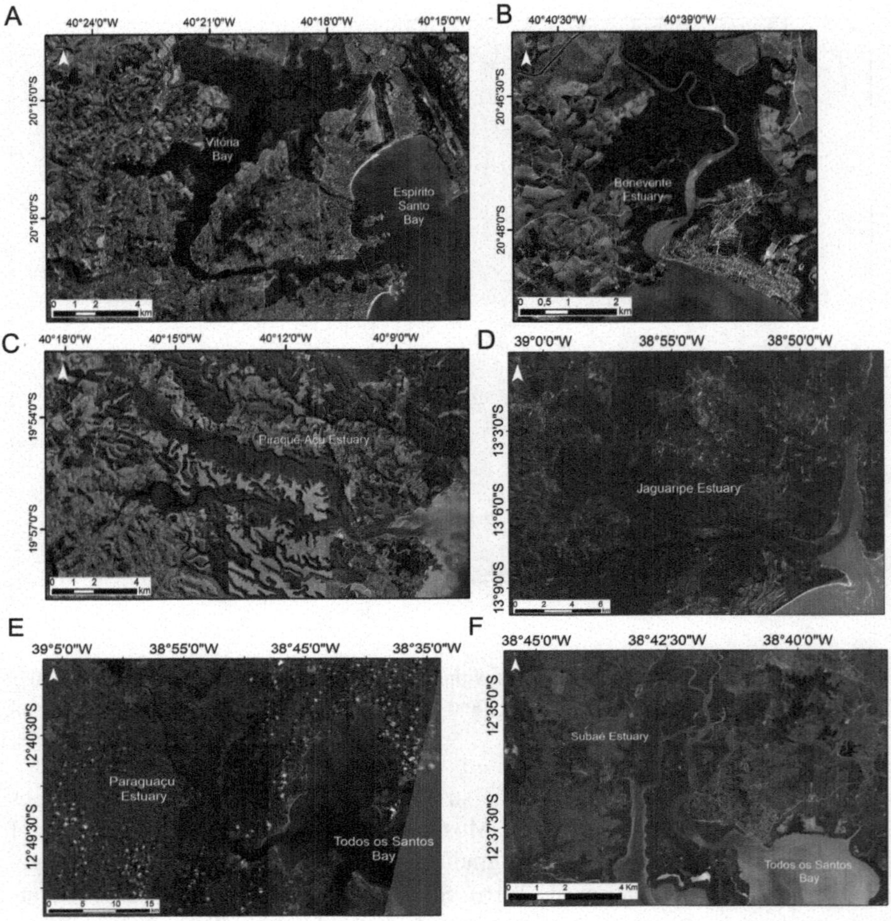

Fig. 4.2 Satellite image of six estuaries within the Eastern Brazil Marine Ecoregion (EME). A. Vitoria Bay (ES); B. Benevente estuary (ES); C. Piraquê-Açu-Mirim estuary (ES); D. Jaguaripe estuary (BA); E. Paraguaçu estuary (BA); F. Subaé estuary (BA). Images A, B, and C kindly provided by IEMA-ES. Images D, E and F from LandSat 7 and 8

wastewater treatment. Habitat loss through mangrove removal or pollution is a major problem within most estuaries and even "visually preserved" ecosystems show some degree of contamination by sewage or industrial pollutants (Grilo et al. 2013; Ribeiro et al. 2016; Bernardino et al. 2018). These pressures impact the biodiversity associated with estuarine ecosystems (Barros et al. 2014) and likely put additional pressures on the services they provide (Lotze et al. 2006), but those effects have not been quantified in most estuaries within the EME (Servino et al. 2018).

A variety of habitats within estuarine ecosystems, including mangrove and seagrass habitats, mudflats, and sublittoral unvegetated channels, provide support to a wide variety of benthic assemblages. These habitats have many species in common, but their assemblages are structured by a range of gradients such as salinity and sediment grain size, organic matter availability and processes of disturbance, species interactions and recolonization (Edgar et al. 1999; Ysebaert et al. 2003; Barros et al. 2012; Whitfield et al. 2012). This chapter will review local (i.e., within estuary) and regional (between estuaries) spatial patterns of benthic estuarine assemblages in the EME in an attempt to provide a broader understanding of their biodiversity and ecological patterns. Here we analyze published and unpublished data from the most dominant habitats within each estuary in order to compare their assemblages along the EME. We also review available published information on the impacts affecting benthic assemblages in some estuaries, and make some theoretical predictions on their ecosystem health and vulnerability to future climatic change.

4.2 Mangrove Habitats

As in many other estuaries of Brazil, mangroves are a dominant habitat along estuaries of the EME. The general pattern of mangrove species distribution is presented in Fig. 4.3, there is a substitution pattern along the salinity gradient (Costa et al. 2015). *Rhizophora mangle* occurs mostly on the euhaline and polyhaline sector of TSB estuaries. *Laguncularia racemosa* has a wide distribution, dominating the central part of Jaguaripe estuary (from euhaline until oligohaline regions) and the oligohaline sector of Paraguaçu and Subaé estuaries.

Avicennia schaueriana occurs in all sectors of Jaguaripe and Subaé estuaries and from euhaline until mesohaline sectors of Paraguaçu occurring as co-dominant with *Rhizophora mangle* on the euhaline. *Avicennia germinans* was found only on the oligohaline sector of Jaguaripe and Subaé estuaries (Costa et al. 2015) although species from this genus are recognized for supporting high salinities. TSB mangrove species distribution pattern are different from those found by Chen and Twilley (Chen and Twilley 1999) in Florida, where *L. racemosa* dominates the euhaline region and *R. mangle* dominates the oligohaline.

The southern region of the EME along the Espírito Santo state has about 70 Km^2 of mangrove forests with highest mangrove coverage within Vitória Bay (18 Km^2) and on the Piraquê-Açu-Mirim and São Mateus estuaries (12–19 Km^2; Schaeffer-Novelli et al. 1990; Servino et al. 2018). Mangroves along the inner VB estuary are well developed, with dominance of *R. mangle*, *A. schaueriana*, and *L. racemosa*. Phenotypic responses of *A. schaueriana* in response to latitudinal salinity patterns and to local sewage contamination from estuaries within metropolitan areas have been detected (Arrivabene et al. 2014). The same species of mangroves are distributed along salinity sectors in estuaries on the southern region of the Espírito Santo state, but the structure of most forests has only been studied within a few estuaries.

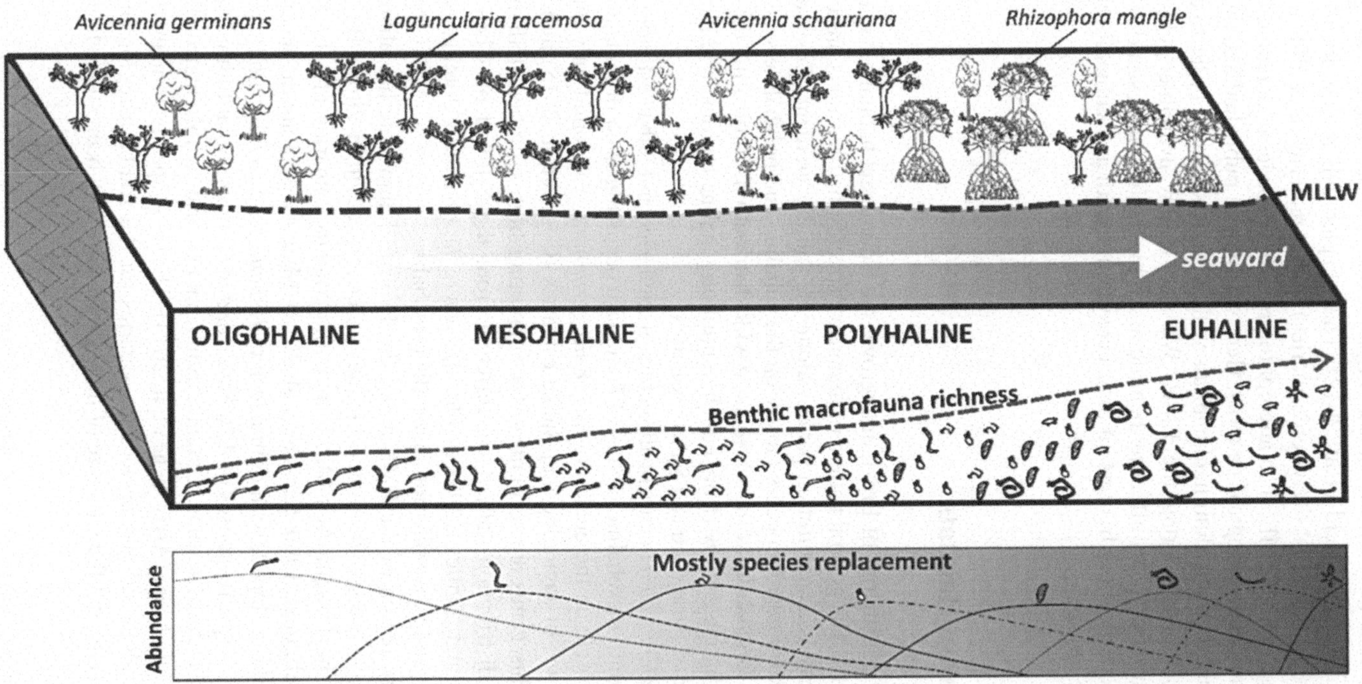

Fig. 4.3 Schematic diagram indicating mangrove and benthic species replacement along estuarine sectors (oligohaline, mesohaline, polyhaline, and euhaline) of Eastern Brazil Marine Ecoregion (EME) estuaries. MLLW: Mean Lower Low Water. Different benthic macrofauna species are represented by different symbols (e.g., worms, crustacean, mollusk, sea star)

4.3 Estuarine Benthic Assemblages

4.3.1 Patterns of Abundance and Biomass

The abundance and biomass of macrofaunal benthic assemblages are highly associated with spatial and temporal patterns of salinity variability, mud content, and estuarine production (Ysebaert and Herman 2002; Gilberto et al. 2004). Most estuaries within the EME can be classified as estuarine bays or floodplain estuaries, with intense input of continental particulate organic matter in short periods of high rainfall (Lessa et al. 2018). The seasonal changes in rainfall along the northern EME therefore can lead to latitudinal changes in community abundance and biomass in estuaries, whereas salinity regimes and sediment content may determine regional differences. However, multiple human impacts additionally alter the abundance and biomass of macrofaunal assemblages at variable spatial scales. For example, local impacts associated with organic inputs (sewage), disturbance (e.g., trampling, fishing), or forest clearing (mangrove) will lead to changes in benthic assemblage abundance and biomass (Bernardino et al. 2018).

Natural variability in salinity clearly changes macrofaunal benthic abundances in tropical estuaries within the EME (e.g., Barros et al. 2008; Mariano and Barros 2015). Higher densities have been observed in sites with higher mud content, presumably in response to higher organic content at mesohaline sectors. There is also spatial and seasonal evidence for changes in macrofaunal abundance along the continental-marine gradient in response to environmental change within estuaries in the EME but it seems that variability in space explains more than variability in time (Barros et al. 2012). Higher macrofaunal densities in areas with high organic availability within impacted estuaries also correspond to a local effect of organic enrichment from pollution. In general, tropical estuaries from the EME exhibit changes in macrofaunal abundance and biomass between estuarine habitats, along salinity sectors, and in response to changes in sediment particle size and other covariables, including sediment organic matter.

4.3.1.1 Intertidal Habitats

Spatial patterns of macrofaunal abundance in intertidal habitats have been studied along several estuaries of the EME, with salinity gradient and habitat type being the most conspicuous factors in the changes observed. For example, mudflats and mangrove habitats in the mesohaline sector of three estuaries of the EME have been compared in respect to macrofaunal abundance, biomass, and secondary production. In the Benevente estuary (BE), mudflats had higher macrofaunal secondary production (138 ± 79 mg/m^2/day), biomass (7677 ± 5051 mg^{AFDW}/m^2), and abundance (33,023 ± 14,709 ind/m^2), if compared to mangrove habitats in the same sector (42.5 ± 46.4 mg/m^2/day, Bissoli and Bernardino 2018). The same study revealed that macrofaunal biomass and secondary production were significantly higher in mangrove habitats at the other two estuaries sampled, Vitória Bay and Piraquê-Açu-Mirim, suggesting that mangrove forests are an important source of organic material to the benthic assemblages in some estuaries but not in all sites studied. When multiple estuaries are compared, there is no clear latitudinal difference in macrofaunal abundance, biomass, or secondary production within the southern sector of the EME. Differences in macrofaunal assemblages observed between mangroves and mudflats are evident, but regional heterogeneity between estuaries makes these differences inconsistent across estuaries.

Changes in abundance of benthic intertidal assemblages along salinity gradients have also been investigated in the southern region of the EME. There are no significant changes in macrofaunal abundance or biomass on mudflat habitats between euhaline and polyhaline sectors in the Piraquê-Açu-Mirim estuary (1560 ± 1170 ind/m^2 and 1221 ± 700 ind/m^2, respectively). In general, macrofaunal abundance, biomass, and production of intertidal benthic assemblages seem to be less influenced by salinity gradients if compared to subtidal habitats (Bissoli and Bernardino 2018).

In the estuaries of Todos os Santos Bay, there is evidence that different macrofaunal assemblages inhabit intertidal mudflats at different haline sectors (Krull et al. 2014; Mariano and Barros 2015). The structure of the benthic assemblages was clearly related to salinity and peaks of abundance of different taxa were reported on different estuaries. For instance, in euhaline sectors there is high abundance of the mollusk *Neritina virginea* on Jaguaripe estuary and Pilargid polychaetes in the Subaé estuary. In the oligohaline sector, Nereididae, Spionidae, Capitellidae and Ampharetidae polychaetes were more abundant (Mariano and Barros 2015). Also, greater intertidal abundances were found on Jaguaripe estuary when compared to Subaé estuary and such pattern is likely to occur due to greater anthropogenic impacts on the latter (Krull et al. 2014).

In general, small-scale patchiness is greatly responsible for the spatial variability in the structure of benthic macrofaunal assemblages within estuarine intertidal habitats. Habitat structure (i.e., heterogeneity and complexity) in estuarine tidal flats is created by a number of environmental factors including shear stress, organic matter availability, sediment grain size, and larval-adult interactions (Ysebaert and Herman 2002; see Carvalho and Barros 2017, for habitat structure conceptual clarification). Nevertheless, consistent patterns of abundance in some taxa have been observed in mudflats along the salinity gradient of different estuaries in the EME, salinity is a major driver, but specific responses to different environmental variables surely needs to be further investigated.

4.3.1.2 Subtidal Habitats

Subtidal habitats of VB and the PAM estuaries have higher macrofaunal abundances within the polyhaline sectors, which contrasts to the spatial patterns observed on intertidal habitats. Higher macrofaunal abundances along intermediate sectors of both estuaries may reflect the fine sediment grain size at those areas and a higher sewage contamination in the case of VB. For example, macrofaunal abundance in areas contaminated by sewage discharge in VB may reach up to 7465 to 12,800 ind/m^2, which is 2–3 orders of magnitude higher than unpolluted sites. Oligohaline sectors of VB and PAM estuaries have macrofaunal densities similar to those encountered in the euhaline sectors, with values varying from 170 to 5760 ind/m^2. In general, there is high spatial variability on total abundances of benthic macrofauna within each sector of all estuaries studied, and as observed on intertidal habitats, multiple environmental factors including current velocities, sediment composition, salinity gradients, and human impacts likely act in concern to structure those assemblages (e.g., Cortelezzi et al. 2007).

4.3.2 Assemblage Composition, Richness, and Diversity

4.3.2.1 Intertidal Habitats

Intertidal mudflats in the euhaline sector of Piraquê-Açu-Mirim estuary have higher taxa richness (identified to family level) and diversity if compared to the mesohaline sector. A total of 55 taxa were identified in the euhaline sector, whereas a total of 37 taxa were sampled in the mesohaline sector of Vitória Bay, Benevente, and Piraquê-Açu-Mirim estuaries.

Intertidal mudflat habitats in the Subaé, Paraguaçu, and Jaguaripe estuaries in the Todos os Santos Bay had similar richness than others estuaries in the EME, with 26–28 macrofaunal taxa (Mariano and Barros 2015). Overall, benthic richness of intertidal mudflat assemblages on the estuaries of the Todos os Santos Bay is lower towards the oligohaline sector (Krull et al. 2014; Mariano and Barros 2015). Species richness in intertidal habitats is lower than on subtidal habitats at the Todos os Santos Bay estuaries, but those differences may be associated with a higher heterogeneity in sediment grain size within estuarine channels. The lower richness on intertidal areas also may be partially associated with stress caused by air exposure time, i.e., high temperature and interstitial salinity during low tide.

Habitat comparisons between mangrove and mudflat habitats revealed higher macrofaunal richness within vegetated habitats on the southern sector of the EME. Overall, infaunal benthic richness was higher within mangrove habitats (18–25 taxa in Vitória Bay, Benevente, and Piraquê-Açu-Mirim estuaries) when compared to nearby mudflats in the same estuaries (14–17 taxa). Diversity patterns were variable between habitats, with higher diversity in mudflats at the Piraquê-Açu-Mirim estuary but the opposite trend, with higher diversity in mangrove sediments, in Vitória Bay and Benevente estuaries.

The estuarine salinity sectors (euhaline, polyhaline, mesohaline, oligohaline, from greater to lesser salinities) are typically recognized in the estuaries of the EME. Some benthic species are accordingly distributed along these gradients with either a constrained (less tolerant species) or a broad distribution, as observed in temperate estuaries. For example, the crustacean isopod genus *Excirolana* and the mollusk *Neritina virginea* seem to occur intertidally only on the euhaline sector of Paraguaçu and Jaguaripe estuaries, respectively (Mariano and Barros 2015). The polychaetes from the family Sternaspidae share both euhaline and polyhaline sectors of TSB estuaries, whereas Paraonidae and Magelonidae families occur only on polyhaline sector and Ampharetidae is apparently adapted to the oligohaline sector (Krull et al. 2014). Other families show a wide distribution along the estuarine salinity gradient. That is the case of the polychaetes from the families Goniadidae, Nereididae, Spionidae, and Capitellidae (Krull et al. 2014; Mariano and Barros 2015). Nevertheless, goniadid polychaetes have higher abundances in euhaline and polyhaline sectors, whereas Nereididae, Spionidae, and Capitellidae are more abundant in the oligohaline sector of TSB estuaries (Krull et al. 2014; Mariano and Barros 2015).

Benthic assemblage composition also changes markedly between vegetated (mangroves) and unvegetated (mudflats) habitats in estuaries within the EME. Macrofaunal composition at family level is more similar between vegetated habitats among the estuaries of Vitória Bay, Piraquê-Açu-Mirim, and Benevente, if compared to mudflats at the same estuaries, suggesting that mudflats have higher spatial variability and dissimilarity. Within mangrove forests, macrofaunal composition was related to higher sedimentary detritus biomass and total sedimentary organic matter, supporting that forests have an important role as ecosystems engineers for the benthic fauna. The most abundant macrofauna in the mangrove forests, the annelids Oligochaeta and Capitellidae, are common in sediments with high organic content and are widely present at other tropical and subtropical mangroves (Netto and Galluci 2003; Mariano and Barros 2015). In general, Oligochaeta and Capitellidae dominated sediments in mangrove habitats (>60%) in all estuaries, whereas mudflat habitats exhibited different dominances among the estuaries, which were likely related to salinity changes as also observed in estuaries of the Todos os Santos Bay (Mariano and Barros 2015). In mudflats within mesohaline sectors, peracarid crustaceans (Kalliapseudidae) and annelids (Oligochaeta) were more abundant in the Benevente estuary, whereas annelids (Spionidae and Capitellidae) dominated (>80% relative abundance) macrofaunal assemblages in Vitória Bay. Annelida is also a dominant group along intertidal mudflats in the Piraquê-Açu-Mirim estuary, but cirolanid isopods occur in low abundances at the euhaline sector, whereas several annelids, including amphinomids and sabellarids, mostly occur at the polyhaline sector and the annelids Capitellidae and Oligochaeta are more abundant in mesohaline sector.

4.3.2.2 Subtidal Habitats

Salinity tolerance is one great natural stressor that dictates most of species distribution patterns in estuaries (Elliott and Quintino 2007; Mariano and Barros 2015). These spatial patterns are observed in EME estuaries, with subtidal benthic richness on the TSB showing a reduction of species richness towards the oligohaline sector (Barros et al. 2012, 2014). This pattern is clearly different from Remane's model as pointed previously (e.g., Barros et al. 2012). In all estuaries studied, there is a strong pattern for high macrofaunal richness on euhaline sectors, a variable richness on polyhaline and mesohaline sectors, and a relatively low richness on oligohaline sector (Barros et al. 2012). Comparatively, species richness on oligohaline sector of Paraguaçu estuary might be higher than in other estuaries of Todos os Santos Bay. The presence of a dam on Paraguaçu estuary allows eventual entrance of water with greater salinity, which could explain these observations. Also, a higher species richness is found inside Iguape Bay, a particularly large and shallow section of Paraguaçu estuarine system (Barros et al. 2012).

Estuarine biological diversity is composed by alpha, gamma, and beta diversity (Barros et al. 2014). Alpha diversity is the small-scale diversity (e.g., stations, sites), gamma diversity refers to the region or ecosystems of interest (e.g., estuary), and beta diversity is the variation in species composition among sites (or places or stations) in the area of interest (Legendre et al. 2005). Alpha, gamma, and beta diversity were estimated for several estuaries in Todos os Santos Bay (Barros et al. 2014). Alpha diversity, (i.e., diversity at each station or site), decreases along the estuarine salinity gradient from the euhaline towards the oligohaline sector (see Fig. 4.3). Estuaries like Paraguaçu and São Paulo presented high gamma diversity, but values are still lower than estuaries from temperate regions in France and in the USA (Barros et al. 2014). When considering gamma diversity as the simple sum of beta and alpha diversities (additive formula from Whittaker's work; e.g., Whittaker 1960), the beta diversity in Todos os Santos Bay estuaries contributes more to gamma diversity than to alpha diversity. The processes behind this pattern still need to be investigated, not only at EME estuaries but also in other regions of the world. Clearly, beta diversity shows greater values on lower estuarine sites (Barros et al. 2014) and beta diversity on Todos os Santos Bay estuaries is mostly driven by turnover (i.e., species replacement) than by nestedness (i.e., subsets of a large group of species). High turnover is usually associated with low impacted estuaries, such as the Jaguaripe estuary. On the other hand, São Paulo estuary shows beta diversity mostly driven by nestedness, which is associated with highly impacted estuaries. In this case, small sets of a large group of species are present in different regions along the salinity gradient due to an increase in the disturbance, pollution for instance, and a decrease in the number of species that are able to persist. Nevertheless, other impacted estuaries like Subaé can show different beta diversity phenomena (nestedness or turnover) at different sampling times. Both Subaé and São Paulo estuaries are known by its high degree of anthropogenic impact (Barros et al. 2014).

Benthic assemblage composition from estuaries in the EME shows typical changes within salinity. In Vitória Bay, subtidal sediments are dominated by a number of annelids, including Oligochaeta, Cirratulidae, Capitellidae, Spionidae, and Magelonidae. Capitellidae (*Capitella* sp.), Cirratulidae (*Mediomastus* sp.), and Onuphidae (*Mooreonuphis lineata*) dominate oligohaline sectors of Vitória Bay estuary, whereas the polyhaline and euhaline sectors are dominated by a range of other species with higher overall diversity. On the Piraquê-Açu-Mirim estuary, a range of different species dominate subtidal sediments, including *Lumbrineris* sp., *Magelona papilicornis*, and *Gimnonereis crosslandi*, which may be associated with a less impacted ecosystem if compared to Vitória Bay. In the Rio Doce estuary located in the central sector of the EME, there is a high diurnal variability in salinity due to tidal and freshwater mixing resulting in strong spatial heterogeneity in sediment composition within the estuary. In sites within the polyhaline sector with sandy sediments, macrofaunal assemblages were dominated by the annelids *Laeonereis* sp., *Neanthes succinea* (Nereididae), *Isolda pulchella* and *Amphicteis* sp. (Ampharetidae). Temporal variability in rainfall volumes within the Rio Doce basin can lead to strong changes in the structure of benthic assemblages, and events of high freshwater input to the estuarine region lead to a 2–3 times higher abundance of Chironomid larvae.

Regarding estuaries on Todos os Santos Bay, several published studies (e.g., Hatje et al. 2006; Barros et al. 2008, 2012, 2014a; Magalhães and Barros 2011; Krull et al. 2014; Mariano and Barros 2015) showed that subtidal patterns of macrofaunal assemblage composition are similar to those found on the intertidal habitats. In spite of all variability, clearly polychaetes from the families Cirratulidae and Orbiniidae are frequently observed on the euhaline and polyhaline sectors of estuaries. Nereididae polychaetes and Tellinidae gastropods have a wider distribution but Nereididae have a higher abundance on the oligohaline sector of the Todos os Santos Bay estuaries.

4.4 Impacts and Conservation Issues

Human development leads to eutrophication, chemical pollution, chronic oil spilling, and clearance of vegetated habitats of estuaries for construction of ports, marinas, housing, and shrimp farms (Lotze et al. 2006). These impacts threaten the biodiversity and ecosystem services provided by estuaries and specially by mangrove ecosystems, which are experiencing fast declines worldwide (Pendleton et al. 2012). On the EME of Brazil there are multiple sources of impact to estuarine ecosystems and their associated habitats, most of them occurring at densely populated areas. In general, studies in the EME showed that benthic estuarine assemblages respond to acute and chronic impacts through changes in community abundance, composition, and diversity, which may lead to functional and important ecological changes that have been less studied in the region (but see Magalhães and Barros 2011).

The climate variability along the wide latitudinal variation within the EME and local impacts on estuarine ecosystems may lead to unique, cumulative or synergistic process between preserved and impacted areas (Vinebrooke et al. 2004; Hoegh-Guldberg and Bruno 2010). Abiotic and biotic stressors may interact to produce combined impacts on estuarine biodiversity as observed in temperate estuaries (France; Pezy et al. 2017). Estuaries previously impacted are expected to be less resilient to climate variability, likely as a result of biodiversity and functional loss (Dolbeth et al. 2007), which is the case of several estuaries in the EME (Krull et al. 2014).

4.4.1 *Organic and Inorganic Pollution*

The Vitória Bay estuary is situated within the largest metropolitan area on the southern region of the EME and receives large inputs of untreated urban and industrial sewage (Jesus et al. 2004; Grilo et al. 2013). As a result, some areas of Vitória Bay are contaminated with high concentrations (> 1.0 μg/g) of fecal lipids (e.g., coprostanol). In general, over 30% of the studied subtidal areas of VB are under high or severe contamination from untreated sewage, with negative effects on benthic diversity and assemblage composition. At the Piraquê-Açu-Mirim estuary, sewage contamination was also detected but with concentrations 10 times lower than observed in Vitória Bay. Trace metals including Fe, Al, Zn, Cr, Co, Cu, and Pb were also detected in higher concentrations within the intermediate (mesohaline) and euhaline sectors of Vitória Bay. These areas receive most untreated sewage and are under direct impact of port activities (Jesus et al. 2004). However, most trace metals exhibited concentrations below the Threshold Effect Levels (TEL) and therefore are not considered dangerous to the estuarine biota. In general, trace metals are carried with organic particles along the river basins and accumulate in areas with high organic deposition (Hatje et al. 2006). Impacts from sewage input on benthic assemblages in Vitória Bay and in the Piraquê-Açu-Mirim estuaries are typical altered patterns of faunal abundance, assemblage composition and diversity which are widely reported in other estuarine and marine ecosystems. However, Vitória Bay is under a much stronger degree of contamination along a larger spatial scale, if compared to local alterations of the benthic assemblages observed in the Piraquê-Açu-Mirim estuary. Contamination of several estuaries within the Todos os Santos Bay has been extensively investigated and exhibits variable levels of contamination and biological degradation (e.g., Hatje et al. 2006; Krull et al. 2014). The Subaé estuary is heavily contaminated by trace metals (Pb, Cd, Co, Mn, Zn, Cr, Cu) due to industrial activities with adverse effects on benthic assemblages (Hatje et al. 2006; Krull et al. 2014). Other estuaries within the TSB typically exhibit lower contamination by heavy metals (Hatje and Barros 2012), but impact sources from agriculture and sewage may also impact benthic assemblages at smaller spatial scales.

4.4.2 *Mine Tailing Impacts*

The collapse of a mine tailing dam in 2015 led to the contamination of the Rio Doce river and the estuary with tailings containing high concentrations of trace metals. A short-term (period of 15 days) assessment of contamination effects on the benthic estuarine assemblages in the Rio Doce estuary revealed changes in macrofaunal composition and trophic guild dominance (Gomes et al. 2017). Macrofaunal composition markedly changed after the mine tailing impacts on the Rio Doce estuary, with tolerant taxa persisting after the impact. Some annelids including *Amphicteis* sp., *Isolda pulchella*, *Laeonereis* sp., *Alitta succinea* and *Heleobia australis* represented over 80% of macrofaunal taxa after the impact. Other species sampled before the impact including *Glycera* sp., *Magelona* sp., *Heteromastus similis*, *Pholoe* sp., *Sigambra grubii*, *Polydora* sp. and *Sthenelais* sp. were not detected after the impact, with an overall 30% reduction on taxa richness. Overall, surface deposit feeders dominated estuarine benthic assemblages after the impact, with a significant decrease in the relative abundances of detritivores and carnivore/omnivores (Gomes et al. 2017). The initial impact of mine tailings deposition (up to 3 days after the impact) revealed a burial impact on the benthic assemblages and the increase in sedimentary trace metal concentrations suggests future impacts due to chronic contamination (Hatje et al. 2017; Queiroz et al. 2018).

4.4.3 Mangrove Clearing and Shrimp Farms

Mangrove deforestation is widespread around the globe and occurs at fast rates on many coastal areas (Alongi 2002; Donato et al. 2011). Mangrove removal for urban development is also common in Brazil despite being their protected status. Mangrove forests are typically removed by cutting down of the above-sediment vegetation with the below-ground root structures left intact, although roots may be completely removed or buried in urban areas. A number of studies revealed that the removal of mangroves significantly change particle sedimentation, carbon stocks, hydrodynamics, and sediment organic matter profiles, which in turn may lead to changes in benthic biodiversity and sediment respiration (Sjoling et al. 2005; Granek and Ruttemberg 2008; Sweetman et al. 2010; Sabeel et al. 2015; Kauffman et al. 2017; Kauffman et al. 2018; Bernardino et al. 2018). These changes occur mainly through modification of sedimentation and changes in mean particle size and on sources of organic matter to the benthic fauna after forest clearing (Granek and Ruttemberg 2008; Bernardino et al. 2018). Although complete removal of epifaunal organisms will follow mangrove clearance, infaunal invertebrates may quickly colonize sediments after initial disturbance (Faraco and Lana 2004). In areas where mangrove forests were introduced, benthic assemblages also show marked differences in macrofaunal composition, dominance, diversity, and food webs when compared to non-vegetated areas or to native ecosystems (Demopoulos et al. 2007; Demopoulos and Smith 2010; Sabeel et al. 2015).

Most Brazilian mangroves and estuaries near urban centers have been impacted by forest clearing. In small estuaries not bordering large urban centers, deforested areas are typically cleared to support local fishermen (Faraco et al. 2010), or they are completely removed and buried for construction of local marinas. In the Piraquê-Açu-Mirim estuary (EME), local effects (1–1000 m scale) of mangrove clearing have been quantified and revealed marked changes in benthic assemblage composition, with changes on benthic food webs in removal areas if compared to natural mangrove forests nearby (Bernardino et al. 2018). The changes observed on benthic assemblages at the deforested site were also related to alteration of the sedimentary habitats and distinct dominance of food sources. The alteration of sedimentary habitats led to a higher similarity between typical estuarine tidal flat assemblages with ones located at the former mangrove forests, evidencing the importance of forests to estuarine benthic diversity (Netto and Lana 1999; Netto and Galluci 2003). The presence of forests fringing tidal flat estuarine assemblages allows a higher amplitude of food sources available to benthic organisms near natural sites, whereas areas cleared of forests exhibit lower trophic levels indicative of lower organic availability (Bernardino et al. 2018). Although there is only limited spatial and temporal assessment of mangrove clearing effects on estuarine benthic assemblages, the functional links between these assemblages and important estuarine ecological processes (Kristensen et al. 2014) indicate that these are important topics for study along Brazilian estuaries. The removal of mangroves over larger spatial scales, including for shrimp farming, potentially will have marked impacts on coastal food webs that have not yet been considered.

Shrimp farming impacts were assessed on Todos os Santos Bay estuaries through multiple lines of evidence. Macrobenthic assemblages in subtidal and intertidal areas were negatively affected by shrimp farm activities with markedly lower species richness and abundance in impacted areas (Hatje et al. 2016; Ribeiro et al. 2016). Also, benthic assemblages from farming impacted areas were consistently different from control areas. Results from those studies alert for the necessity of regulation and management measures to reduce present and future impacts on mangrove forests in the EME, which can deliver important ecosystem services (Alongi 2002; Kauffman et al. 2018; Servino et al. 2018).

4.4.4 Invasive Species

The introduction of invasive species is a main concern on marine coastal areas, mainly due to port activities. In the Todos os Santos Bay estuaries, three benthic invasive species are well known and the invasive marine coral *Tubastraea* (*T. tagusensis* and *T. coccinea*) was found on 12 sites of Paraguaçu estuary. The specimens were observed on intertidal areas, rocky reefs, and consolidated substrata, mostly artificial, such as piers and platforms (Miranda et al. 2016a). The sources of invasion include oil platforms anchorages at nearby sites and the higher water salinities within that estuary. Paraguaçu estuary has a reduced freshwater flow caused by a dam, which is also facilitating the establishment of those species in the estuary. A marked reduction of native species abundances (e.g., native corals, oysters, and barnacles) and changes in benthic assemblage structure were observed in consolidated substrates on areas invaded by *Tubastraea* (Miranda et al. 2016b). The Asian swimming crab *Charybdis hellerii* is one of the most widespread invasive benthic species in Brazil and is also a very frequent species in Todos os Santos Bay (Silva and Barros 2011; Silva 2016). This crab does not have commercial value and is not under fishing pressure like native species (Lopes 2009). Additionally, this species may compete for habitat and food with commercially important species such as crabs *Callinectes* causing a possible abundance reduction of the latter (Silva 2016). Some species such as the bryozoan *Triphyllozoon arcuatum* (Almeida et al. 2015) and the ascidians *Ascidia tapuni* and *Cnemidocarpa irene* (Rocha et al. 2012) were also recognized as exotic species in the Todos os Santos Bay and further studies are necessary to assess their role as invasive species. In general, invasive species within estuaries of the EME have similar deleterious effects on native biodiversity, but there is still a large gap understanding the degree of infestation by invasive species and their ecological impacts at other estuaries of the EME.

4.4.5 Climate Change

Climate change is altering Brazil's estuarine ecosystems at a regional scale due to increases in mean atmospheric temperatures (Bernardino et al. 2015). Most Marine Ecoregions in Brazil have been under increased high temperature anomalies (i.e., higher annual temperatures above historic means) in the last decades. In the EME, there is a decadal warming trend of 0.1 °C and positive annual temperature anomalies above 1 °C were common during the last decade (2000–2010; Bernardino et al. 2015). The warming atmospheric trends observed near estuaries of the EME are within the range expected from regional climate model forecasts (Marengo et al. 2010) and agree with the observed national atmospheric dataset (Alvares et al. 2014). Although the warmer anomalies observed in the EME have smaller ranges if compared to other Marine Ecoregions of Brazil (Bernardino et al. 2015), estuarine communities within this region, under the typical tropical and dry climate, may show biological stress due to higher temperatures and lower rainfall. We expect that climate-associated impacts may lead to impacts on benthic estuarine organisms at some estuaries within the EME through physiological stress, changes in assemblage composition and productivity (Doney et al. 2012; Semeniuk 2013). Although these climate-driven impacts have been observed in temperate estuaries (Cardoso et al. 2008; Dolbeth et al. 2011), there is already preliminary data from estuarine benthic assemblages in the EME indicating impacts of a warmer and drier climate.

In the EME, a 14-month assessment of benthic assemblages in the Piraquê-Açu-Mirim estuary indicated changes in community structure across an exceptionally dry season associated with a strong El-Niño year (2015–2016 Golden Gate Weather Services Comparative Climatic Data 2016). The drought season recorded in the Piraquê-Açu-Mirim estuary between 2015 and 2016 was one of the strongest in the last 60 years, with rainfall volumes 3–8 times lower than average (Servino et al. 2018). During the drought period, significant decrease in rainfall volumes and increase in water salinity and temperature were observed in the Piraquê-Açu-Mirim estuary. Low dissolved oxygen concentrations were also observed during the drought period, suggesting that short periods of hypoxia (<2 mg/L) may also have impacted benthic assemblages in this estuary (Brown et al. 2004; Bishop et al. 2006; Harris et al. 2006; Vaquer-Sunyer and Duarte 2008). During droughts periods, sediment silt fraction increased towards the lower estuary likely as a result of aggregation processes and flocculation by salinization and pH increase (Eisma 1986; MacKay et al. 2010; Mari et al. 2012). The higher mud content was correlated to decreased macrofaunal density, species richness and changes species composition, which resulted in altered ecological functions provided by benthic assemblages (Anderson 2008; Pratt et al. 2014).

4.5 Final Considerations

The Brazilian EME includes over 50 estuaries with broad geomorphology and environmental dynamics. Small estuaries are located within reserves and are relatively preserved if compared to large ecosystems located near heavily urbanized areas and with multiple impacts due to pollution and habitat loss. Benthic estuarine biodiversity within the EME is structured by typical estuarine environmental dynamics such as salinity, sediment grain size, organic matter availability, and habitat types. Estuaries within the EME exhibit comparable spatial and temporal changes in abundance and diversity, but local disturbance effects and specific environmental drivers of macrofaunal composition and diversity alter large scale patterns. Although benthic assemblages have been successfully used as environmental indicators of impact, their taxonomy, function, and importance to estuarine services such as organic carbon sequestration, water quality, and food provision have been understudied within the EME. Common impacts associated with pollution and habitat loss exist within most estuaries and there is initial evidence for climate change effects that also need further study. The remaining ecosystems within urban estuaries will likely see major climatic impacts in the future with sea level rise resulting in loss of habitats and lower precipitation and decreases in water quality. Regional scale assessments of those impacts during long-term studies would clearly benefit understanding these climate impacts and favor conservation of their biodiversity and ecosystem services to local communities.

Acknowledgments AFB was supported by FAPES (52638090/2011; 61847429/2013) and CNPq (470542/2013-6; 441243/2016-9) research grants. FB was supported by CNPq fellowship (239978/2012-9; 306332/2014-0), Projeto BTS, PRONEX and Baías da Baía (PET 0035/2012). LEOG, LBB, LR, and AD were supported by graduate scholarships from FAPESB/FAPES/CAPES/CNPq. This is also a PELD-HCES contribution #003.

References

Almeida ACS, Souza FBC, Gordon DP et al (2015) The non-indigenous bryozoan *Triphyllozoon* (Cheilostomata: Phidoloporidae) in the Atlantic: morphology and dispersion on the Brazilian coast. Zoologia (Curitiba) 32:476–484. https://doi.org/10.1590/S1984-46702015000600007

Alongi DM (2002) Present state and future of the world's mangrove forests. Environ Conserv 29:331–349. https://doi.org/10.1017/S0376892902000231

Alvares CA, Stape JL, Sentelhas PC et al (2014) Koppen's climate classification map for Brazil. Meteorol Z 22:711–728. https://doi.org/10.1127/0941-2948/2013/0507

Anderson MJ (2008) Animal-sediment relationships re-visited: characterising species' distributions along an environmental gradient using canonical analysis and quantile regression splines. J Exp Mar Biol Ecol 366:16–27. https://doi.org/10.1016/j.jembe.2008.07.006

Arrivabene HP, Souza I, Có WLO et al (2014) Functional traits of selected mangrove species in Brazil as biological indicators of different environmental conditions. Sci Total Environ 476–477:496–504. https://doi.org/10.1016/j.scitotenv.2014.01.032

Barros F, Blanchet H, Hammerstrom K et al (2014) A framework for investigating general patterns of benthic β -diversity along estuaries. Estuar Coast Shelf Sci 149:223–231. https://doi.org/10.1016/j.ecss.2014.08.025

Barros F, Carvalho GC, Costa Y et al (2012) Subtidal benthic macroinfaunal assemblages in tropical estuaries: generality amongst highly variable gradients. Mar Environ Res 81:43–52. https://doi.org/10.1016/j.marenvres.2012.08.006

Barros F, Hatje V, Figueiredo MB et al (2008) The structure of the benthic macrofaunal assemblages and sediments characteristics of the Paraguaçu estuarine system, NE, Brazil. Estuar Coast Shelf Sci 78:753–762

Bernardino AF, Netto SA, Pagliosa PR et al (2015) Predicting ecological changes on benthic estuarine assemblages through decadal climate trends along Brazilian marine ecoregions. Estuar Coast Shelf Sci 166:74–82. https://doi.org/10.1016/j.ecss.2015.05.021

Bernardino AF, Pagliosa PR, Christofoletti RA et al (2016) Benthic estuarine communities in Brazil: moving forward to long term studies to assess climate change impacts. Braz J Oceanogr 64:81–96. https://doi.org/10.1590/S1679-875920160849064sp2

Bernardino AF, Gomes LEO, Hadlich HL et al (2018) Mangrove clearing impacts on macrofaunal assemblages and benthic food webs in a tropical estuary. Mar Poll Bull 126:228–235. https://doi.org/10.1016/j.marpolbul.2017.11.008

Bishop MJ, Powers SP, Porter HJ et al (2006) Benthic biological effects of seasonal hypoxia in a eutrophic estuary predate rapid coastal development. Estuar Coast Shelf Sci 70:415–422. https://doi.org/10.1016/j.ecss.2006.06.031

Bissoli LB, Bernardino AF (2018) Benthic macrofaunal structure and secondary production in tropical estuaries on the Eastern Marine Ecoregion of Brazil. PeerJ 6:e4441. https://doi.org/10.7717/peerj.4441

Boone Kauffman J, Arifanti VB, Bernardino AF, Ferreira TO, Murdiyarso D, Cifuentes M, Norfolk J (2018) And details for land-use carbon footprints arise from quantitative and replicated studies. Front Ecol Environ 16(1):12–13

Brown JH, Gillooly JF, Allen AP et al (2004) Toward a metabolic theory of ecology. Ecology 85:1771–1789. https://doi.org/10.1890/03-9000

Cardoso PG, Raffaelli D, Pardal MA (2008) The impact of extreme weather events on the seagrass *Zostera noltii* and related *Hydrobia ulvae* population. Mar Pollut Bull 56:483–492. https://doi.org/10.1016/j.marpolbul.2007.11.006

Carvalho LRS, Barros F (2017) Physical habitat structure in marine ecosystems: the meaning of complexity and heterogeneity. Hydrobiol 797:1–9. https://doi.org/10.1007/s10750-017-3160-0

Chen R, Twilley RR (1999) Patterns of mangrove forest structure and soil nutrient dynamics along the Shark River Estuary, Florida. Estuaries 22:955–970. https://doi.org/10.2307/1353075

Costa P, Dórea A, Mariano-Neto E et al (2015) Are there general spatial patterns of mangrove structure and composition along estuarine salinity gradients in Todos os Santos Bay? Estuar Coast Shelf Sci 166:83–91. https://doi.org/10.1016/j.ecss.2015.08.014

Cortelezzi A, Capítulo AR, Boccardi L et al (2007) Benthic assemblages of a temperate estuarine system in South America: transition from a freshwater to an estuarine zone. J Mar Syst 68(3–4):569–580

Demopoulos AWJ, Fry B, Smith CR (2007) Food web structure in exotic and native mangroves: a Hawaii-Puerto Rico comparison. Oecologia 153:675–686. https://doi.org/10.1007/s00442-007-0751-x

Demopoulos AWJ, Smith CR (2010) Invasive mangroves alter macrofaunal community structure and facilitate opportunistic exotics. Mar Ecol Prog Ser 404:51–67. https://doi.org/10.3354/meps08483

Dolbeth M, Cardoso PG, Ferreira SM et al (2007) Anthropogenic and natural disturbance effects on a macrobenthic estuarine community over a 10-year period. Mar Pollut Bull 54:576–585. https://doi.org/10.1016/j.marpolbul.2006.12.005

Dolbeth M, Cardoso PG, Grilo TF et al (2011) Long-term changes in the production by estuarine macrobenthos affected by multiple stressors. Estuar Coast Shelf Sci 92:10–18. https://doi.org/10.1016/j.ecss.2010.12.00

Donato DC, Kauffman JB, Murdiyarso D et al (2011) Mangroves among the most carbon-rich forests in the tropics. Nat Geosci 4:293–297

Doney SC, Ruckelshaus M, Duffy JE (2012) Climate change impacts on marine ecosystems. Annu Rev Mar Sci 4:11–37. https://doi.org/10.1146/annurev-marine-041911-111611

Edgar GJ, Barrett NS, Last PR (1999) The distribution of macroinvertebrates and fishes in Tasmanian estuaries. J Biogeogr 26:1169–1189. https://doi.org/10.1046/j.1365-2699.1999.00365.x

Eisma D (1986) Flocculation and deflocculation of suspended matter in estuaries. Neth J Sea Res 20:183–199. https://doi.org/10.1016/0077-7579(86)90041-4

Elliott M, Quintino V (2007) The estuarine quality paradox, environmental homeostasis and the difficulty of detecting anthropogenic stress in naturally stressed areas. Mar Pollut Bull 54:640–645. https://doi.org/10.1016/j.marpolbul.2007.02.003

Faraco LFD, Andriguetto Filho JM, Lana PC (2010) A methodology for assessing the vulnerability of mangroves and fisherfolk to climate change. Panam. J Aquat Sci 5:205–223

Faraco LFD, Lana PC (2004) Leaf-consumption levels in subtropical mangroves of Paranaguá Bay (SE Brazil). Wetl Ecol Manag 12:115–122. https://doi.org/10.1023/B:WETL.0000021666.42546.c2

Gilberto DA, Bremec CS, Acha EM et al (2004) Large-scale spatial patterns of benthic assemblages in the SW Atlantic: the Río de la Plata estuary and adjacent shelf waters. Estuar Coast Shelf Sci 61:1–13. https://doi.org/10.1016/j.ecss.2004.03.015

Golden Gate Weather Services Comparative Climatic Data (2016) http://ggweather.com/enso/oni.htm. Accessed 15 Mar 2017

Granek E, Ruttemberg BI (2008) Changes in biotic and abiotic processes following mangrove clearing. Estuar Coast Shelf Sci 80:555–562. https://doi.org/10.1016/j.ecss.2008.09.012

Grilo CF, Neto RR, Vicente MA et al (2013) Evaluation of the influence of urbanization processes using mangrove and fecal markers in recent organic matter in a tropical tidal flat estuary. Appl Geochem 38:82–91. https://doi.org/10.1016/j.apgeochem.2013.08.009

Gomes LEO, Correa LB, Sá F et al (2017) The impacts of the Samarco mine tailing spill on the Rio Doce estuary, eastern Brazil. Mar Pollut Bull 120:28–36

Harris LA, Duarte CM, Nixon SW (2006) Allometric laws and prediction in estuarine and coastal ecology. Estuar Coasts 29:340–344. https://doi.org/10.1007/BF02782002

Hatje V, Barros F (2012) Overview of the 20th century impact of trace metal contamination in the estuaries of Todos os Santos Bay: past, present and future scenarios. Mar Pollut Bull 64:2603–2614. https://doi.org/10.1016/j.marpolbul.2012.07.009

Hatje V, Barros F, Figueiredo DG et al (2006) Trace metal contamination and benthic assemblages in Subaé estuarine system. Brazil. Mar Pollut Bull 52:969–987. https://doi.org/10.1016/j.marpolbul.2006.04.016

Hatje V, Barros F, Magalhaes W et al (2008) Trace metals and benthic macrofauna distributions in Camamu Bay, Brazil: sediment quality prior oil and gas exploration. Mar Pollut Bull 56:363–370. https://doi.org/10.1016/j.marpolbul.2007.10.029

Hatje V, de Souza MM, Ribeiro LF et al (2016) Detection of environmental impacts of shrimp farming through multiple lines of evidence. Environ Pollut 219:1–13. https://doi.org/10.1016/j.envpol.2016.06.056

Hatje V, Pedreira RMA, de Rezende CE, Schettini CAF, de Souza GC, Marin DC, Hackspacher PC (2017) The environmental impacts of one of the largest tailing dam failures worldwide. Sci Rep 7(1):10706

Hoegh-Guldberg O, Bruno JF (2010) The impact of climate change on the world's marine ecosystems. Science 328(5985):1523–1528. https://doi.org/10.1126/science.1189930

Jesus H, Costa EA, Mendonça ASF et al (2004) Distribuição de metais pesados em sedimentos do sistema estuarino da ilha de Vitória-ES. Quim Nova 27:378–386. https://doi.org/10.1590/S0100-40422004000300004

Kauffman JB, Arifanti VB, Trejo HH et al (2017) The jumbo carbon footprint of a shrimp: carbon losses from mangrove deforestation. Front Ecol Environ 15(4):183–188. https://doi.org/10.1002/fee.1482

Kauffman JB, Bernardino AF, Ferreira TO, Bolton NW, de Gomes LE, O, Nobrega GN (2018) Shrimp ponds lead to massive loss of soil carbon and greenhouse gas emissions in northeastern Brazilian mangroves. Ecol Evol. doi.org/10.1002/ece3.4079

Kristensen E, Delefosse M, Quintana CO et al (2014) Influence of benthic macrofauna community shifts on ecosystem functioning in shallow estuaries. Front Mar Sci 1:1–14. https://doi.org/10.3389/fmars.2014.00041

Krull M, Abessa DMS, Hatje V et al (2014) Integrated assessment of metal contamination in sediments from two tropical estuaries. Ecotoxicol Environ Saf 106:195–203. https://doi.org/10.1016/j.ecoenv.2014.04.038

Legendre P, Borcard D, Peres-Neto PR (2005) Analyzing beta diversity: partitioning the spatial variation of community composition data. Ecol Monogr 75:435–450. https://doi.org/10.1890/05-0549

Lessa GC, Santos FM, Souza Filho PW et al (2018) Brazilian estuaries: a geomorphologic and oceanographic perspective. In: Lana PC, Bernardino AF (eds) Brazilian Estuaries. Springer International Publishing, Brazilian Marine Biodiversity. https://doi.org/10.1007/978-3-319-77779-5_4

Lopes RM (2009) Informe sobre as Espécies Exóticas Invasoras Marinhas no Brasil, Brasília, Ministério do Meio Ambiente, Série Biodiversidade doi: https://doi.org/10.1017/CBO9781107415324.004

Lotze HK, Lenihan HS, Bourque BJ et al (2006) Depletion, degradation and recovery potential of estuaries and coastal seas. Science 312:1806–1809. https://doi.org/10.1126/science.1128035

MacKay F, Cyrus D, Russell KL (2010) Macrobenthic invertebrate responses to prolonged drought in South Africa's largest estuarine lake complex. Estuar Coast Shelf Sci 86:553–567. https://doi.org/10.1016/j.ecss.2009.11.011

Magalhães W, Barros F (2011) Structural and functional approaches to describe polychaete assemblages: ecological implications for estuarine ecosystems. Mar Fresh Res 62:918–922. https://doi.org/10.1071/MF10277

Marengo JA, Ambrizzi T, da Rocha RP et al (2010) Future change of climate in Sotuh America in the late twenty-first century: intercomparison of scenarios from three regional climate models. Clim Dyn 35:1073–1097. https://doi.org/10.1007/s00382-009-0721-6

Mari X, Torréton JP, Trinh CBT et al (2012) Aggregation dynamics along a salinity gradient in the Bach dang estuary, North Vietnam. Estuar Coast Shelf Sci 96:151–158. https://doi.org/10.1016/j.ecss.2011.10.028

Mariano DLS, Barros F (2015) Intertidal benthic macrofaunal assemblages: changes in structure along entire tropical estuarine salinity gradients. J Mar Biol Assoc UK 95:5–15. https://doi.org/10.1017/S002531541400126X

Mcleod E, Chmura GL, Bouillon S et al (2011) A blueprint for blue carbon: toward an improved understanding of the role of vegetated coastal habitats in sequestering CO2. Front Ecol Environ 9:552–560. https://doi.org/10.1890/110004

Miranda RJ, Costa Y, Lordes FL et al (2016a) New records of the alien cup-corals (*Tubastraea spp.*) within estuarine and reef systems in Todos os Santos Bay, southwestern Atlantic. Mar. Biodivers Rec 9:1–6. https://doi.org/10.1186/s41200-016-0053-2

Miranda RJ, Cruz IC, Barros F (2016b) Effects of the alien coral *Tubastraea tagusensis* on native coral assemblages in a southwestern Atlantic coral reef. Mar Biol 163:1–12. https://doi.org/10.1007/s00227-016-2819-9

Netto SA, Galluci F (2003) Meiofauna and macrofauna communities in a mangrove from the island of Santa Catarina, South Brazil. Hydrobiologia 505:159–170. https://doi.org/10.1023/B:HYDR.0000007304.22992.b2

Netto SA, Lana PC (1999) The role of above-and belowground components of *Spartina alterniflora* (Loisel) and detritus biomass in structuring macrobenthic associations of Paranaguá Bay (SE, Brazil). Hydrobiologia 400:167–177. https://doi.org/10.1023/A:1003753001411

Pendleton L, Donato DC, Murray BC et al (2012) Estimating global "blue carbon" emissions from conversion and degradation of vegetated coastal ecosystems. PLoS One 7:e43542. https://doi.org/10.1371/journal.pone.0043542

Pezy JP, Baffreau A, Dauvin JC (2017) What are the factors driving long-term changes of the suprabenthos in the seine estuary? Mar Pollut Bull 118:307. https://doi.org/10.1016/j.marpolbul.2017.03.008

Pratt DR, Lohrer AM, Pilditch CA et al (2014) Changes in ecosystem function across sedimentary gradients in estuaries. Ecosystems 17:182–194. https://doi.org/10.1007/s10021-013-9716-6

Queiroz HM, Nóbrega GN, Ferreira TO, Almeida LS, Romero TB, Santaella ST, Bernardino AF, Otero XL (2018) The Samarco mine tailing disaster: a possible time-bomb for heavy metals contamination? Sci Total Environ 637-638:498–506

Ribeiro LF, Eça GF, Barros F et al (2016) Impacts of shrimp farming cultivation cycles on macrobenthic assemblages and chemistry of sediments. Environ Pollut 211:307–315. https://doi.org/10.1016/j.envpol.2015.12.031

Rigo D (2004) Análise do escoamento em regiões estuarinas com manguezais – Medições e modelagem na Baía de Vitória, ES. Thesis, Federal University of Rio de Janeiro

RMA P, Barros F, Farias CO, Wagener AL, Hatje V et al (2017) Tropical bar as a reference área defined by multiple lines of evidence. Mar Pollut Bull 123:291–303. https://doi.org/10.1016/j.marpolbul.2017.08.041

Rocha RM, Bonnet NY, Baptista MS et al (2012) Introduced and native Phlebobranch and Stolidobranch solitary ascidians (Tunicata: Ascidiacea) around Salvador, Bahia, Brazil. Zoologia (Curitiba) 29:39–53. https://doi.org/10.1590/S1984-46702012000100005

Rondinelli SF, Barros F (2010) Evaluating shellfish gathering (*Lucina pectinata*) in a tropical mangrove system. J Sea Res 64:401–407. https://doi.org/10.1016/j.seares.2010.06.002

Sabeel RAO, Ingels J, Pape E et al (2015) Macrofauna along the Sudanese Red Sea coast: potential effect of mangrove clearance on community and trophic structure. Mar Ecol 36:794–809. https://doi.org/10.1111/maec.12184

Schaeffer-Novelli Y, Cintrón-Molero G, Adaime RR et al (1990) Variability of mangrove ecosystems along the Brazilian coast. Estuaries 13:204–218. https://doi.org/10.2307/1351590

Semeniuk V (2013) Predicted response of coastal wetlands to climate changes: a western Australian model. Hydrobiologia 708:23–43. https://doi.org/10.1007/s10750-012-1159-0

Servino RN, Gomes LEO, Bernardino AF (2018) Extreme weather impacts on mangrove forests in the eastern Brazil marine ecoregion. Sci Total Environ 628-629:233–240. https://doi.org/10.1016/j.scitotenv.2018.02.068

Silva EC (2016). Efeito das espécies invasoras Charybdis hellerii (Crustacea; Portunidae) e Melanoides tuberculatus (Gastropoda; Thiaridae) sobre a estrutura das assembleias macrobentônicas em dois ambientes aquático, Bahia, Brasil. Thesis, Federal University of Bahia

Silva EC, Barros F (2011) Macrofauna bentônica introduzida no brasil: lista de espécies marinhas e dulcícolas e distribuição atual. Oecol Aust 15:326–344. https://doi.org/10.4257/oeco.2011.1502.10

Sjoling S, Mohammed SM, Lyimo TJ et al (2005) Benthic bacterial diversity and nutrient processes in mangroves: impact of deforestation. Estuar Coast Shelf Sci 63:397–406. https://doi.org/10.1016/j.ecss.2004.12.002

Spalding MD, Fox HE, Allen GR et al (2007) Marine ecoregions of the world: a bioregionalization of coastal and shelf areas. Bioscience 57:573–583. https://doi.org/10.1641/B570707

Sweetman AK, Middelburg JJ, Berle AM et al (2010) Impacts of exotic mangrove forests and mangrove deforestation on carbon remineralization and ecosystem functioning in marine sediments. Biogeosciences 7:2129–2145. https://doi.org/10.5194/bg-7-2129-2010

Vaquer-Sunyer R, Duarte CM (2008) Thresholds of hypoxia for marine biodiversity. Proc Natl Acad Sci U S A 105:15452–15457. https://doi.org/10.1073/pnas.0803833105

Vinebrooke RD, Cottingham KL, Norberg J et al (2004) Implications of multiple stressors on biodiversity and ecosystem functioning. Oikos 104:451–451. https://doi.org/10.1111/j.0030-1299.2004.13255.x

Whitfield AK, Elliott M, Basset A et al (2012) Paradigms in estuarine ecology - a review of the Remane diagram with a suggested revised model for estuaries. Estuar Coast Shelf Sci 97:78–90. https://doi.org/10.1016/j.ecss.2011.11.026

Whittaker RH (1960) Vegetation of the Siskiyou Mountains, Oregon and California. Ecol Monogr 30:279e338. https://doi.org/10.2307/1943563

Ysebaert T, Herman PMJ (2002) Spatial and temporal variation in benthic macrofauna and relationships with environmental variables in an estuarine, intertidal soft-sediment environment. Mar Ecol Prog Ser 244:105–124. https://doi.org/10.3354/meps244105

Ysebaert T, Herman PMJ, Meire P et al (2003) Large-scale spatial patterns in estuaries: estuarine macrobenthic communities in the Schelde estuary, NW Europe. Estuar Coast Shelf Sci 57:335–555. https://doi.org/10.1016/S0272-7714(02)00359-1

Chapter 5
Benthic Estuarine Assemblages of the Southeastern Brazil Marine Ecoregion (SBME)

Paulo da Cunha Lana, Ronaldo Christofoletti, João Bosco Gusmão Jr, Thayanne Lima Barros, Daphne Spier, Tânia Marcia Costa, Abílio Soares Gomes, and Cinthya Simone Gomes dos Santos

Abstract We assess the current knowledge of the benthic assemblages in the Southeastern Brazil Marine Ecoregion (SBME), which extends for approximately 1200 km of coastline and includes seven major estuarine systems from Guanabara Bay in Rio de Janeiro to Babitonga Bay (or São Francisco do Sul) in Santa Catarina. The high ecosystem diversity of SBME putatively accounts for the high levels of endemism of the regional marine invertebrate fauna. However, until more taxonomical and biogeographical evidence is available, the SBME should be treated as a working biogeographical hypothesis rather than a cohesive unit identified by endemic fauna. As a consequence of urban, agricultural, and industrial development, the coastal areas from the SBME have been the most altered in the country over the last 500 years. Some of the largest cities and busiest harbors of the country are in or near the regional estuarine areas. The rapid environmental changes over the last several decades do not allow for the assessment if current similarities and

P. C. Lana (✉) · T. L. Barros · D. Spier
Universidade Federal do Paraná, Centro de Estudos do Mar, Paraná, Brazil

R. Christofoletti
Universidade Federal de São Paulo, Instituto do Mar, Santos, São Paulo, Brazil
e-mail: christofoletti@unifesp.br

J. B. Gusmão Jr
Universidad Católica del Norte, Facultad de Ciencias del Mar, Departamento de Biología Marina, Millennium Nucleus for Ecology and Sustainable Management of Oceanic Islands (ESMOI), Coquimbo, Chile

T. M. Costa
Universidade Estadual Paulista, Instituto de Biociências, Campus do Litoral Paulista, São Vicente, São Paulo, Brazil
e-mail: costatm@clp.unesp.br

A. S. Gomes · C. S. G. dos Santos
Universidade Federal Fluminense, Departamento de Biologia Marinha, Campus do Valonguinho, Niterói, Rio de Janeiro, Brazil
e-mail: abiliosg@id.uff.br; cinthyasantos@id.uff.br

P. C. Lana, A. F. Bernardino (eds.), *Brazilian Estuaries*, Brazilian Marine Biodiversity, https://doi.org/10.1007/978-3-319-77779-5_5

dissimilarities in the benthic assemblages express pristine conditions or if they are already the result of major human interventions, especially in the case of the Guanabara, Sepetiba, and Santos estuaries.

Keywords Southeastern Brazil · Benthic assemblages · Estuarine gradients

5.1 Introduction

This chapter assesses, mainly from a benthic perspective, the current knowledge of the estuaries from the Southeastern Brazil Marine Ecoregion (hereon SBME), as defined by Spalding et al. (2007). The name itself and the limits of the SBME, the most developed region in the country, do not coincide with the legal and geopolitical divisions in Brazil. In a legal context, Southeastern Brazil includes the coasts of Espírito Santo, Rio de Janeiro, and São Paulo states, whereas Southern Brazil includes Paraná, Santa Catarina, and Rio Grande do Sul states. For the purposes of this review, SBME will include estuaries that occur along the coastline for approximately 1200 km from Guanabara Bay in Rio de Janeiro to Babitonga Bay (or São Francisco do Sul) in Santa Catarina (Fig. 5.1). Although the main drainage systems of Southeastern and Southern Brazil are directed westwards to the La Plata Basin, the SBME still includes dozens of small rivers that drain directly to the coast and will not be evaluated here. We will focus on the seven major estuaries, as defined by Lessa et al. (2018, Chap. 1), namely, Guanabara and Sepetiba (in Rio de Janeiro State), Santos and Iguape-Cananéia (in São Paulo State), Paranaguá and Guaratuba (in Paraná State), and Babitonga (in Santa Catarina State).

The regional subdivisions of Spalding et al. (2007) may not truly express the latitudinal gradients of the meteo-oceanographic drivers and associated biota along the southwestern Atlantic Ocean, especially in the case of the inner shelf and other shallow-water marine environments. The Southeastern Brazil Marine Ecoregion, although included in the Warm Temperate Southwestern Atlantic province of Spalding et al. (2007), is in fact subtropical, with a markedly wet summer season, high summer temperatures, and high humidity levels. In this context, the consistency and validity of Spalding's ecoregions need to be empirically tested in coastal and shallow-water systems. Differently from the Eastern Marine Brazilian Ecoregion, which encompasses a large latitudinal variation, the SBME is oriented southwestwards along only 3° of latitudinal coverage. The SBME is hydrologically and topographically complex, and its dominant estuarine and coastal ecosystems alternate among mangroves, mudflats, dunes, sand banks, oceanic and estuarine beaches, rocky shores, lagoons, estuaries, salt marshes, transitional freshwater plant formations, and seagrasses (Fig. 5.2). This ecosystem diversity putatively accounts for the high levels of endemism of the marine invertebrate fauna. However, until more taxonomical and biogeographical evidence is available, the SBME should be

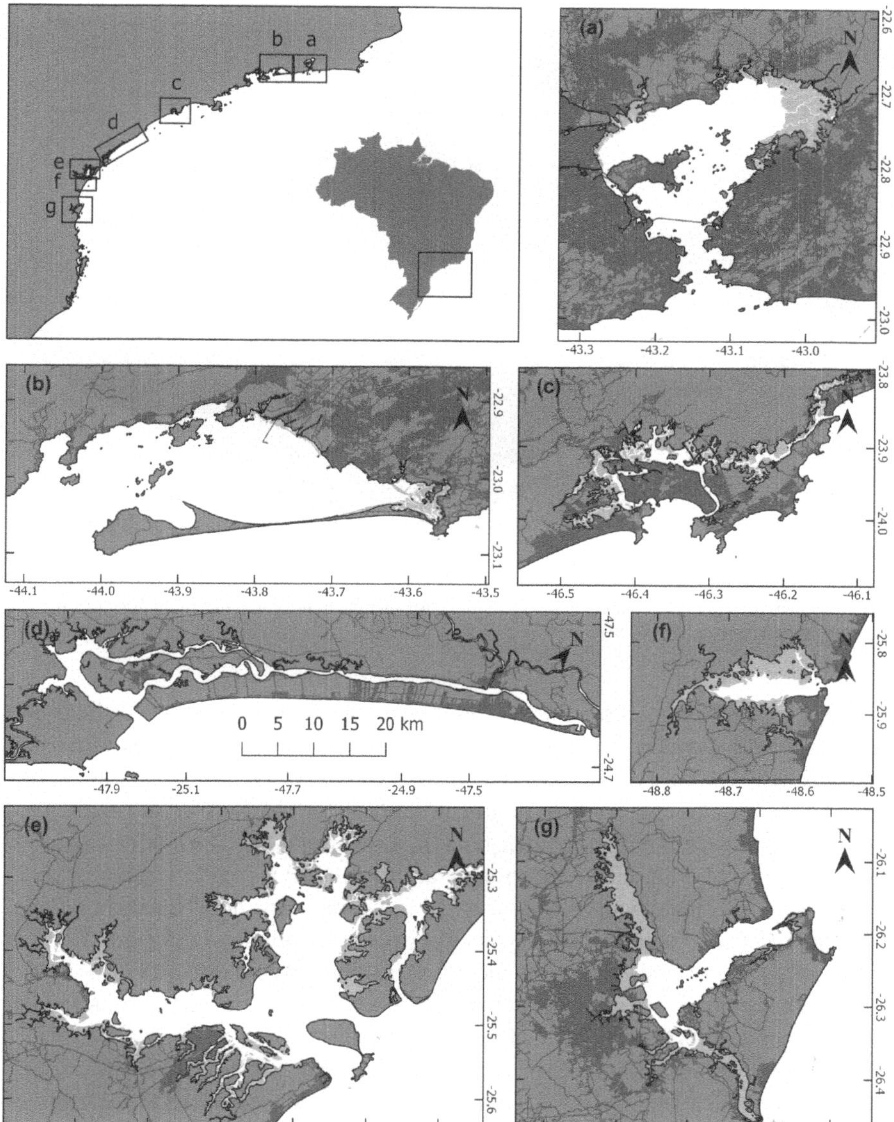

Fig. 5.1 The seven major estuaries of the Southeastern Brazilian Marine Ecoregion (SBME) all at the same scale: (**a**) Guanabara Bay; (**b**) Sepetiba Bay; (**c**) Santos Bay; (**d**) Iguape-Cananéia Bay; (**e**) Paranaguá Bay; (**f**) Guaratuba Bay; (**g**) Babitonga Bay

treated as a working biogeographical hypothesis rather than a cohesive unit identified by endemic fauna (Floeter and Soares-Gomes 1999; Caires 2014).

The estuaries of the SBME generally display small (less than 4000 km^2), steep catchment areas, with a dense river drainage. The estuaries share the ubiquitous

Fig. 5.2 Schematic representation of the diversity of estuarine habitats and plant formations in a subtropical estuary of the SBME

presence of the Atlantic Rain Forest ("Mata Atlântica") along their watersheds and have typically developed mangrove forests along their estuarine margins. Salt marshes, which appear as marginal mangroves fringes in the estuaries of the northern SBME, become increasingly dominant in the intertidal flats in the south, until they totally replace mangroves at Laguna, in Santa Catarina. Thus, with respect to the dominance of estuarine vegetated habitats, the SBME is clearly a biogeographic transitional zone that changes from tropical to warm temperate coastal conditions, as observed by Palacio (1982). Typically, tropical benthic assemblages become increasingly impoverished towards the colder and less humid climates of Southern Brazil. This trend in estuarine assemblages closely follows the parallel rarefaction of the Atlantic Rain Forest and mangrove vegetation.

Benthic diversity and dominance patterns along the SBME reflect the varying degrees of connection of the local estuaries with the adjacent shelfs, the rates of nutrient and freshwater influxes, the temperature regimes, the exposure degree of inlets, the influence of upwelling, the sediment sources, and the bathymetric or coastal complexity (Noernberg et al. 2006). Such differences in the annual rainfall, mean temperatures and freshwater input are among the main drivers of the benthic assemblages in the SBME, even at local scales.

Bernardino et al. (2015) used polychaetes as surrogates for benthic assemblages to show that the Southeastern and Rio Grande ecoregions (sensu Spalding et al. 2007) are more similar to each other than to the other Brazilian ecoregions. The numerically dominant species in the southeastern and southern estuaries are apparently restricted to these ecoregions (Bernardino et al. 2015). The estuarine seascapes are monopolized by mangroves and salt marshes in the Amazon and Rio Grande ecoregions (Chaps. 2 and 6). This condition gives rise to a type of unimodal distribution of the diversity of estuarine habitats or compartments along the Brazilian coast, which reaches a maximum in the SBME and the Northeastern and Eastern ecoregions and decreases towards the Amazon and the Rio Grande ecoregions. Along the SBME, greater habitat complexity is related to, among other factors, the close proximity of the Serra do Mar and the Atlantic Rain Forest, the greater climatic variability expected in subtropical transition zone, and the prevalence of upwelling and frontal systems. As a result, the large estuarine systems of the SBME display diverse habitats, bringing together tropical, subtropical and warm temperate systems, such as mangroves, mudflats, dunes, sand banks, oceanic and estuarine beaches, rocky shores, lagoons, estuaries, salt marshes, transitional freshwater plant formations, and seagrasses (Fig. 5.2). In general, these estuarine and coastal habitats tend to occur from Guanabara Bay to Babitonga Bay. However, the benthic assemblages associated with these habitats are only partially recurrent, as shown in the following sections. The rapid environmental changes over the last five to six decades do not allow for the assessment if current similarities and dissimilarities in the regional benthic assemblages express pristine conditions or if they are already the result of major human interventions, especially in the case of the Guanabara, Sepetiba, and Santos estuaries.

The taxonomic information on the benthic fauna of the SBME is the broadest available for the Brazilian coast, despite the inconsistencies and asymmetries in the knowledge of the minor phyla and meiofaunal groups, which are due to the lack or shortage of specialists. Unfortunately, only a few of the major oceanographic expeditions from the second half of the nineteenth century collected biological samples along the Brazilian coast. Nonato (1964) listed the major contributions resulting from these expeditions, which sampled mainly the northern and northeastern coasts. Hansen (1882), who was an occasional polychaetologist that is much better known as the discoverer of the leprosy bacillus, was meritoriously the first researcher to recognize endemism in the fauna of southeastern Brazil. Although the descriptions of South American marine species peaked around 1850, 1900, and 1970 (Miloslavich et al. 2011), the rate of discovery of new species, even among the best-known taxonomic groups, is still significant. Only cnidarians, echinoderms, and macroalgae seem to have reached a relatively stable number, with few recent additions (Miloslavich et al. 2011). The increasing application of molecular methods to less explored, cryptic groups or environments is also contributing to the current information on regional marine biodiversity (Prantoni et al. 2017). However, attempts to understand the ecological processes underlying the distribution patterns and function of benthic assemblages are scarce. Quantitative data, mainly on biomass and production, are missing or unsatisfactory, making it difficult to assess energy flow and energy budgets.

Southeastern Brazil holds approximately 45% (around approximately 85 million people) of Brazil's population, who live in 10% of the territory and are mainly concentrated in the coastal zone. The region accounts for a large percentage of the Brazilian GDP due to its resources (e.g., fishing, agriculture, mining, tourism), shipping and trade as well as tourism (Copertino et al. 2016). As a consequence of urban, agricultural, and industrial development, the coastal areas from the SBME have been the most altered in the country over the last 500 years. Some of the largest cities and busiest harbors of the country are in or near the regional estuarine areas (Rio de Janeiro, Sepetiba, Santos, and Paranaguá). Since the 1940s, urban development, iron and steel industries, exports from agrobusinesses (soy, sugar, cellulose, and bovine meat), and other industrial activities have greatly expanded in the ports of Santos (SP), Rio de Janeiro (RJ), and Paranaguá (PR). More recently, since the 1970s, a tourism boom has heavily impacted the southeastern Brazilian coast (Costa et al. 2006; Mello 2008). The degree of urban growth and habitat loss has increased over the recent years (Galvão and Nolasco 2013; Sousa et al. 2013). As a result of these accelerated changes, land use has already led to heavy contamination conditions in Guanabara, Sepetiba, Santos, and Babitonga bays (Molisani et al. 2004; Martins et al. 2011; Bordon et al. 2011; Fistarol et al. 2015; Soares-Gomes et al. 2016). The recent changes in pollutant concentrations suggest that the environmental conditions have worsened (Sousa et al. 2014; Fistarol et al. 2015). Some of the estuaries in the SBME have also suffered from environmental changes caused by the diversion of river waters from adjacent basins. The loss of environmental quality has compromised the estuarine biodiversity and affected the fisheries and tourism in most of the SBME estuaries. The concentrations of trace metals in algae, mollusks, crustaceans, and oysters are often well above the limits allowed by Brazilian legislation in Guanabara, Sepetiba, and Santos Bays (Sousa et al. 2014). Despite the persistence of quasi-pristine conditions in some sectors of Iguape-Cananéia, Paranaguá, and Guaratuba bays, unplanned land-based activities have resulted in pessimistic scenarios for the other major estuarine systems of the SBME.

5.2 Environmental Settings, Recent Decadal Oscillations and Future Scenarios

The coastline along the SBME stretches for approximately 1200 km from Guanabara Bay (~23 °S) to Babitonga Bay (~26 °S). The coastline is bordered by the "Serra do Mar" continuous mountain range ("sea mountain range"), which originated from Precambrian crystalline rocks. The Serra do Mar has a scarped coastal slope that creates embayments with wide strandplains (~50 km) and large estuarine systems of up to 600 km^2 (Angulo et al. 2016). The coastline looks like a submerged landscape from Guanabara to Santos Bay, and displays a sequence of high headland, small coves and beaches interconnected by rocky shores (Dominguez 2006). From Santos Bay to northern Santa Catarina, including the coast of Paraná, the coastline is formed by long beaches and well developed coastal plains with wide estuaries, such as Santos and Cananéia in São Paulo, Paranaguá and Guaratuba in Paraná and Babitonga in Santa Catarina.

Lessa et al. (2018) (Chap. 1) indicated that the seven major estuaries between the States of Rio de Janeiro and Santa Catarina are in grabens that run parallel to the coast and bear the same names (Fig. 5.1). There is evidence of Quaternary tectonic movements for at least the Guanabara, Santos-Bertioga, and Cananéia grabens (Lessa et al. 2018, Chap. 1). Local subsidence is likely accentuating the rates of ongoing sea level rise, which may impact the long term urban and industrial development. It is controversial if there was Late Quaternary tectonism in Paranaguá Bay, as suggested by the alluvium deposits, anomalous drainage lines, and uplifted catchments (Nascimento et al. 2013). There is even less evidence for tectonic control in Guaratuba and Babitonga Bays. Paleo-mangrove plains are present in the southern margin of Guaratuba Bay, whereas the present mangrove vegetation thrives on the northern side (Lessa et al. 2018, Chap. 1).

The development of Holocene barriers and the paleogeographic evolution of the coastal plains are closely related to the morphodynamics of the estuarine systems of the SBME (Angulo et al. 2009). There were at least three major cycles of marine regression and transgression during the interglacial periods of the Quaternary in southeastern Brazil, but precise dating is still lacking (Angulo et al. 2006). The sea level was approximately 8 m higher than present during the last interglacial highstand (MIS 5e) and approximately 3 m higher than present during the last postglacial maximum some 5100 year ago (Angulo et al. 2006). These cyclic sea-level changes may have caused the correspondingly isolation and reconnection of the coastal hydrographic basins. There is indeed evidence of population expansion of coastal and estuarine species over the past 200,000 years. The resulting reconnections and subsequent fragmentation and isolation between the estuarine and freshwater bodies were putatively relevant to the historical demography of the estuarine species (Tschá et al. 2017a, b). However, an alternative hypothesis suggests that the origins of endemic marine taxa in the SBME, at least in the case of the mobile fish fauna, are older and resulted from the dispersal from the eastern South Pacific Ocean (Caires 2014).

The climate conditions along the coast display complex temporal variability and are controlled by local and remote land–ocean–atmosphere processes (Marengo 2001; Liebmann and Marengo 2001). The climate is humid subtropical, with wet summers (Cfa according to the Köeppen classification, Alvares et al. 2014). The seasonal cycle of precipitation shares general characteristics with typical monsoon climates in other parts of the world (Nieto-Ferreira et al. 2011). Strong western winds in the limits of the subtropical and subpolar regions of South America (mainly in the eastern sector of the Andes) promote the incursion of mid-latitude cold fronts well into the southeastern Brazilian coast (Lupo et al. 2001). Thus, the Subtropical High Pressure (SHP) center of the South Atlantic and its interactions with the Subpolar Pressure govern the atmospheric circulation along the coastal plains and estuaries (Harari et al. 2008). Complex topography also influences the wind regime along the SBME. The most frequent winds in all seasons blow from the east, with average velocities of 1.5 m/s. Cold fronts force eastern winds to rotate to the north and west in only a few hours, with wind velocities of up to 10 m/s. This rotation is followed by winds blowing from the south for 1–3 days, with velocities between 5 and 10 m/s. As the air temperature drops and the atmospheric pressure rises, the winds turn back from the east until the temperatures and pressures return to normal values (Harari et al. 2008; Nieto-Ferreira et al. 2011). The wind regime seems to be a major driver of the intertidal flats dynamics in the outer estuarine sectors of the SBME, which act by generating short waves that produce turbulent bottom shear stress in shallow-water environments. Cold fronts may also be frequent in the summer when the South Atlantic Convergence zone (SACZ) becomes more intense (Barreiro et al. 2002). Stronger southerly winds greater than 6 m/s may cause meteorological tides (Campos et al. 2010), increasing the height of the sea surface and confining the water in the estuarine areas (Camargo and Harari 1994).

Interannual climate variability is strongly driven by the El Niño-Southern Oscillation (ENSO), which leads to below-normal rainfall and a decrease in extreme precipitation during El Niño years (Grimm and Tedeschi 2009). Both El-Niño (positive ENSO) and La-Niña (negative ENSO) affect the rainfall patterns and alter estuarine dynamics (García-Rodríguez et al. 2014). The ENSO is considered the main source of rainfall variability in South America (Grimm and Tedeschi 2009) increasing rainfall in SE Brazil and Rio Grande and decreasing it in the Northeastern ecoregions.

The SBME displays a regressive sea level fall, tropical to subtropical (21.5 °C annual mean temperature and 1400–5300 mm/year of rainfall), humid (over 85%), microtidal (1.7 m spring tide range) coast that is exposed to wave heights and wave periods of approximately 1.0 m and 8 s, respectively (Angulo et al. 2006; Noernberg et al. 2006). The tidal amplitudes increase towards the inner estuarine sectors, and are amplified less than two times. The tidal phases and amplitudes indicate that the tidal wave propagates in a mixed form, with a progressive form at the outer region and a standing wave form in the inner sectors. During neap cycles, strong nonlinear interactions allow for the formation of up to six high and low tides per day (Lana et al. 2001). Following cold front forcing, storm surges elevate the water levels up to 80 cm above astronomical tides (Marone and Camargo 1994). As exemplified by Paranaguá Bay, the tidal current velocities increase upstream, with maxima of 0.8–0.85 m/s at ebb and 1–1.4 m/s at flood (Marone and Camargo 1994). The mean precipitation during the rainy season in the austral summer, from December to March, can be up to three to six times higher than during the dry season in the winter, from June to August (Kjerfve et al. 1997; Noernberg et al. 2014; Soares et al. 2014). Riverine discharge reaching the ocean is generally low, as small drainage basins (<4000 km^2) discharge at rates less than 200 m^3/s into the estuaries (Angulo et al. 2009). The fluvial bedload yield to the nearshore areas is therefore small and large ebb-tidal deltas at the estuarine mouths are mostly fed by littoral drift (Angulo et al. 2016).

Stratification and mixing processes inside the estuaries of the SBME are primarily regulated by tidal currents but are heavily influenced by freshwater discharge, which causes seasonal variations in the magnitude of the vertical salinity stratification. The seven major bays of the SBME can exhibit highly stratified (summer neap tides), partially mixed (summer spring tides), and well-mixed (winter spring tides) conditions within a single season or fortnight (Mantovanelli et al. 2004; Soares et al. 2014). The salinity variations are usually in phase with the water level, and the tidal propagation is well represented by a standing wave (Noernberg et al. 2014). The optical environment has a mixed dominance of optically active substances, as indicated by the absorption coefficients of dissolved and particulate matter. However, due to the influence of freshwater discharge and the resuspension of bottom sediments induced by physical processes, the concentrations of the optical components in the water column do not generally display simple relationships (Noernberg et al. 2014). The phytoplanktonic contribution increases as a function of saline intrusion (Noernberg et al. 2014).

The seven major estuaries of the SBME differ widely in size and spatial complexity (Table 5.1). At least three distinct spatial configurations can be recognized, depending on the inlet characteristics, degree of development, compartmentalization, and interconnection of the inner water bodies. Increasing complexity in the estuarine geomorphological and hydrodynamic conditions leads to a corresponding heterogeneity of the benthic habitats and associated assemblages. In such cases, the linear estuarine gradients that are expected in less complex estuarine bodies are broken by the presence of inner sub-estuaries and the progressive compartmentalization and channelization of inner water bodies.

The linear estuarine pattern is exemplified by Guanabara, Sepetiba, and Guaratuba says (Fig. 5.1a, b, f), which have unique, usually narrow and shallow inlets, a main central natural channel, and only one large inner water body, that is either elliptical or semicircular. Well-marked gradients in environmental energy and salinity extends from their euhaline sectors near the inlets to the inner, mesohaline or oligohaline sectors, which display greater riverine influence (Molisani et al. 2004; Morais et al. 2016; Soares-Gomes et al. 2016). The surface sediment distributions and sedimentary facies are well known for both Guanabara (Kjerfve et al. 1997; Soares-Gomes et al. 2016) and Guaratuba bays (Barbosa and Suguio 1999). A tri-partite facies distribution that is typical of wave-dominated estuaries is well defined in these estuarine patterns. Riverine sandy sediments, that form bay-head deltas, accumulate near the river mouths. In the low energy, central estuarine section, sandy-mud and muddy-sand deposits prevail (Angulo et al. 2009). Marine sands predominate in the lower half of the estuary, which is subjected to marine influence from the inner shelf.

The multi-inlet and multi-compartment pattern is exemplified by the large Paranaguá Bay (Fig. 5.1e), which has three inlets and well-developed, interconnected inner water bodies. Paranaguá Bay should be better referred to as an estuarine complex, made up by different water bodies, composed of Paranaguá, Antonina, Laranjeiras, and Pinheiros bays. Each of these water bodies has unique linear salinity and energy gradients, although they are interconnected. All individual compartments have main channels that extend from the entrance to the inner sectors. The salinity displays a spatial linear gradient from the outermost to the innermost sectors of the individual bays, but this pattern is often broken by the presence of sub-estuaries (Noernberg et al. 2006). The individual bays display a tripartite zonation similar to other barrier and coastal-plain estuaries, as demonstrated by Lessa et al. (1998). The bay-mouth zone is located in the lower part of the estuarine system, where the main channel is deeper and under the influence of inner-shelf dynamics. The bottom sediments are characterized by well-sorted fine to very fine sand, and

Table 5.1 Dimensions of the estuaries of the Southeastern Brazilian Marine Ecoregion, along-channel extension of mangrove vegetation, fluvial catchment area and mean annual river discharge to the estuaries (modified from Lessa et al. 2018, Chap. 1)

	Area (km^2)							
Estuary	Intertidal vegetated	Supratidal	Water surface + non-vegetated intertidal	Total	Length (km)	Catchment area (km^2)	Mangrove extension (km)	River discharge (m^3/s)
Guanabara	84.9	1.3	362.6	448.8	48.2	3768	47.1	105.8
Sepetiba	30.5	10.1	507.1	547.7	48.0	2388	48.0	32.2
Santos	85.5	2.0	59.5	147.0	23.6	1052	21.1	42.8
Iguape-Cananeia	90.0	[a]	[a]	115.0	110	1339	72.0	43.0
Paranaguá	191.0	15.3	591.48	797.5	56.2	3272	89.7	230.4
Guaratuba	59.2	4.6	54.5	118.3	30.6	1810	21.8	80.0
Babitonga	68.2	12.0	177.4	257.6	55.0	1198	54.4	22.2

[a]Not measured

silt and clay contents vary from 0% (seaward end) to 40% (landward end). The estuary funnels have the widest cross sections and their bottoms are characterized by muddy sediments, with less than 50% sand (Bigarella et al. 1978). The estuary funnel zones of the individual bays gradually narrow upstream and form low-tide bars composed of shelly, dark mud. The meander zones are the narrowest stretches of the bays, with shallow channels and greater fluvial influence. The bottom sediments are mostly poorly sorted fine to medium sand, with less than 40% silt and clay and common granule and gravel due to the riverine inputs from the Serra do Mar. Riverine deltas develop in the meander zone.

The multi-entry channelized pattern is exemplified by Santos, Iguape-Cananéia, and Babitonga bays (Fig. 5.1c, d, g), which display at least two or more inlets at their southern and northern ends. Those estuaries present very narrow, confined, and shallow (6–12 m) inner bodies that are channels behind an extensive barrier island (Schaeffer-Novelli et al. 1990; Harari et al. 2008). Santos estuary is more complex than the other estuaries and composed of three main channels (São Vicente, Santos, and Bertioga channels), with dozens of fluvial inputs within the system. The Santos and São Vicente channels flow into Santos Bay which is open bay to the sea, while Bertioga channel flows straight to the platform. Coarser sediments can be found in areas under marine influence, where a considerable fraction of fine sediment (up to 80% at the entrance of the port) is also found. Sediments of the inner sector are mainly silt (Speranzini 2017). The Iguape-Cananéia estuary has been controversially treated as a lagoon system in the literature, probably due to its lower geomorphological complexity, with long channel that runs parallel to the coast.

Bernardino et al. (2015) showed that the decadal oscillations in temperature and rainfall regimes display a clear latitudinal trend along the Brazilian ecoregions that is consistent with global and regional climate forecasts (Marengo et al. 2010). At the Southeastern Brazil and Rio Grande ecoregions, the summer months (December and February) are characterized by higher temperature increment than in the winter, and this pattern has been consistent over the last four decades. All marine ecoregions have also displayed an increase in positive temperature anomalies in the last decade. Warmer years were more frequent and were marked in the Eastern Brazil, Southeastern Brazil, and Rio Grande regions, with yearly peaks of 1.5 °C above the mean decadal temperatures (Bernardino et al. 2015). The only significant increase in rainfall was detected in the SBME, where there was evidence of higher summer peaks over the last two decades. According to Bernardino et al. (2015), the estuaries from the SBME may experience higher than present salinity oscillations and sediment disturbances over short periods in the future decades if the extreme summer storm events observed during the last few decades continue. Extreme flooding events and changes in the benthic habitats within estuaries of the Rio Grande ecoregion are well documented (Costa et al. 2003; Colling et al. 2007), but consistent records of decadal changes in benthic assemblages are still missing for the SBME. Apparently, a larger influx of fluvial sediments is causing an expansion of the tidal sand bars or bay-head deltas, as already noted by Lessa et al. (1998) approximately 30 years ago, probably due to an increase in sediment input associated to vegetation cover loss in the Atlantic Rain Forest, and accelerated urban and industrial development. Soares et al. (2014) investigated the vulnerabilities of the Brazilian Large Marine Ecosystems (LMEs) to local and remote forcing, based on the total and partial correlations between climate indices and oceanic and atmospheric variables including sea surface temperatures (SST), wind stress, Ekman transport, sea level pressure, and outgoing longwave radiation. They indicated that future ecosystem-based management actions aimed at the conservation of regional marine resources need to consider the high complexity of the basin-scale interactions between local and remote climate forcing, including their effects on the ocean-atmosphere system of the southwestern Atlantic Ocean.

In short, the decadal record indicates that the estuaries of the SBME are currently influenced by global warming effects, and these effects overlap with the accelerating anthropogenic changes. Shifts in species distribution and assemblage compositions associated with climate change and habitat loss are expected to increase in frequency over the next few decades (Miloslavich et al. 2011; Bernardino et al. 2016).

5.3 Benthic Assemblages

5.3.1 Overview

An exhaustive revision of the extensive literature concerning the benthic assemblages of the SBME is beyond the scope of this chapter. Knowledge of the Brazilian benthic marine fauna throughout time have been presented by Lana et al. (1996), Amaral et al. (2003), Amaral and Jablonski (2005), Neves and Valentin (2011), Miloslavich et al. (2011), and Longo and Amado Filho (2014). Rather than presenting a chronological or geographical survey of the regional assemblages, we emphasize the shared patterns and the faunistic differences among the seven major estuarine systems of the SBME, organized by benthic compartments and exemplified by representative cases. The lagoonar systems of Rio de Janeiro were previously treated by Esteves (1998), Esteves et al. (2008) and Laut et al. (2016). We tried to update species names following the World Registry of Marine Species (WoRMS, online access in December 2017).

Pioneering work on the benthic assemblages of the SBME estuaries was mainly led at the Fundação Osvaldo Cruz and Universidade Federal do Rio de Janeiro, in Rio de Janeiro, and the Instituto Paulista de Oceanografia, later called the Instituto Oceanográfico of the Universidade de São Paulo, during the first half of the 1900s. However, until the 1960s, there were no benthic ecology schools or traditions in the country. From the 1970s on, there was a diversification and specialization of benthic surveys, which are still mainly concentrated in Guanabara, Santos and Paranaguá bays (Tommasi 1970; Lana 1986; Jorcin 1997; Soares-Gomes et al. 2016). Studies have only more recently addressed the spatiotemporal variations of benthic assemblages and have formally tested the role played by environmental drivers. Macroecological approaches are also very recent (Bernardino et al. 2015; Lana et al. 2017).

Collections of the benthic invertebrate fauna of the SBME are currently deposited at the Museu Nacional and Instituto de Biologia at the Universidade Federal do Rio de Janeiro; Museu de Zoologia, Departamento de Ecologia Geral (Instituto de Biociências); Instituto Oceanográfico at the Universidade de São Paulo; Museu de Zoologia at the Universidade Estadual de Campinas "Adão Jose Cardoso"; Departamento de Zoologia at Universidade Federal do Paraná, and Museu Oceanográfico at Universidade Federal de Rio Grande. Illustrated guides and reviews of the marine biodiversity of the SBME are also available (Lana et al. 1996; Amaral et al. 2003; Amaral and Jablonski 2005; Miloslavich et al. 2011).

Bernardino et al. (2016) suggested that the benthic assemblages are more diverse in the eastern and southeastern ecoregions of Brazil than in the Amazonian and Rio Grande ecoregions, probably as a result of the higher mean temperatures and increased habitat complexity, which is mostly associated with carbonate bottoms. This hypothesis still needs to be consistently tested, as it may result from a taxonomic artifact, due to the limited taxonomic knowledge of the eastern, northeastern and northern Brazilian ecoregions. This taxonomic artifact was recently demonstrated for polychaete species, which were treated as surrogates for the benthic fauna (Lana et al. 2017). The accumulation curves by marine ecoregion highlighted the regional differences in species diversity but also the large variance in sampling efforts along the Brazilian coast (Fig. 5.3). The published and publicly available data are mostly concentrated in the Southeastern and Rio Grande ecoregions (~60% of the records available for Brazil), due to the greater number of locally active benthic taxonomists and ecologists during the last few decades. Species accumulation curves tend to stabilize in the SBME, but asymptotic curves indicate that more data are necessary for the Northern, Northeastern, and Eastern ecoregions. Species richness seems to follow a trend opposite to the expended effort when interpolated data are used (continuous lines in Fig. 5.3). Conversely, when searching for extrapolated data (dashed lines in Fig. 5.3), the Southeastern ecoregion has the largest species diversity. But the larger and more comprehensive dataset in this region clearly biases the analysis, and these results should be considered with caution for the moment.

5.3.2 Benthic Assemblages of Mangroves and Salt Marshes

Mangroves in the SBME are close to their southern distribution limit at the Lagoa de Santo Antônio in Santa Catarina (28°28.5′S), where *Laguncularia racemosa* and *Avicennia schaueriana* dominate. The southernmost limit of mangroves in Brazil has not changed over the last 36 years and *L. racemosa* is most likely limited by the temperature at its southernmost extent (Soares et al. 2012). However, as the mean annual air temperatures are likely to increase by 3–5 °C by 2100 as a result of climate change, the latitudinal mangrove extents are expected to increase southwards.

The mangrove composition in this region is similar to that in the northern ecoregions (Chaps. 2, 3, and 4), where *Rhizophora mangle*, *Avicennia shaueriana*, and

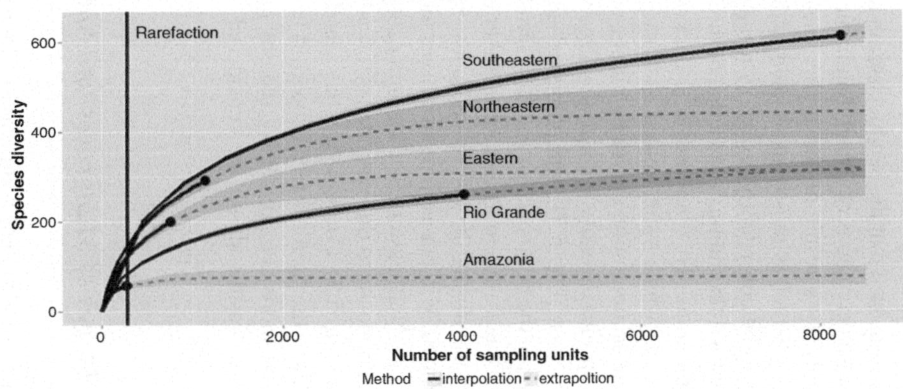

Fig. 5.3 Accumulation curves of number of polychaete species for the five Brazilian ecoregions, sensu Spalding et al. (2007), adapted from Lana et al. (2017)

Laguncularia racemosa are the main species. However, the mangroves in the SBME are less developed, less structured and less productive than those in the northern, tropical ecoregions. The tidal range in the SBME (approximately 1.7 m during the spring tides) and the presence of the Serra do Mar mountains near the coast, do not allow for wider mangrove expansions. The lower salinity areas contain transitional vegetation that integrates mangrove and Atlantic Rain Forest species. In these regions, some transitional species are frequently found, such as *Acrostichum aureum*, *Hibiscus tiliaceus*, *Typha domingensis*, *Scirpus californicus*, *Crinum salsum*, *Conocarpus erecta*, and monospecific beds of *Spartina alterniflora*. In the intermediate and high-salinity regions, the dominance of mangrove forest depends on the distance to the Serra do Mar mountain range, salinity, soil composition and height. In these regions, the mangrove forest can be monospecific for any of the mangrove species or present a mixed composition. The zonation is not always clear, especially in the more geomorphological complex estuaries with multiple inlets. Therefore, in this transitional region of the SBME, a mosaic of mangrove forests may be found to be closely associated with the Atlantic Rain Forest, salt marshes, mudflats, estuarine beaches, and estuarine rocky shores.

The composition, abundance, distribution, and zonation patterns of benthic fauna along the SBME are closely related to the presence of plant cover in the intertidal estuarine habitats and to the tidal regime. Spier et al. (2016) showed that at large spatial scales (>10 s km), flood regime is the leading factor that determines spatial distribution of animals in mangrove and salt marsh habitats across gradients of tidal energy in Paranaguá Bay. Secondary factors might be related to salinity, sediment composition, and nutrient flow. These authors indicated that the distribution of crabs is centered around mean high tide. In addition, during a 2-year period, inundations rates were low. Bivalves were not continuously exposed for periods longer than 10 days, providing an empirical estimate of the order of magnitude of their maximum desiccation tolerance.

As in other tropical and subtropical regions, the benthic epifauna and infauna in the SBME mangroves are composed of and dominated by a few animal groups, mainly polychaetes, crabs, bivalves, and gastropods. The benthic biodiversity of the fauna associated with sediments differs from that associated with prop roots, but both show low species richness. Polychaetes are the dominant group in unconsolidated bottoms, whereas isopods and tanaidaceans are the dominant groups on the prop roots (Silva-Camacho et al. 2017). The benthic community associated with the mangroves exhibits spatiotemporal variability, with higher infauna species richness and diversity in the innermost channel areas and a higher abundance of epifaunal organisms during the summer (Silva-Camacho et al. 2017). Differently from the vagile epifauna such as crabs, bivalves and mollusks, infaunal species are not exclusive to the mangroves and are frequently found in nearby habitats such as salt marshes and mudflats (Lana and Guiss 1991; Netto and Lana 1994, 1995; Lana et al. 1997a; Lorenzi 1998).

Crustaceans are the most studied benthic group in the mangroves of the SBME. Independent of the size of the estuaries, the mangroves seem to support crab and shrimp populations of commercial interest all along the SBME (*Farfantepenaeus brasiliensis* and *Farfantepenaeus paulensis*) where juveniles develop in the estuarine system while adults grow in the open bays (Costa et al. 2008). In an overall view, 28 species of crabs are found in the intertidal zones of the estuaries in the SBME, with a dominance of Grapsoidea and Ocypodoidea (Colpo et al. 2011), including common species such as *Goniopsis cruentata, Callinectes* spp., and *Aratus pisonii.*

The semiterrestrial crab *Cardisoma guanhumi* plays a relevant ecological role in the transitional zone from mangrove to terrestrial areas. The strong fishery pressure on this species has decreased the population size in the SBME and the species is probably threatened. *Ucides cordatus* is a semiterrestrial large-bodied crab of economic and ecological importance. This crab is one of the main crustacean fishery resource in the mangroves in Brazil. In the SBME, traditional fishing communities explore *U. cordatus*, but in a less intense way than in the northern and northeastern Brazilian regions. *Ucides cordatus* is a burrowing crab and plays an important role in the ecosystem functioning by bioturbation, affecting the biogeochemistry of the sediment and carbon cycling. Galleries are essential for the recruitment of *U. cordatus*, and their conservation seems mandatory for recovery programs for *Ucides* populations (Kassuga and Masunari 2015).

Fiddler crabs of the genus *Uca* represent other benthic components that are relevant for ecosystem functioning in mangrove areas. Studies carried out in the estuaries in the central area of the São Paulo coast have stressed the relevant role of these crabs in ecosystem functioning, as they increase the meiofaunal diversity and microphytobenthic biomass (Citadin et al. 2016). The distribution of different species of fiddler crabs across the intertidal zone is correlated with soil characteristics and plant cover. In monospecific patches, smaller crabs occupy the lower level of the intertidal zone, which contains more organic matter, whereas larger crabs are dominant in upper levels (Grande 2016). Grande (2016) suggested that organic food matter in the environment indicates the amount of food available for fiddler crabs. Sites with high organic content support higher-density populations (Grande 2016). However, it is still unclear how fiddler crabs influence organic matter contents by bioturbation, as this influence has been proven in controlled laboratory conditions but was overcome by other environmental drivers in field experiments (Natálio et al. 2017).

The literature on salt marsh plant communities in South America was reviewed by Costa and Davy (1992) and Isacch et al. (2006). Research concerning the fauna of the SBME salt marshes peaked in the 1990s but has almost been discontinued since then. Salt marshes occur regionally as narrow belts of *Spartina alterniflora* that border mangrove stands, and they may rarely dominate in intertidal flats to the detriment of mangrove stands (Lana et al. 1991). *Spartina alterniflora* belongs, as a pioneer colonizer, to the successional series that leads to mangrove development (Cunha-Lignon et al. 2009). As a result, salt marshes have been considered by local marine ecologists and the Brazilian environmental legislation (Lana 2003) as structural and functional components of mangroves. One should expect structural and functional similarities in the benthic assemblages of mangroves and salt marshes, considering that they are adjacent to each other, both develop in similar environments, are subject to similar ecological drivers, and face similar evolutionary pressures. Indeed, the macrofauna of the upper euhaline zone in sandy marshes can be quite similar to that in the adjacent mangroves, with the numerical dominance of oligochaetes, the polychaete *Capitella* sp. and the fiddler crab *Uca leptodactyla* (Couto et al. 1995; Netto and Lana 1995). However, there is unequivocal evidence that the species richness and abundance of macrobenthic infauna are significantly lower in mangroves than in adjacent salt marshes or unvegetated flats (Couto et al. 1995; Netto and Lana 1995). Between-habitat variation is more pronounced in high-energy euhaline sectors where a more abundant and diversified group of species occurs in marsh sites, in opposition to the impoverished fauna in mangrove sites (Lana et al. 1997b). Faunal differences among vegetated habitats seem to be less significant in inner oligohaline sectors, where the annelids *Laeonereis culveri* (also referred to in the regional literature as *L. acuta*) and *Nephthys fluviatilis* may be dominant, irrespective of tidal elevation or plant cover type, although they are less common in mangrove sediments (Lana 1986; Lana et al. 1997a).

The species composition is similar in the salt marshes along the SBME (Tararam and Wakabara 1987; Lana and Guiss 1992; Flynn et al. 1996, 1998; Netto and Lana 1997; Attolini et al. 1997). Macroinfauna are numerically dominated by the annelids *Isolda pulchella, Nereis oligohalina, Sigambra grubii*, and deposit-feeding capitellids. The tanaid *Monokalliapseudes schubarti* may display high population densities in the winter–spring, but it is less frequent during the rest of the year (Lana and Guiss 1991). On the other hand, the epifaunal gastropods *Heleobia australis* and *Littorina angulifera* and the epifaunal isopods *Tholozodium rhombofrontalis* and *Sphaeromopsis mourei* peak during the late spring and summer when the aboveground biomass is higher (Flynn et al. 1996). Species composition and abundance may vary with salt marsh elevation, even when grass stands are narrow (Netto and Lana 1997; Flynn et al. 1998). The abundance of the local macrobenthic species is also significantly correlated with rainfall and tends to decline after intense freshwater input in the summer, as shown by intense population shifts in *Monokalliapseudes schubarti* and the amphipod *Monocorophium acherusicum* (Pagliosa and Lana 2005).

The marked seasonal differences in the primary production cycles and litter dynamics (Domingos and Lana 2017) may explain the corresponding variation in the dominant benthic assemblages of coexisting mangroves and salt marshes. Winter and early spring starch stocking in subsurface rhizomes is very noticeable in the regional salt marshes (Lana et al. 1991), whereas there is no evidence of similar processes in mangroves. Since the plant biomass and litter produced in mangroves and salt marshes can serve as potential food or refuge for detritivores, pronounced seasonal differences should be expected in the abundance and diversity of the associated fauna. Although this hypothesis awaits experimental testing, the numerically dominant polychaetes *Isolda pulchella* and *Nereis oligohalina* are known to peak at the end of winter and the beginning of spring, when there is an up to fivefold increase in the belowground biomass of *Spartina* (Tararam and Wakabara 1987; Lana and Guiss 1991, 1992).

Aerial plant cover clearly plays a key role in the distribution patterns of salt marsh macrofauna, as experimentally shown by Pagliosa and Lana (2005), who followed the faunal responses after removing leaves and culms in *Spartina alterniflora* marshes. The responses of the infaunal or epifaunal species clearly depended on their abilities to discriminate between vegetated or unvegetated habitats. Epifaunal habitat specialists associated with leaves and culms, such as the gastropod *Neritina virginea,* the isopod *Sphaeromopsis mourei* and an unidentified gammarid amphipod, were expectedly affected by plant removal, whereas the habitat generalist epifauna *Monokalliapseudes schubarti* did not respond to it. The infaunal densities of the polychaetes *Isolda pulchella, Nereis oligohalina, Laeonereis acuta,* and *Capitella* sp. did not change significantly after plant removal.

Significant variation also occurs between the epifauna and infauna of salt marshes and adjacent unvegetated flats (Lana and Guiss 1992; Flynn et al. 1996). Salt marsh areas display significantly higher species richness and mean density; only a few species dominate faunistic composition. The presence of plant cover obviously affects epifaunal grazers, such as the gastropods *Neritina virginea* and *Littorina flava* and the isopod *Sphaeromopsis mourei*, in addition to a number of infaunal dominant species that are directly associated with the root-rhizome mat, such as the polychaetes *Isolda pulchella* and *Nereis oligohalina*. Benthic populations also seem to be more persistent in vegetated areas than in the adjacent flats, despite the marked short-term or seasonal variations (Pagliosa and Lana 2005).

5.3.3 Benthic Assemblages from Tidal Flats

The tidal flats of the SBME, which are locally called "baixios" (not to be confounded with the estuarine beaches, which are addressed in the next section), are usually well-developed and may represent up to 20% of total estuarine areas (Lessa et al. 2018; Chap. 1). The upper zones of the tidal flats are almost always bordered by salt marshes and mangrove stands. The tidal flat dynamics is strongly influenced by tidal currents, which surpass the effect of fluvial fluxes (Bigarella et al. 1978). In the winter, strong southern winds may enhance the tidal flat erosion and sediment resuspension, mainly near the estuarine inlets (Marone and Camargo 1994; Fonseca et al. 2013).

A sediment distribution model for tidal flats, which may prove relevant to explain the zonation of macrobenthic assemblages, was proposed by Fonseca et al. (2013). The sediments at the lower zones are often characterized by high gravel and carbonate content, the latter consisting mainly of shell fragments. In contrast, silt, clay, and organic matter content tend to accumulate in the more sheltered middle and upper zones, closer to the vegetated habitats. During high tides, the waves broke and dissipate their energy in the salt marshes and mangroves, beyond the higher tidal flat zone. During the ebb tides, the lower zone is also subjected to breaking waves, which prevents the deposition of silt-clay. The lower hydrodynamics at the middle and upper flat zones favor the deposition of fine suspended matter and may lead to sediment stabilization, both of them essential for microphytobenthic and macrobenthic development. Microphytobenthic primary production is indeed higher at the middle and upper zones as a function of incident radiation (Fonseca et al. 2013).

The microphytobenthic biomass content also displays a conspicuous seasonal gradient. Seasonality is determined mainly by turbulence of the water column, which, in turn, is influenced by atmospheric events (Fonseca et al. 2013). As in other tropical and subtropical areas (Alongi 1990), the seasonality of the benthic assemblages from tidal flats is also strongly influenced by the rainfall regime and by its correspondingly effects on microphytobenthic production (Murolo et al. 2006). In the absence of the algal cover (Couto 1996), the microphytobenthic biomass seems essential to support the trophic webs of idal flats, because it is more palatable and accessible than organic debris produced by adjacent salt marsh and mangrove plants (Domingos and Lana 2017).

The tidal flat fauna is numerically dominated by deposit-feeding or filter-feeding species, favored by the deposition of fine and particulate organic matter. Even in the absence of macrophytes, the biogenic structures may confer spatial heterogeneity to tidal flats, thus generating correspondingly complex macrofaunal distribution patterns (Couto et al. 1995; Sandrini-Neto and Lana 2014). The tanaid *Monokalliapseudes schubarti* (referred to in the literature prior to 2006 as *Kalliapseudes schubarti*), the bivalve *Anomalocardia flexuosa* (previously referred to as *A. brasiliana*), capitellid polychaetes, tubificids, and the polychaete *Nephtys fluviatilis* (mostly in oligohaline flats) are usually the dominant macroinfaunal species in tidal flats all along the SBME.

Morais (2015) and Morais et al. (2016) assessed the variability of the intertidal macrofauna of Guaratuba Bay at nested spatiotemporal scales. The numerically dominant species present highly variable responses to large- and small-scale drivers in space and time. The variance associated with the large-scale salinity gradient was always greater than the variance associated with time or spatiotemporal interactions for the amphipod *Monocorophium acherusicum*, unidentified ostracods, the polychaetes *Nephtys fluviatilis* and *Aricidea* sp., and the mollusks *Heleobia australis* and *Mytella guayanensis*. Conversely, unidentified oligochaetes, and the annelids *Heteromastus similis* and *Streblospio benedicti* occurred in highly variable and temporary population patches, and were consistently responsive to small-scale drivers, such as sediment texture, organic matter content and the presence of biogenic structures. Although the total abundance and assemblage composition varied significantly from hundreds of meters to kilometers, the variations in species numbers could not be explained by the large-scale estuarine gradient.

Boehs et al. (2004) showed that the recruits and juveniles of the mollusks *Lucina pectinata*, *Tagelus divisus*, and *T. plebeius* are more abundant in the upper levels of tidal flats, and suggested that these areas are preferential for settlement, followed by a later horizontal migration of adults along the intertidal region. A similar recruitment pattern was suggested for the bivalve *Anomalocardia flexuosa* (Guiss 1993).

5.3.4 Benthic Assemblages of the Estuarine Beaches

The dynamics of the estuarine beaches from the southern sector of SBME was reviewed by Angulo et al. (2016). Their length may vary from less than 100 m to more than 5 km. The dominant process of sediment reworking is by locally generated wind waves in a fetch-limited, low-height environment, although refracted ocean waves may be more important near the estuarine inlets. The tide-modified beaches close to inlets have both a beach face and low-tide terrace composed of well-sorted fine sands, while the inner more tide-dominated beaches have beach faces composed of poorly sorted coarser sands while the low-tide terrace to tidal flats become muddy (Rosa and Borzone 2008). Variations in grain texture between beach face and low tide terrace denotes a lack of sedimentary exchange between them, as is typical of tide-dominated beaches (Short 2006).

There are similarities in the composition of the benthic assemblages of estuarine beaches along the SBME, but human impacts may have already altered pristine or quasi-pristine conditions. The structure of polychaete assemblages, as exemplified by Omena et al. (2012) for 20 estuarine beaches in Guanabara Bay, is clearly related to beach slope, beach exposition, and to water and sediment quality. *Scolelepis chilensis* dominated the bay entrance sectors, while *Streblospio gynobranchiata* and a species from the *Capitella capitata* complex dominated the inner estuarine beaches. The highest densities and lowest specific richness were found in contaminated sites, near Governador Island and Ilha do Fundão, in the inner sector, with codominance of *Saccocirrus* sp., *Capitella capitata*, and *Polydora* sp., indicators of organic enrichment.

Santos et al. (2014) assessed the effect of morphodynamics and season on the macroinfauna assemblages of two sandy beaches inside Guanabara Bay during winter 2005 and summer 2006. They showed that macroinfaunal structure differed between the beaches and strata, with the more dissipative beach displaying higher richness but lower density than the reflective beach. Lower densities at the more dissipative beach was associated with high anthropogenic impact, while no major seasonal variation was detected.

The distribution of polychaete annelids was also assessed on 12 sandy beaches in Sepetiba Bay (Mattos et al. 2013). It was also shown that the more protected, inner beaches presented a greater diversity of trophic guilds than beaches close to the estuarine inlet. The diversity and composition of the trophic guilds of polychaetes are mainly driven by the average grain size and wave-exposure levels. These results agree with the reflective-dissipative morphodynamic gradient, with dissipative beaches supporting a more complex food web with more trophic links in response to a combination of abiotic and biotic variables (Mattos et al. 2013).

Corbisier (1991) analyzed species composition, abundance, species diversity and zonation patterns of the benthic macrofauna in three transitional beaches of the Santos estuary. Polychaetes were dominant in both number of species and number of individuals representing 92.2% and 98.5% of the total fauna, respectively. There was a lower faunal diversity in the upper level of all the beaches. Assemblages of the more exposed beaches were characterized by high abundance, and low diversity and species richness, due to the numerical dominance of the spionid *Scolelepis squamata* (probably *Scolelepis goodbodyi*). Inner, more protected beaches were numerically dominated by the polychaetes *Laeonereis acuta* and a species from the *Capitella capitata* complex. The main faunal differences among beaches were again attributed to sediment texture, and to the degree of wave protection. The reduced number of crustaceans and mollusks was attributed to high pollution levels all over the studied area. Such a dominance pattern was also indicated by Baroni et al. (2015), who also showed a dominance of polychaetes (>90% of species number and number of individuals) and a lower abundance of crustaceans and mollusks in the polyhaline sectors.

Rosa (2009) surveyed the benthic fauna of 13 estuaries distributed along the two main axes of Paranaguá Bay, during summer and winter conditions. The beaches were characterized by a narrow and steep upper portion followed by a generally extensive plain with low slope in the lower portion, which was absent in some cases. Grain size increased towards the outer estuarine sectors, contrasting with the muddy facies in the inner estuarine beaches. Reduction in width and increase in face slope, as well as reduction in the wave regime and increase of tidal influence, were the main drivers of faunal change between oceanic and estuarine beaches. Only the beaches near the estuary inlet displayed benthic assemblages similar to those of adjacent oceanic, exposed beaches, whereas typically estuarine animals dominated the fauna of beaches from the inner sector. Crustaceans and polychaetes were the dominant taxonomic groups, although oligochaetes and mollusks could be numerically dominant in the inner beaches. Of the four species recorded of talitrid amphipods in Brazil, three (*Atlantorchestoidea brasiliensis, Talorchestia tucurauna,* and *Platorchestia monodi*) occurred in 11 of the 13 estuarine beaches, although *A. brasiliensis* was represented by only one individual. *Talorchestia tucurauna* presented the highest densities, and its positive correlation with salinity values was attributed to its low osmotic tolerance. *Platorchestia monodi* was positively correlated with the debris biomass, which probably serves as a shelter for the species against predation and desiccation and as an alternative food source. The densities of the ghost crab *Ocypode quadrata* crab burrows in regional estuarine beaches were similar to those recorded for oceanic beaches. However, the absence of burrows in the inner estuarine beaches suggested that salinity plays a major role in the distribution of *O. quadrata.* Low density of crab burrows in winter was probably related to a decrease in crab activity under low temperatures.

Aluizio (2007) presented a detailed analysis of macrobenthic assemblages of Brasilia and Coroazinha beaches, in the euhaline sector of Paranagua Bay, both chosen because of the great abundance of plant debris, a common feature of estuarine beaches. Faunal patters were markedly influenced by the debris type, composed of varying amounts of mangrove or macroalgae detritus, which varied significantly among strata, beaches, tidal periods, and seasons. As a result, diverse faunal assemblages were associated with varying debris drift-lines in summer and winter periods. The fauna was composed of the crustaceans *Talorchestia tucurauna, Platorchestia monodi, Atlantorchestoidea brasiliensis, Bathyporeiapus ruffoi,*

Metamysidopsis neritica, Excirolana armata, Tholozodium rhombofrontalis and by the insects *Bledius bonariensis, Bledius fernandezi*, and *Bledius* sp. Non-metric multidimensional scaling and redundancy analyzes showed that there was a fauna typically associated to the detritus lines in both beaches, consisting of the talitrid species *Talorchestia tucurauna* and *Platorchestia monodi* that probably use the detritus for food source and refuge. Both species presented higher densities in the beach of Brasilia, possibly due to the greater contribution of algae in the debris lines. The isopods *Excirolana armata* and *Tholozodium rhombofrontalis*, although quite abundant, showed a weak relationship with the debris lines and were strongly related to the moisture content of the sediment. In contrast, the three species of *Bledius* correlated negatively with sediment moisture, and were more abundant above the debris line.

Pagliosa (2006) assessed the spatial variation of benthic assemblages along the intertidal–subtidal gradient of an estuarine beach located at the inlet of Babitonga Bay. Sediments with high concentrations of silt-clay were densely colonized (mean of 22,400 inds.m^2) by the polychaete *Clymenella brasiliensis*. Benthic assemblages subjected to wave action and erosion–accretion dynamics were impoverished at landward stations. Seaward macrofauna (10 and 20 m depths) inhabiting sediments with higher organic content and poorly sorted sands were more diverse and numerically dominated by ophiuroids and by cirratulid (*Tharyx* sp.) and maldanid polychaetes (*Clymenella brasiliensis*). There was a marked breakdown in the zonation patterns due to the presence of a creek outlet, with high densities of the anthozoan *Edwardsia* sp. and the polychaete *Lumbrineris atlantica*.

5.3.5 Benthic Assemblages of Seagrass Meadows

We summarize herein part of the extensive reviews presented by Marques and Creed (2008) and Copertino et al. (2016), focusing on seagrasses from the SBME. Whereas shallow rocky shores and reefs are dominated by macroalgal beds, the seagrass meadows are regionally restricted to shallow sandy bottoms of outer estuarine sectors and coastal lagoons, and are less frequent in exposed sites. The sporadic patches of *Halodule wrightii*, *H. decipiens*, and *Ruppia maritima*, with reduced biomass, depend on suitable estuarine conditions, which include shallow unconsolidated substrate, wave-protected conditions, and less turbid waters.

There is some evidence that regional seagrass meadows have declined along the last decades (Copertino et al. 2016). In 2013, as part of an ongoing National Seagrass Mapping Project, nine sites previously known from the well-studied northern coast of Rio de Janeiro, were revisited and yielded six recordings (Copertino et al. 2016). In São Paulo, seagrasses were recorded at 12 beaches of the inner shelf, mostly in protected coves, during the 1980s (Oliveira-Filho et al. 1983). The sites were revisited in 2014 and meadows were found in only three of them (Copertino et al. 2016).

While many studies have been published on the seagrass biology and ecology of the Rio de Janeiro coast (Marques and Creed 2008 for a review), almost nothing is known about seagrasses in São Paulo and Paraná States. Not considering the observed decreased biomass and densities (Copertino et al. 2016), there is just a few published systematic characterizations of benthic structure or population dynamics in the seagrass beds of this SBME sector.

The short-term temporal dynamics and morphology of regional seagrass beds were first assessed by Creed (1997, 1999). Macroalgae, mainly *Jania adhaerens*, *Hypnea* spp., and *Acanthophora spicifera*, may be abundant inside the meadows. The gastropod *Cerithium atratum* is numerically dominant reaching densities up to 2000 ind/m^2 (Creed 2000). Fish, shrimp and blue crab from the seagrass beds are commercially exploited.

Corbisier (1994) described the benthic macrofauna of *H. wrightii* meadows of Codó Beach, Ubatuba (São Paulo) in both summer and winter conditions. Assemblage structure depended on sediment type, plant presence, and shoot density. Deposit-feeding infaunal polychaetes numerically dominated vegetated areas, whereas carnivores, filter feeders, and omnivore polychaetes were more common in unvegetated areas. Benthic density was not affected by plant biomass, probably due to the small size and biomass of the plants, which do not favor a diversified epifauna. However, species numbers and diversity were higher in vegetated than in unvegetated areas.

Omena and Creed (2004) investigated the polychaete assemblages of six beds of *H. wrightii* along the coast of the State of Rio de Janeiro and found 69 species of polychaetes belonging to 24 families. The dominant species, as well as the species composition, varied substantially between sites. Species number was positively correlated to the percentage of silt-clay and to the sediment classification coefficient, while the composition of the polychaete communities was influenced by plant morphology. Plants with larger axes mainly harbored surface deposit-feeding species, such as the polychaete *M. papillicornis*, and plants with a higher ratio of root–axis length harbored mostly suspension feeders like the sabellid *Fabricia filamentosa*.

Oigman and Omena (1999) also assessed the effects of *H. wrightii* on the distribution of the polychaete annelids from Ilha do Japonês, Rio de Janeiro, at the inlet of Araruama Lagoon. Of the 12 recorded species, only *Magelona papillicornis* Muller was numerically dominant, comprising more than 50% of the total number of polychaetes recorded in most cases. Low density of polychaetes (less than 239 inds/m^2) was attributed to strong tidal currents, intense predation, and/or low grass biomass. Using the same sampling effort, Ribeiro and Junqueira (1999) compared the abundance and diversity of mollusks in vegetated and non-vegetated habitats. Fifteen species of gastropods and ten of bivalves were recorded, mainly *Cerithium atratum, Divaricella quadrisulcata, Tellina* sp., *Anomalcardia brasiliana, Codakia costata*, and *Bittium varium*. There was a positive correlation between the abundance of gastropods and vegetation, which may be related to the feeding habits of *C. atratum*, which is a herbivore gastropod. Variation in abundance and diversity of crustaceans was also assessed in Ilha do Japonês by Rumjaneck and Lavrado (1999). They observed a higher species number and greater abundance of the dominant amphipods *Melita orgasms* and *Cymadusa filosa* in vegetated areas. Crustacean abundance was also dependent on the morphology and biomass of marine grass, which is related to herbivory and protection against predation. Grass biomass in this region (mean of 30.83 g/m^2, dry weight) was low compared to that found in similar systems, which could explain the low density of crustaceans. This was not observed for the polychaetes (Oigman and Omena 1999) and for the bivalves (Ribeiro and Junqueira 1999), which were more related to the granulometric characteristics of the sediment, indicating that the presence of marine grass affects the main groups differently of macrofauna.

Halodule wrightii meadows occur in shallow subtidal sandy bottoms in the euhaline high-energy sector of Paranaguá. Near their southernmost limit in the South Atlantic, the local plants display shorter and narrower leaves, shorter leaf sheaths, thinner rhizomes, a lower number of leaves per shoot, and higher internodal distances when compared with those from other *Halodule* meadows along the Atlantic coast (Sordo 2008). Because of its great adaptability as a pioneer and short-lived species, *H. wrightii* can grow and reproduce vegetatively under stressful conditions such as high turbidity and lower salinities and temperatures. As a result, the local seagrass populations are patchy, unstable, and infrequently sexually reproductive. Sordo (2008) and Sordo et al. (2011) assessed the seasonal morphological and biomass variations of *H. wrightii* meadow on Rasa da Cotinga (from a healthy state to its subsequent decline and die-off), and Mel Island, near the inlet of Paranaguá Bay, from 2005 to 2008. The highest values for all plant variables were found in the summer. Thereon, there was an important decrease in shoot density and in belowground and aboveground biomass in Cotinga Island, with the clearance of the internal areas of the meadow, in colder months. The complete regression of the Rasa da Cotinga meadow by the beginning of 2008 was followed by an overgrowth of the epiphytic brown alga *Hincksia mitchelliae*. The differences between sites increased with the progressive degradation of the Rasa da Cotinga meadow. Such marked temporal variations in morphology and plant biomass, which lead to local meadow regression and disappearance, had a clear effect on local benthic assemblages. With an unexpected boost of epiphytic biomass, plant growth was suppressed and the numbers of burrowing and opportunistic macrobenthic species increased in the Cotinga meadow. Epifaunal and infaunal populations responded to such a variation in plant density and biomass with corresponding increase in the density of the amphipod *Caprella scaura* and a significant decrease in the numbers of the polychaete *Magelona papillicornis*. At the same time, the stable Mel Island meadow presented minor seasonal or internal variations, with the exception of a high-density patch of the polychaete *Magelona papillicornis* in May 2007. The biomass and number of leaves increased with time and macrobenthic abundance and species richness tended to decrease without any other major changes in the structure of the community. This study indicated that the first signs of degradation or regression of local seagrasses were rather reflected on biological variables than on chemical ones. As a result, variations in plant biomass and numbers of animals were proposed as early indicators of water quality degradation in events of seagrass regression.

Short-term or interannual shifts in shoot density, average canopy, and belowground biomass of seagrasses along the SBME may thus prove usual for a better understanding of the regional effects of climate change (Marques et al. 2015;

Copertino et al. 2016). Increases in spatial differences in plant cover and abundance within stable seagrass meadows may be an early indicator of future regressions and can be used in assessing and monitoring environmental impacts (Sordo et al. 2011).

5.3.6 The Subtidal Benthic Assemblages

The subtidal benthic assemblages display similar distribution patterns in the SBME, despite marked differences in geomorphology, hydrodynamics, and contamination levels among estuaries. The species composition and the patterns of numerical dominance may vary along the 1200 km coastline of the SBME, but the macrofaunal assemblages are always more diversified and stable at the outer, more saline sectors, and get impoverished or even disappear in the hypoxic bottoms of the inner sectors, as first shown by Tommasi (1970) and later confirmed by Lana (1986), Santi et al. (2006), Mendes et al. (2006), Santi and Tavares (2009), and Soares-Gomes et al. (2012). The outer estuarine bottoms are mostly composed of sandy, oxidized or moderately reduced sediments with normoxic conditions in the water column. Higher diversity and biomass of deposit- and suspension-feeding species are usually found in the mesohaline zones, with sand-muddy bottoms. Low diversified assemblages or even azoic conditions may develop in the low-energy inner zones, which are dominated by muddy bottoms, with reduced sediment surface conditions, high organic matter content, and the frequent development of hypoxic conditions in the water column near the bottom.

The expected linear benthic gradients may be disrupted by local changes in tidal circulation, by the presence of river and tidal creek mouths and by local changes in the intensity and direction of tidal flow. Lateral estuarine gradients may originate from the freshwater input of rivers and tidal creeks, which may create several "micro-estuaries" or sub-estuaries in the euhaline and polyhaline sectors (Knoppers et al. 1987; Noernberg et al. 2006). The changes in freshwater input, with daily, seasonal, and interannual components, may also overlap and confound the expected salinity gradient (Lana et al. 2001).

The benthic distribution model pioneered by Tommasi (1970) probably reflects the pristine or quasi-pristine configurations of the SBME estuaries, eventually altered by human changes over the last few decades. Tommasi (1970) recognized five faunal zones, which closely follow the salinity and sediment texture gradients across the estuarine system of Cananéia estuary: (a) outer sector of Trapandé Bay, near the estuarine inlet and deltaic bars, with direct influence of oceanic waters, sandy bottoms with low organic matter content and dominated by the octocorallians *Renilla reniformis* and *Virgularia* sp.; (b) the transitional zone of Mar de Cananéia, with mixed sediment bottoms, dominated by deposit-feeding bivalves, such as *Macoma* sp. and *Tellina alternata*; (c) Mar de Cubatão, already under the influence of riverine inputs, with higher values of particulate organic matter than the Cananéia Sea, and dominated by the tanaid *Monokalliapseudes schubarti* and the phoronid *Phoronis* sp.; (d) northern region of Cananéia Island, from Ilha de Graças to the mouth of the Buguassu River, with high organic content, little particulate suspended matter and low dissolved oxygen, numerically dominated by the gastropod *Heleobia australis*, which is probably favored by the "*pessimus*" conditions for other macrofaunal animals; (e) hard-bottom areas of the Ilha da Casca region, where the most abundant organisms were the hydrozoans *Eudendrium carneum*, the ophiuroid *Ophiothrix angulata*, and a diversified set of amphipods. The presence of a diverse epifauna was attributed to tidal currents strong enough to avoid intense sediment deposition.

Later studies carried out in Iguape-Cananéia estuary also indicated a marked gradient in macrofaunal distribution (Carcamo 1980; Varoli 1990; Souza 2009). The local macrofaunal assemblages are composed by approximately 170 species, with polychaetes as the dominant group both in species numbers and abundance, followed by mollusks (reaching almost 50% of fauna), mainly represented by *Heleobia australis* sp. and *Turbonilla* sp. (Tommasi 1970; Carcamo 1980; Varoli 1990; Souza 2009). Echinoids and ophiuroids are found mainly at the polyhaline region (Tommasi 1971). Ostracods are a low abundant but diverse group in the region, with distribution patterns also influenced by the salinity gradient (Coimbra et al. 2007). Almost 40 years after the pioneer work of Tommasi (1970), Souza et al. (2013) assessed the spatial and temporal patterns of the subtidal benthic macrofauna and its relationship with environmental factors in Trapandé Bay. The greatest number of species and the highest densities were again found near the estuary inlet, decreasing towards inner sectors. However, the temporal and spatial changes observed at the estuary inlet strongly reflected seasonal variations in nutrient and freshwater input, attributed to increased rainfall. The increased flow in the Cananéia Sea, coming from the drainage basin, produces major changes in sediment and faunal composition. In late spring, the polychaete *Aricidea* sp. was predominantly found at the shallow waters near the Cananéia Sea, associated with higher values of total phosphorus contents organic matter concentrations. However, the densities of *Aricidea* sp. decreased at the end of summer, most likely due to the increased rainfall and hydrodynamic flows at estuarine bottoms, which lead to a corresponding decrease in total phosphorus contents organic matter concentrations. Higher densities of magelonid polychaetes were usually found in sediments with approximately 5–10% of organic matter. The distribution pattern of the omnivorous lumbrinerid *Scoletoma tetraura* was intrinsically linked to the total density and number of species. Ophiuroids were restricted to the outer estuarine area in the dry season (October). At the end of the southern hemisphere summer, in March, most likely due to the freshwater inflow that originates from the North through the Cananéia Sea, their occurrence was restricted to the southern bank of the outer sector.

Similar macrofaunal patterns in subtidal bottoms of the SBME were also shown by Santi and Tavares (2009) and Cardoso (2010) for Guanabara Bay: (1) euhaline and polyhaline zones, with salinity from 33 to 30, dominated by the polychaetes *Magelona posterelongata*, *M. papillicornis*, and *Mediomastus* sp., with high species numbers; (2) a transitional mesohaline zone, with salinity from 33 to 22, with decreased macrofaunal diversity; (3) inner oligohaline zones, with salinities from 19 to 8, dominated by the gastropod *Heleobia australis*, the polychaete *Nephtys fluviatilis* and the tainadacean *Monokalliapseudes schubartti*. There is also a change in functional groups with a dominance of polychaete surface deposit feeders in the polyhaline and transition zones to carnivore and subsurface deposit feeders in the inner zones (Cardoso 2010). Similar diversity gradients are also evident for mollusks (Mendes et al. 2006; Neves et al. 2013) and decapod crabs (Lavrado et al. 2000; Calil et al. 2006), The commercially explored portunid crabs *Callinectes ornatus, Callinectes danae*, and the penaeid shrimps *Farfantepenaeus brasiliensis* and *Farfantepenaeus paulensis* are common in subtidal bottoms all along the SBME estuaries. They tend to be more abundant in the outer estuarine sectors, but may be also dominant in inner sectors during the dry winter season (Lavrado et al. 2000).

Pieper (2007) carried out the first systematic survey of the benthic sublittoral associations of the Laranjeiras Bay in the Paranaguá estuarine system, whose benthic assemblages were previously assessed by Lana (1986). The survey was based on two seasonal sampling campaigns in winter 2006 and summer 2007 looking for possible differences between the eastern margins (greater marine contribution) and western margin (largest continental contribution). 197 taxa were identified up to the species or morphotype level, being granulometry and the concentration of organic matter the main drivers of the distribution of organisms. There was a clear differentiation in benthic assemblages between the margins, but not between seasons.

Blankensteyn and Moura (2002) presented a preliminary list of macrofauna of non-consolidated subtidal bottoms of Guaratuba Bay, certainly the least studied of the great estuarine systems along the southeastern coast of Brazil. A total of 69 taxa were identified, of which only 22 (31.9%) had been previously recorded, with dominance of polychaetes and bivalve mollusks. Silva (2008) investigated Guaratuba Bay subtidal in winter of 2006 and summer 2007 campaigns, trying to correlate benthic distribution patterns with the salinity and hydrodynamics gradients. 152 taxa were found, with mollusks representing 93% of the total macrofauna, followed by annelids (3.26%) and crustaceans (2.82%). Three large sectors were defined by different faunistic structures, hydrographic and granulometric properties, where sediment texture and organic matter content were the main drivers of macrobenthic composition and distribution. There was a spatial difference between sectors, but not between seasons.

Haponiuk (2007) investigated the spatial distribution patterns of the benthic community and their relationships with environmental variables in Babitonga Bay. He identified three major sectors defined by the faunistic structure and by different hydrographic and granulometric properties. The general distribution of the organisms is closely related to sediment texture as in the other estuaries of the SBME. The first sector comprised the outermost sampling stations, with high hydrodynamics and the presence of the lancelet *Branchiostoma caribaeum*. The second sector, located in the intermediate zone of the estuary, displays lower environmental energy and is dominated mainly by the polychaete *Laonice branchiata*. The third sector, composed of the very low-energy inner bottoms, is dominated by the polychaete *Magelona papillicornis*. Some taxa were frequent and persistent along the estuary gradient, such as sipunculids and the polychaete *Capitella* sp. The abundance and number of species also followed the patterns already described for the other SBME estuaries, decreasing from the mouth to the inner sectors of the bay. Vieira (2007) also described the spatial variation of the macrobenthic assemblages in subtidal sites of Babitonga Bay. A total of 104 taxa were listed, and polychaetes were the dominant group, with *Magelona papillicornis* and *Polydora websteri* being the most abundant species. In general, species numbers followed a salinity gradient, increasing in the outer, intermediate and near the Canal do Linguado, which presented higher salinity and falling towards the inner sectors near the Palmital Canal where the salinity decreased. The number of taxa was higher in the outer sector, decreasing towards the inner ones. The composition and dominance patterns of benthic macrofauna were related to variations in the sand, silt, organic matter and calcium carbonate contents. Particulate organic matter and carbonate contents were low in the outer and inner areas, but increased in the intermediate sectors with fine sediments. Contrary to what happens in other estuarine systems of the SBME, the local hydrodynamics seems to be the conditioning factor for the dominance of fine sediments in the intermediate sector, a reflection of the low hydrodynamics and the deposition of material of continental origin. On the other hand, in the outer and inner sectors, the sediments were characteristic of high-energy dynamics, due to the mouths of the Cubatão, Palmital, and Garuva rivers.

5.3.7 Benthic Assemblages of Tidal Creeks

Tidal creeks, locally called *gamboas*, are common features of coastal plains dominated by tidal flow all along the SBME. They are routes for the exchange of water and dissolved particulate matter between mangroves, salt marshes and adjacent coastal systems (Lana et al. 1989). The alternation of depositional environments in such meandering water courses creates conditions for the formation of salt marshes and mangroves which typically develop along the meander bars (Lana et al. 1989). Tidal creeks also drain areas of continental vegetation, such as the sand-dune vegetation or *restingas* (Ovalle et al. 1990). Their capacity to carry sediment, conditioned by rainfall and tidal regimes, is low, due to correspondingly low slope, meandering contours and reduced flow rates (Lana et al. 1989). The marked decrease of pH in tidal creeks is associated to surface and porewater migration from the mangroves and adjacent *restingas*, mainly during low tides (Ovalle et al. 1990). The daily variations of temperature and dissolved oxygen tend to be more pronounced near their catchment areas than in their mouths, subjected to greater marine influence.

Despite their relevance to estuarine metabolism, there are just a few studies dealing with the composition and distribution patterns of the benthic macrofauna of tidal creeks along the southeastern coast of Brazil. Por et al. (1984) were the first to assess the biota of the so-called *gamboas* or blackwater rivers of the SBME. Later, Lana et al. (1989) described the macrofaunal distribution patterns along the Gamboa Perequê in the euhaline sector of Paranaguá Bay. Benthic macrofauna of *gamboas* is clearly conditioned by the marked physical, chemical gradients and by the alternance of erosion and accretion environments.

The most detailed survey on tidal creek macrofauna of the SBME is still unpublished. To assess macrobenthic distribution patterns, Lorenzi (1998) sampled five tidal creeks along a gradient of salinity and hydrodynamics of Paranaguá Bay: (a) Gamboa Papagaios (euhaline sector); (b) Gamboa das Ostras (polyhaline sector); (c) Gamboa Pinheirinho (mesohaline sector), and (d) Gamboa Xaxim (oligohaline sector). Sediment texture patterns of the *gamboas* followed, in general lines, the same pattern previously described by Netto and Lana (1997) for intertidal flats. The macrofauna was dominated by a few numerically dominant species, mainly of polychaetes, as previously shown by Por et al. (1984) for the Juréia region, in the central coast of São Paulo state. Recurrent sets of species were recognized in distinct salinity sectors. The association made up by *Monokalliapseudes schubarti*, *Nephtys fluviatilis*, *Erodona mactroides*, dipteran larvae, and enchytraeids was typical of the oligohaline and mesohaline tidal creeks. The association of *Heteromastus similis*, *Laeonereis acuta*, and *Nereis oligohalina* was common in creeks from the euhaline and polyhaline sectors. The bivalve *Mytella guyanensis* and the barnacle *Fistulobalanus citerosum* were typical of hard substrates in the euhaline sector. The polychaete *Aricidea albatrossae*, also common in euhaline subtidal bottoms of other estuaries from the SBME, was recorded only in the euhaline sector. Total macrofaunal density, species richness and density of dominant species did not seem to differ significantly from the catchment areas to the mouth of the creeks, regardless of the considered salinity sectors. However, the polychaetes *Aricidea albatrossae, Glycera americana*, and *Polydora websteri*, and the commercially exploited bivalve *Mytella guyanensis* may be dominant at the mouths of tidal creeks in polyhaline and mesohaline sectors. Likewise, *Monokalliapseudes schubarti,* a species common to all gamboas, had significantly higher densities in the innermost sectors of the bay, near the catchment area of the *gamboas*, where fine sediments predominate. A similar distribution pattern was detected by Lana et al. (1989) in the Gamboa Perequê in the euhaline sector of Paranagua Bay, where higher densities of this tanaid were recorded in the vicinity of the catchment area.

The species diversity of tidal creeks is rather low in comparison to that of other adjacent environments, such as unvegetated tidal flats, salt marshes, and mangroves (Lana et al. 1997a, b; Lorenzi 1998), probably as a result of fast salinity changes and low pH values, which may exclude estenohaline freshwater and marine species. In contrast to mangrove and subtidal bottoms from the inner estuarine sectors, the availability of dissolved oxygen does not appear to be a limiting factor for the macrofauna, as saturation is always relatively high due to tidal circulation (Lorenzi 1998). Nutrient availability also does not appear to be a limiting factor of primary production, since the recorded values recorded are similar to those of the adjacent estuarine water masses (Brandini and Thamm 1994; Machado et al. 1997). Por et al. (1984) and Lorenzi (1998) have hypothesized that the low diversity and low abundance of tidal creek macrofauna reflects the dominance of acid waters with a high concentration of humic substances.

In summary, the benthic associations found in tidal creeks are also recurrent in other estuarine environments, as adjacent unvegetated intertidal flats and shallow subtidal bottoms. Although they drain areas of mangroves and salt marshes, tidal creeks should not be seen as ecotones or transitional environments between these ecosystems and the adjacent estuary. Its fauna should be considered as a less diversified and less abundant continuum of the benthic assemblages from adjacent estuarine compartments.

5.3.8 Benthic Assemblages of Estuarine Rocky Shores

Estuarine rocky shores are present in the SBME due to the intrusion of the Serra do Mar ("sea mountain range") in the coastal areas. There are a few studies on the biodiversity and dynamics of the estuarine rocky shores, and most of the studies have focused on specific taxonomic groups rather the benthic assemblages as a whole. Previous studies were mainly conducted in the polyhaline estuarine regions, with a gap of knowledge on the biodiversity patterns and processes in the mesohaline and oligohaline regions. In general, the lower the salinity along the salinity gradient, the lower the species richness and the percent cover of the substrate.

Studies from the Guanabará, Santos and Paranaguá estuaries have indicated the presence of macroalgae, barnacles, decapod crustaceans, mollusks, polychaetes, ascidians, hydroids, and ophiuroids (Lotufo 1997; Junqueira et al. 2000; Lavrado et al. 2000; Santos 2004; Breves-Ramos et al. 2005; Van der Ven 2005; Grohmann 2009; Santi and Tavares 2009; Rocha 2010; Omena et al. 2012; Soares-Gomes et al. 2012; Yoneshigue-Valentin et al. 2012; Mattos et al. 2014). Macroalgae is the most diverse group, with 173 taxa in Guanabara Bay, and the taxon is dominated by red and brown macroalgae (Yoneshigue-Valentin et al. 2012). The species richness in the consolidated estuarine substrata increases when the fauna associated with the secondary mussel bed (*Perna perna*) substrate, that is found in the polyhaline regions, is considered, with dominance by polychaetes and amphipods (Jacobi 1987).

5.3.9 Meiofaunal Patterns of Diversity

The ecological and functional role played by meiofauna in the subtropical estuaries of the SBME has been little studied. The first taxonomical and ecological studies on Brazilian meiofauna were carried out in the Cananéia-Iguape system (see Gerlach 1958 for a synthesis). However, only during the last two decades there has been an increase in the number of meiofaunal researchers in southeastern Brazil. Meiofauna of the Rio Grande ecoregion is better known than that of SBME (see Netto et al. 2018; Chap. 6, for a review on composition and anthropogenic disturbances). Studies along the SBME are still sporadic and ecological studies with taxonomic resolution higher than phylum or class are still scarce. The scarcity of data about the meiofauna of estuarine beaches of the SBME and other intertidal compartments were recently evidenced by an extensive review on the meiofauna of sandy beaches along the Brazilian coast (Maria et al. 2016). Corgosinho et al. (2003) compared meiofaunal patterns in two sandy beaches under varying hydrodynamical conditions in Ilha do Mel, Paraná state (25°29' S and 48°17' W), Brazil. Total meiofaunal diversity and nematode densities were higher at the exposed site, a pattern attributed to the high vertical migratory capacity of nematodes in comparison with other meiofaunal taxa, and the almost complete absence of other interstitial meiofaunal groups. The low Nematoda–Copepoda ratio at the sheltered estuarine beach could indicate moderate hydrodynamic stress, since copepods are more sensitive to environmental disturbances than nematodes. Copepod densities, Shannon diversity, and evenness were also higher at the more eutrophic sheltered site.

Corgosinho et al. (2003) and Vicente (2008) carried out the only two studies on subtidal meiofaunal assemblages in the estuarine system of Paranaguá Bay. They indicated that salinity and sediment characteristics, conditioned by hydrodynamic conditions, are the main drivers of species composition and meiofaunal abundance. Vicente (2008) suggested that meiofaunal assemblages along the eastern–western more-contaminated axis of Paranaguá Bay differed from those of the southern–northern less-impacted axis, Laranjeiras Bay, mainly as a function of variation in nematode densities. Spatial gradients in meiofaunal distribution were detected in both bays, but without evidence of marked seasonal variation.

Thomas and Lana (2011) evaluated the efficacy of vital Neutral Red for the experimental analysis of nematode dispersion rates in intertidal flats from the polyhaline sector of Paranagua Bay. Dispersion rates were expressed as the number of stained captured nematodes by comparison to average densities of nematodes at stained sediment in source sites. The local association was numerically dominated by the nematode genera *Comesa, Terschellingia, Microlaimus, Metachromadora, Sabatieria*, and *Viscosia*. Stained individuals of only 4 (*Terschellingia, Metachromadora, Sabatieria*, and *Viscosia*) of the 23 identified genera were recaptured in the sediment. Tidal currents with average velocities of 9 cm/s were able to ressuspend the numerically dominant nematode taxa, dispersing them to distances up to 180 cm from the stained corers during a single tidal event. Thomas (2011) suggested that the morphology and lifestyles of nematodes themselves, besides local hydrodynamics, were determinant of their dispersal processes, by defining which genera are more prone to be ressuspended and which tend to stay in the sediment during ebb tides or flood tides.

In an unpublished thesis, Thomas (2011) categorized free-living marine nematodes of an intertidal flat of Paranaguá Bay into functional dispersal groups by combining biological data (mobility strategies, locomotion patterns, morphology, and life strategies), later used to assess genetic flow. The validity of functional dispersal groups was tested through a field experiment, carried out during single tidal events replicated in winter and summer conditions. A vagile-swimmer functional dispersal group displayed active behavior, swimming capacity and senoidal omega locomotion patterns, facilitated by a slender body and conical/cylindrical tail. Dispersal of vagile-swimmer species was mainly related to algal rafting events, suggested as relevant transport vectors during ebb tides and flood tides. Unlike vagile-swimmer species, vagile-crawlers showed circular senoidal locomotion pattern and could be grouped into a selective deposit-feeding trophic group. They avoided being resuspended by actively moving to deeper subsurface sediment layers. A sedentary dispersal group was made up by relatively small and robust species, with restricted mobility or completely stationary. The presence of such sedentary species in the water column was a function of resuspension by tidal currents and their subsequent passive transport. Populations of *Metachromadora chandleri* showed high gene flow, and no evidence of genetically structured populations. On the other hand, the vagile-swimmer *Viscosia cobbi* was able to avoid being carried in the water column and adapted to the classical metapopulation concept, with regional populations with restricted gene flow.

Leite et al. (2014) assessed if the changes in the structure of nematode assemblages could be used as indicators of moderate petroleum contamination in salt marshes of the euhaline sector of Paranaguá Bay. They did not observe significant differences in nematode total density, number of taxa and the overall assemblage structure between control and impact treatments from before to after experimental diesel oil spills. They claimed that, despite being good indicators of environmental stress, free-living marine nematodes are able to tolerate low concentrations of hydrocarbons and to survive in moderately contaminated areas.

5.4 Variation of Benthic Assemblages of the SBME in Time and Space

The understanding of community dynamics depends on the comprehension of their variations at different spatial and temporal scales. The latitudinal gradient might be the main driver of the spatial variability in estuaries, and this gradient is addressed as different ecoregions in this book. Salinity gradients across individual estuaries, often in the scale of dozens of kilometers, are also major drivers of benthic changes, whereas sediment texture, organic matter content, local hydrodynamics, geomorphological complexity, and biological interactions may influence the fauna at smaller spatial scales. At the temporal scale, estuarine dynamics vary from hours, due to the tidal influence, to seasons due to the rainy/dry periods and frontal systems. Recently, the impacts at larger temporal scales associated to climate changes, have been one of the main concerns for the management and conservation of estuaries, whose impacts require a clear understanding of the integration of the different spatiotemporal scales (Bernardino et al. 2016).

Disturbance is the main driver of variations at small spatial and temporal scales in estuaries, and its impacts depend on the intensity and origin of the disturbance (Xavier et al. 2008; Barboza 2010). Short-term disturbances are associated with several environmental changes, such as thermal stress during low tides, desalinization, erosion, and sediment resuspension. Pereira et al. (2013) showed that artificial disturbance was one of the main causes for the high diversity in Guanabara Bay. By evaluating the recovery process in soft-bottom macrobenthic communities after sterilization of the sediment, these authors indicated that disturbance could be responsible for the diversity found in polluted bays, probably due to the increase in the amount of organic matter in the sediment due to high mortality rates of the fauna. In addition, the authors found that the recolonization of defaunated sediments was rapid and was essentially complete after 27 days, with similar composition and abundance between control and impacted treatments.

Seasonal variation is an important natural driver in this dynamic environment. In the SBME, higher rainfall occurs in the summer (Bernardino et al. 2015) and, as a consequence, the freshwater input is also higher, decreasing the salinity in the estuary. On the other hand, during the winter, frequent meteorological-oceanographic events (frontal systems) increase seawater height and pull water from the coast (Mazzuco et al. 2015), which further increases the salinity in the estuaries. A seasonal pattern in the macrobenthic communities was described in the pristine estuaries in the Juréia-Itatins region (São Paulo State), in both salt marshes and mudflats, where the abundance and species richness was normally lower during the rainy and warm season than during the cold and dry season (Laurino 2016). In this region, the months with the highest degree of climate variability demonstrated negative effects on species richness and abundance, and sediment properties were the main factors responsible for the variations in richness. The macrofauna community was dominated by the tainad *Monokalliapseudes schubarti*, oligochaetes, and the polychaete *Capitella* sp., with greater macrobenthic richness and diversity in the higher salinity zones than in the lower salinity zones (Laurino 2016).

Seasonal cycles also have pronounced effects on species dynamics all along the subtropical SBME. Several species have a markedly seasonal reproductive cycle, as the commercially exploited crab *Ucides cordatus* (Hattori and Pinheiro 2003; Wunderlich et al. 2008), polychaetes (Lana and Guiss 1992; Pagliosa and Lana 2000), and bivalves (Christo et al. 2016). Marked seasonal variation in abundance and diversity of associated fauna has been attributed to correspondingly seasonal differences in primary production cycles and litter dynamics in local salt marshes and mangroves (Lana and Guiss 1992; Domingos and Lana 2017). Winter and early spring starch stocking in subsurface rhizomes is very noticeable in regional salt marshes (Lana et al. 1991), whereas litter production in mangroves is significantly higher in summer months (Bernini and Rezende 2010).

Short-scale meteorological-oceanographic events in the Paranaguá region did not cause significant and recurrent reductions in the number of species and individuals in the estuarine macrofauna in shallow subtidal estuarine flats (Negrello-Filho and Lana 2013). However, frontal systems significantly impacted exposed and estuarine beaches from the outer estuarine sectors (Brauko 2008). Although frontal systems may not be a major or regular driver of the composition and abundance of shallow-water estuarine macrofauna, extreme meteorological events, such as subtropical storms and cyclones over the Western South Atlantic Ocean, have relevant impacts on the benthic assemblages (see details in Sect. 5.6.4). The increase in the frequency and intensity of frontal events in different climate change scenarios may accelerate interannual and decadal changes that have been seen in the estuaries of this transitional ecoregion (Bernardino et al. 2015). However, the assessment and prediction of decadal changes is still fragmentary due to the lack of continuous monitoring and long-term studies, coupled with the increasing influence of human activities all around the SBME estuaries (Bernardino et al. 2016).

5.5 Processes Mediated by Benthic Organisms

Only more recently, from the 2000s on, the traditional descriptive focus on the taxonomy and structure of benthic assemblages of the SBME estuaries and coastal lagoons has been complemented by functional approaches which emphasize the role played by plants and animals in shaping and conditioning their own habitats (Caliman et al. 2013; Gusmão Jr 2017). Benthic systems are largely controlled by hydrodynamic forces that shape habitats and directly affect the biota. Habitat complexity may be also introduced in sedimentary systems at shorter spatial scales by the presence of bioturbators and habitat-forming engineer species. These organisms can alter the environmental conditions and directly or indirectly regulate the availability of food and habitat for other species (Gusmão Jr 2017).

The so-called habitat-forming organisms are autogenic ecosystem engineers that via their physical structure and behavior create new habitats for other species, increasing structural complexity in otherwise more simple and monotonous unconsolidated substrata (Gusmão Jr 2017). Although most of the ecosystem engineers are epifaunal, the relevance of infaunal species should not be underestimated, as they can alter sediment geochemistry, through their feeding, mating, and locomotion strategies at subsurface layers (Fig. 5.4). With the exception of large burrowing crabs, as *Ucides cordatus* (Santos et al. 2009), *Ocypode quadrata* (Rosa and Borzone 2008), and species of *Uca* (Machado et al. 2013), most epifaunal species have a limited capacity to disturb the subsurface sediment matrix. However, they

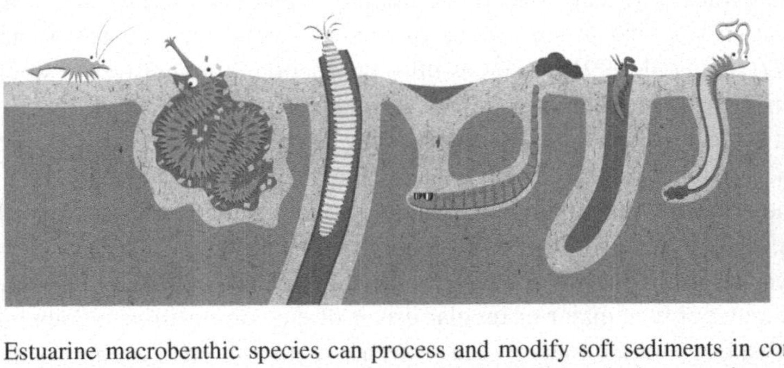

Fig. 5.4 Estuarine macrobenthic species can process and modify soft sediments in contrasting ways. From the left to the right: an epifaunal shrimp with limited bioturbation capacity at the sediment surface; a goniadid polychaete disturbing sediment in all directions (biodiffusing); a tube-dwelling onuphid worm; a lugworm bioirrigating the substrate and depositing fecal material in the surface (upward conveyor); a gallery-dweller shrimp; and a spionid polychaete depositing fecal material under the surface (downward conveyor)

can still affect sediment surface by selective deposit feeding (Colpo and Negreiros-Fransozo 2011). The pure size of burrowing crab species clearly increases their capacity to alter the vertical profile and geochemical properties of unconsolidated bottoms (Sayão-Aguiar et al. 2012).

Tube and gallery dwellers not only promote deeper oxygenation of the sediment but also can change sediment microtopography. Conversely, sediment biodiffusers can disturb sediment matrix in all directions, destabilizing the substrate and increasing nutrient release (Caliman et al. 2013). Bioirrigators promote water flow through the sediment matrix, resulting in increased nutrient release. Upward and downward conveyors transport material out or down into the sediment and may directly regulate the release and burial of organic matter in the sediment matrix (Gusmão Jr 2017).

Considering the differences in bioturbation capacities among macrobenthic species, changes in assemblage structure from subtidal to intertidal compartments have clear implications for sediment-related processes. The main introducers of sediment heterogeneity in shallow and intertidal estuarine habitats along the SBME estuaries are halophyte plants. Checon et al. (2017) tested the prediction that mangrove vegetation would not exert the same positive effects on infaunal organisms that they obviously exert on epifauna. For that, they investigated polychaete assemblages in mangrove stands of northern coast of São Paulo state. Mangrove areas were shown to display lower polychaete density than that of adjacent sandflats, and polychaete density was negatively related to root biomass. Species richness was not affected by the presence of vegetation, but the number of exclusive species was higher inside the mangrove. Changes in composition of polychaete assemblages were mainly attributable to reductions in species density rather than species replacement. Trophic structure was not influenced by mangrove vegetation, as subsurface feeders dominated inside and outside mangrove zones. Marine macrophytes such as *Spartina alterniflora* (Flynn et al. 1998; Netto and Lana 1997), macroalgae such as *Hypnaea musciformis* (Couto et al. 1995), and seagrasses (Marques and Creed 2008; Copertino et al. 2016) may promote flow attenuation and decrease bedload transport, reducing sediment resuspension, increasing sedimentation rates, providing refuge or food, and thus creating a new habitat for diverse faunal assemblages.

Extensive bivalve beds of the native *Mytella guyanensis* and *Crassostraea brasiliana*, also known as *C. gasar* (Westphal and Ostrensky 2016) may also introduce hard substrate into intertidal flats and shallow subtidal bottoms, allowing for the establishment of fouling organisms and other associated biota. Dead shells may also introduce complexity and heterogeneity into soft-sediment habitats of the SBME and may regulate the structure and functioning of benthic assemblages. The effects of high densities of empty shells of the bivalve *Anomalocardia flexuosa* on macrofaunal recolonization were experimentally assessed during a 3-month period by reciprocal transplantation of defaunated sediment between shelled and bare habitats (Sandrini-Neto and Lana 2014). Despite the well-recognized impact of mollusk shell production on soft-sediment habitats, effects were mostly scale-dependent, both spatially and temporally, and might be regulated by the structure of local assemblages and species-specific responses related to dispersal mechanisms.

Fiddler crabs are key species all along the intertidal flats, salt marshes, and mangroves of the SBME. Their varying engineering effects are an important source of habitat heterogeneity and a structuring agent of microfaunal and meiofaunal assemblages, as they manipulate large amounts of sediment, altering the physical and geochemical properties of the substrata. Citadin et al. (2016) investigated whether different types of sediment bioturbation produced by fiddler crabs modulate meiofaunal assemblages and microphytobenthic content. They hypothesized that sedimentary structures produced by burrowing (the burrow itself and the excavation pellets) and feeding (feeding pellets) generate different microenvironments compared with areas without apparent signs of fiddler crab disturbance, affecting both meiofauna and microphytobenthos, independent of the sampling period. Engineering effects of burrow construction and maintenance and the engineering effects of fiddler crab foraging clearly modulated meiofaunal assemblages in different ways. Overall, meiofauna from burrows and excavation pellets was more abundant and diverse than at control sites, whereas feeding pellets contained poor meiofaunal assemblages. By contrast, only foraging effects were detected on microphytobenthos; independent of the sampling period, chlorophyll-a and phaeopigment content were higher in the feeding pellets, but similar among burrows, excavation pellets, and control sites. However, it is still unclear the role played by fiddler crabs on organic matter contents by bioturbation as it is proved in controlled laboratory conditions but overcame by environmental variables in field experiments (Natálio et al. 2017).

Brustolin et al. (2014) assessed the effects of the sand-dollar *Encope emarginata* on subtidal meiofauna and microphytobenthos in shallow subtidal bottoms of Paranaguá Bay. They provided strong evidence that top-down effects related to the occurrence of *E. emarginata* act in synergy with bottom-up structuring related to hydrodynamic processes in determining overall benthic spatial variability. Species richness was mainly influenced by environmental heterogeneity at small spatial scales (centimeters to meters), which creates a mosaic of microhabitats. They evaluated variation patterns at five different spatial scales, from kilometers to centimeters, in sites with patches of living and dead echinoids and in sites without them. The overall structure of microphytobenthos and meiofauna was always less variable at sites without *Encope* and much of variation at the scale of hundreds of meters was related to variability in sites with living or dead *Encope*, due to foraging activities or to the presence of shells. Likewise, increased variability in chlorophyll-a and phaeopigment contents was observed among sites with living *Encope*, although textural parameters of sediment varied mainly at smaller scales.

Bioturbation also produces biogeochemical changes in the sediments as it promotes sediment aeration and stimulates aerobic microbial activity. Bioturbation ultimately leads to changes in sediment redox conditions with major implications for the cycling of nutrients, organic matter and contaminants in the unconsolidated substrata. Caliman et al. (2013) experimentally assessed bioturbation effects on the bottoms of a coastal lagoon in Rio de Janeiro state. They showed that bioturbator species richness had no overall significant effect on sediment methane when comparisons were made among different species richness levels. However, bioturbator species richness significantly reduced sediment methane when species richness levels were compared to control defaunated treatments, effect that was significant only at the deepest sediment layer. Furthermore, bioturbator species composition had significant, but distinct effects on the patterns of reduction in sediment methane depending on the sediment depth and the bioturbator species. They concluded that both the number and composition of bioturbator species are important to determine the effects of benthic bioturbators on spatial patterns of sediment methane, but the strength of these effects depend on species traits that determine interspecific interactions strength across the sediment vertical profile.

Suzuki et al. (2012) estimated that benthic activity on metal diffusion into mangrove sediments was responsible for 32% (Cr), 39% (Co), and 44% (Zn) of total inventories of diffused metals. Benthic activity was quantitatively evidenced as a control on trace metal diffusion into the sediments, contributing to determine the sediment role as a metal sink. Bioturbation was also shown to affect metal potential bioavailability, considering that recently diffused metals can be more readily available to biological uptake. Correia and Guimarães (2016) assessed the effects of *Uca* on mercury (Hg) and methylmercury (MeHg) which are poorly studied in regional mangroves. They found that crab bioturbation favored Hg retention in the sediment but also promoted MeHg formation near the surface. Hg volatilization and water Hg concentrations were higher in treatments without crabs.

Functional diversity, which is high among benthic invertebrate bioturbators, is conditioned by biological traits such as body size, feeding mode, and burrowing capacity (Gusmão Jr 2017). In addition, invertebrate bioturbators themselves can alter their functional attributes in response to environmental change or species interactions (Caliman et al. 2013). The discrepancy in functional traits among benthic invertebrate species is considered a criterion to preserve bioturbator diversity in aquatic ecosystems, given the disparate effects of species belonging to different functional groups may have on benthic–pelagic processes (Caliman et al. 2013).

5.6 Human Impacts and Conservation Issues

Disturbances are among the most important controlling drivers of ecosystem structure and function in marine sedimentary environments. As exposed above, small-scale disturbances that are driven by bioturbators play an important role in maintaining a mosaic of patches in different successional stages, which promotes local heterogeneity (Gusmão Jr 2017). However, natural or human-driven large-scale disturbances can have major effects on regional benthic biota and consequently affect process dynamics in sedimentary systems. The increasing pressures of human-driven large-scale disturbances on coastal seascapes, such as climate change and overfishing, have generated concern among researchers and marine managers about the functional consequences of benthic biodiversity loss. Analytical tools that integrate information about the macrobenthic species to estimate the relationship between species composition and community function are thus valuable and can predict the potential services losses of services in the case of local or broader scale disturbances.

Most of the human population and economy of Brazil are concentrated in the SBME and, as a consequence, several environmental stressors influence the estuaries in this region. In this section, we will cover the main drivers of the impacts to the estuaries of the SBME, including climate change and the goals for conservation. It is important to highlight that the patterns and processes of biological communities are exposed to multiple stressors that may act simultaneously with additive and non-additive (i.e., synergetic or antagonistic) outcomes. Therefore, it is essential to consider the potential interaction of multiple stressors when designing impact studies, as it is difficult to understand the environmental impact by analyzing single stressors on benthic communities, as the real impacts are more than the sum of single stressors. In this context, we provide information based on field studies since, even when addressing single stressors, these studies evaluated the natural complexity of the estuarine system and under other stressors. Therefore, the results from these studies will reflect the responses of the biota under the putative interactions of all local stressors. Laboratory studies usually address single stressors in a controlled system; thus, the applications of the results of laboratory studies to natural ecosystems, although valid, are limited.

5.6.1 The SBME History and the Multiple Stressors of the SBME

The coast of the SBME is strongly influenced by urban sprawl, which has increased both habitat losses and the discharge of pollutants and sewage. The current trends of urbanization and industrialization are schematically represented in Fig. 5.5. The land use types in this area have been impacted for several centuries (see Fistarol et al. 2015 for the case study of Guanabara Bay), and there is no previous knowledge of the fully natural patterns of this region. Thus, it is important to bear in mind that the actual biological patterns that we describe in this chapter are a result of the long-term influence of human activities. The impact of habitat loss in the SBME is clearly represented by the loss of the Atlantic Rain Forest, which covers less than 16% of its original area (Ribeiro et al. 2009).. This huge deforestation gave way to the development of cities, industrial complexes and agricultural activities, and the residuals of these activities are discharged mainly in coastal estuaries. The urbanization of the region began with the first urban settlements on the coast in the sixteenth century. From that moment, the impacts on the SBME estuaries occurred at different times and intensities. In the Guanabara estuary, cities such as Rio de Janeiro and Niteroi developed, and these developments included significant land reclamation and decreases in environmental quality over the last 200 years (Fistarol et al. 2015). More recently (100–150 years), other estuaries in the SBME showed changes of different intensities. More intense pressures occurred in the Santos, Sepetiba, and Babitonga estuaries, where ports and industries rapidly developed. On the other

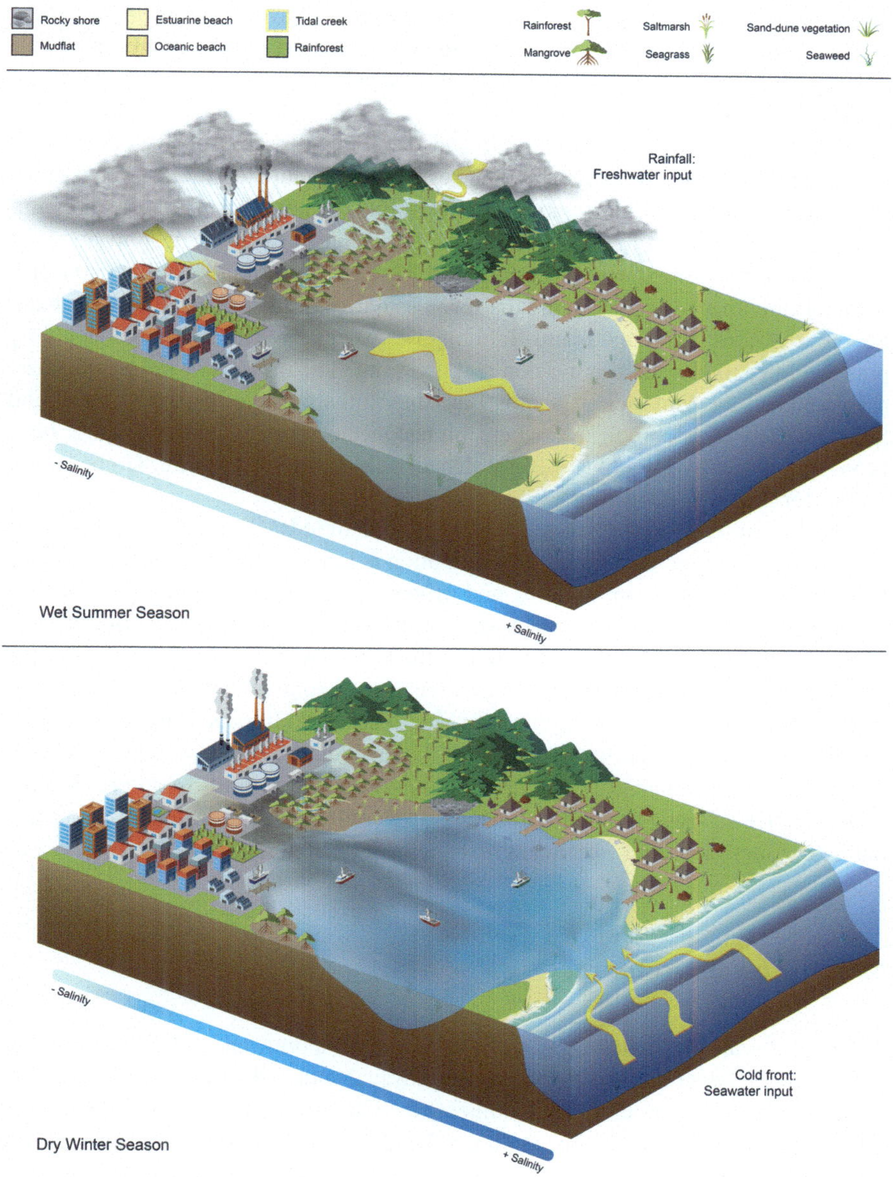

Fig. 5.5 Dynamic behavior of an idealized subtropical estuary under the dry summer and the wet winter seasons

hand, other estuaries (Juréia, Iguape-Cananéia, and Guaratuba) presented less-intense urban development. There are only small cities and local traditional communities in these estuaries where pristine and less-impacted regions are still found, including the longest sector of native Atlantic Rain Forest in the Lagamar region, which harbors different protected areas and the Juréia, Iguape-Cananéia, and

Paranaguá estuaries. Paranaguá Bay is considered to have faced an intermediate level of impact, as the southern side is influenced by a port, contaminants, and habitat loss, while the northern side is occupied by small cities, traditional communities, and the native Atlantic Rain Forest.

The SBME is the most urbanized and economically developed ecoregion in Brazil. The most common stressors in the estuaries are tourism, litter, sewage, fertilizers, agricultural pesticides, and industrial pollutants, and these stressors are consequences of the development of cities. The intensities of their impacts vary according to the closest demography, and the influence of the rivers that carry pollutants from other regions. Among the seven estuaries of the SBME, the Guanabara, Sepetiba, Santos, Paranaguá, and Babitonga estuaries have also been impacted by port activities, including bioinvasion and dredging. The impacts of industrial waste are greater in the estuaries of Santos Bay, due to the presence of the Cubatão and Sepetiba industrial complexes, where high concentrations of contaminants have been discharged over the past several decades (Marques et al. 2006; Gomes et al. 2009; Bordon et al. 2016).

The economic development and urbanization in the SBME has also increased the social inequality, and the vulnerable population is large. Vulnerable populations are from both traditional communities and urbanized low-income employees. In this context, urbanized regions have a mosaic of different social-economic-cultural realities. In this mosaic, there are different vulnerabilities of the impacts of multiple stressors depending on the socioeconomic reality as, for example, slams with lower infrastructure are present along the estuarine border. Additionally, there are different connections between man and nature that are related to the sustainable use of resources by traditional communities and the lower connections to nature in some higher social classes. These differences bring conflicts between urbanization and nature conservation with significant management challenges.

5.6.2 Fishery Resources

Most traditional populations rely on fishing as an economic and cultural activity. In regions with larger human populations, artisanal and industrial fishing are also stressors on the estuaries that combine with other stressors (e.g., pollutants and contaminants) to increase the conflict related to ecosystem services and health implications. The fishing of benthic estuarine organisms is mainly focused on crustaceans and mollusks from mudflats and mangroves (Westphal and Ostrensky 2016; Castilho-Westphal et al. 2014; Duarte et al. 2014). In urban areas, overfishing can be a problem because of population decreases and the potential changes in ecosystem functions. In addition, fishery resources can accumulate contaminants that will be ingested by the human population and will bring about health implications for humans, especially in low-income communities that utilize fisheries as the main sources of protein.

Swimming crabs are an important resource for local communities according to Severino-Rodrigues et al. (2001), who described the artisanal swimming crab fishery in Santos Estuary from 1987 and 1994. *Callinectes danae* was the main resource (84% of the total) consistently over that period; however, the abundance decreased over that period. This decrease was probably a result of the overfishing of juveniles and ovigerous females. Shrimp fisheries are also common in larger estuaries (Natividade et al. 2006; Andriguetto-Filho et al. 2009), and these fisheries have impacts on the benthic community due to the by-catch of fauna with low economic interest (López-Barrera et al. 2012; Stanski et al. 2016).

In mangroves, *Cardisoma guanhumi* is the species that is most impacted by fisheries, and *Ucides cordatus* is the main fishery resource. The density of exploited species seems greater near mangrove trees because roots affect fishing gear (Kassuga and Masunari 2015). Mangrove crabs of commercial interest have also been impacted by contaminants in the estuaries in the SBME (Pinheiro et al. 2013; Oliveira-Neto et al. 2014). Another important fishery resource is the mangrove oyster *Crassostrea* spp., and the sustainability of the extraction in the Iguape-Cananéia estuary is more related to the social, technological, and ethical dimensions than the ecological and economic dimensions (Mendonça and Machado 2010; Machado et al. 2015). This species has been cultured at a commercial scale (Henriques et al. 2010). Although the culture seems to have a low impact on the environment, it causes a positive effect on the biodiversity due to the increase in associated fauna (Monteiro 2007).

5.6.3 Benthic Fauna and Anthropogenic Stressors

When considering the actual impacts on estuaries, benthic macrofauna can be used as bioindicators and can support management and decision-making for estuary conservation. Studies on estuarine benthic macrofauna have shown that both species and assemblages can be used as bioindicators for several natural and anthropogenic drivers (e.g., C, N, P, steroids, sewage indicators, $CaCO_3$, and metals). However, it is still controversial how benthic fauna respond to multiple stressors. This issue needs to be addressed carefully and with more elaborate studies and experimental designs, as the interpretation of benthic macrofauna as bioindicators needs to be specifically addressed together with the covariation among several factors, such as sedimentology, organic matter, salinity gradients, the presence of contaminants and pollutants, and the understanding of the life strategies and resilience of organisms.

Pollutants and contaminants have been shown to be the main drivers that have an influence at the intermediate (km) and short scale (m) in estuaries. Changes in the benthic macrofauna structure in polluted areas have been observed (Barboza et al. 2015; Souza et al. 2013, 2016), while variation patterns have also reflected significant small-scale (centimeters and meters) changes that are not directly related to contamination (Souza et al. 2013, 2016). The influence of sewage at the temporal scale is still unclear. In Guanabara Bay, communities in polluted areas recover faster than those in unpolluted regions, which is probably due to the presence of more resilient species in the polluted environments (Breves-Ramos et al. 2005). For a better understanding of such differences, it is also important to have an understanding of the future spatiotemporal variability integrated in the SBME. For example, Hadlich (2010) integrated spatiotemporal variation by studying polychaetes in a hierarchical design from centimeters and dozens of meters to hundreds of meters and to kilometers in sewage-contaminated gradient during the dry and wet subtropical seasons. She found a seasonal variation in the polychaete community; however, the spatial variation changed over time. In the dry season (winter) the variability was expressed at the smaller spatial scale, while during the wet season (summer), when the pollutant gradients were stronger, the benthic variation appeared from hundreds of meters to kilometers.

Pristine or quasi-pristine benthic gradients may thus be confounded by anthropogenic influence associated with urban development, harbor activities, industrial activities and waste disposal along the major SMBE estuaries (Fistarol et al. 2015; Soares-Gomes et al. 2016; Rodrigues et al. 2017). Due to land-usage dynamics, human environmental impacts are indeed more impacting in the inner estuarine sectors, as in Guanabara (Soares-Gomes et al. 2016), Sepetiba (Rodrigues et al. 2017), and Santos estuaries (Ferreira 2008). Although there is some evidence that contamination levels by trace metals may not be critical for the fauna as yet (Abuchacra et al. 2015; Aguiar et al. 2016 but also Rodrigues et al. 2017), sewage disposal is unequivocally a major driver of benthic assemblage structure in regional estuaries (Souza et al. 2013; Brauko et al. 2016). Short-lived opportunistic species or assemblages seem to be recurrent in contaminated areas including nematodes, the tanaid *Monokalliapseudes schubarti*, the polychaetes *Streblospio* cf. *benedicti*, *Sternaspis* sp., and *Scolelepis* sp., and the gastropods *Bulla striata* and *Heleobia australis* (Rodrigues et al. 2017). In Paranaguá bay, shallow bottoms contaminated by sewage are numerically dominated by tubificids, and by the polychaetes *Laeonereis culveri* and *Heteromastus* sp. (Souza et al. 2013). Ferreira (2008) described the spatial and vertical structure of the population of the polychaete annelids of the contaminated Santos bay and adjacent continental shelf. The lowest values of density, richness, diversity, and equitability were observed in the central portion of Santos bay, with numerical dominance of the polychaetes *Mediomastus capensis*, *Prionospio* sp., and *Aricidea* cf. *catharinae*, mostly concentrated in the 5-cm surface layer of fine bottoms. This pattern was attributed to the greater retention of Pb, Zn, Cr, n-alkane hydrocarbons, HPAs, LABs, PCBs, and fecal sterols arising from urban effluents, regular port dredging, and atmospheric contribution. Ferreira (2008) indicated that these species could be used as indicators of high contamination levels in the region.

Macrobenthic communities with high abundances of polychaetes (e.g., *Edwardsia fusca*, *Magelona papillicornis*, *Aricidea taylori*, *Laeonereis culveri*, and *Owenia* sp.) (Rocha 2006; Hadlich 2010; Soares-Gomes et al. 2012; Souza et al. 2013), ophiuroids (Barboza 2010), and mollusks (the bivalve *Ervilia concentrica* and the gastropod *Heleobia australis*) (Neves et al. 2011) are good indicators of sewage pollution in different estuaries of the SBME. Considering the polychaete communities, Soares-Gomes et al. (2012) showed that an analysis carried out solely at the family level is perfectly adequate to describe the pollution gradient. Therefore, the use of polychaetes at the family level may be a useful tool for a rapid environmental assessment.

The understanding of how estuarine benthic communities respond to sewage has not come to a consensus yet. Some studies have shown that communities under disturbances in polluted areas recover faster than those in less polluted areas, probably due to the existence of more resilient communities in the impacted areas (Breves-Ramos et al. 2005). The intermediate disturbance theory is applied for estuarine benthic macrofauna, as species richness is higher in moderately polluted areas (Breves-Ramos et al. 2005). However, the spatiotemporal variability needs to be integrated in such analyses of the community since experimental outcomes can be related to stochastic processes rather than anthropogenic impacts. Souza et al. (2016) assessed the effects of sewage on the spatial and temporal variation of benthic macrofauna in Paranaguá Bay. Despite the expected influence of sewage input on the macrofaunal structure, the variation in species distribution also reflected significant other small-scale temporal changes not directly related to contamination. In this study, temporal variability patterns were similar between polluted and unpolluted sites and are possibly regulated by simultaneous processes operating under different conditions at scales greater than thousands of meters.

Considering the effects of organic enrichment on community recolonization, Gern and Lana (2013) showed that the macrofauna was resilient to contamination by domestic sewage and rapidly recovered to the background or control conditions after the contaminated sediment was transplanted to non-contaminated areas. The recolonization process was strongly dependent on the migration of adults present in the sediments adjacent to the experimental units. The differences in sediment quality (organic enrichment gradient) did not determine macrofaunal recolonization, at least in the spatial and temporal scales of meters and weeks that were considered. The density of the polychaete *Capitella* sp. was extremely high in the contaminated area, as was the density of the gastropod *Cylichna* sp. in the non-contaminated area. A similar result was observed for the effects of an experimental in situ diesel oil spill on the benthic community. Egres et al. (2012) found an acute effect immediately after the impact, but the recovery to the pre-disturbance population levels was extremely fast. The increase in the total density of the benthic community after the disturbance was caused by increases in the densities of *Heleobia australis*, oligochaetes, and ostracods in both impacted and control treatments, and these increases were a result of background variability and not the presence of the contaminant. The experimental spill had little influence on the biological descriptors of the benthic community, which were resilient or tolerant to oil disturbance at the temporal (147 days) and small spatial (cm^2) scales used in the experiment.

Apart from benthic communities, some species have also been used as bioindicators for detecting the impacts in the SBME estuaries. Due to its opportunistic life strategy, the gastropod *Heleobia australis* is a good indicator of anthropogenic activities such as dredging and sediment discharge, which increase the environmental variability and turbidity (Hostin et al. 2007). The polychaete *Scolelepis chilensis* is sensitive to environments with contaminated sediments and low oxygen supplies

(Rocha et al. 2013). Sandrini-Neto et al. (2016) suggest that the bivalve *Anomalocardia brasiliana* and the polychaete *Laeonereis culveri* are more suitable sentinels due to their greater responses to oil spills and their wider geographical distributions. The studies of Teixeira et al. (1984a, b), Carneiro et al. (1987) and Teixeira et al. (1987) showed that certain taxa from more polluted areas might present changes in population density and body size. For example, the vegetation cover by *Ulva fasciata* was greater in a polluted area than a control area, and *Cladophora vagabunda* dominates certain coasts and prevents the occurrence of other taxa, reducing the diversity in impacted areas. On the Island of Boa Viagem, which is located in the external sector of Guanabara Bay, the disappearance of 30 taxa over an interval of 30 years was verified. Paradoxically, on the Island of Paquetá, which is in the internal sector of the bay, the disappearance of seven taxa was verified; however, 22 other taxa appeared.

Kolm et al. (1999), in an unpublished technical report, suggested the alternative view that *Heleobia australis* is rather an indicator of areas subjected to high environmental variability and maximum turbidity zones than of highly contaminated sites. Populations may reach densities of up to 80,000 inds/m^2 (unpublished observations) in areas subjected to rapid and wide variations in salinity and in the input of organic particulate at the mouths of rivers and tidal creeks. This hypothesis is partially supported by experimental evidence (Echeverría et al. 2010; Neves et al. 2011). Three distinct dispersal strategies were described for *H. australis* in lab assays): (a) mobility on soft sediments; (b) mobility on hard substrata; and (c) the ability to lift from the bottom to the surface, to again sink down. Such abilities may facilitate dispersal by surface currents or floating debris. A hypothesis of source–sink metapopulation dynamics was also tested by Echeverría et al. (2010), who showed that opportunistic life strategies might indeed allow for *Heleobia* to quickly recolonize areas disturbed by natural or anthropic disturbance in various estuarine compartments, from mangroves to subtidal bottoms.

Understanding multiple stressors becomes more important in more impacted estuaries when the contamination of the estuarine benthic fauna and the possible impacts to the ecosystem and human populations are considered. Studies have shown that some contaminants present in the estuarine substrate are not bioavailable for the benthic fauna (Abreu et al. 2016). On the other hand, several studies have shown decreases in the abundance of benthic fauna, species richness or alterations in the functions of assemblages (Donnici et al. 2012; Pereira et al. 2014; Barboza et al. 2015; Capparelli et al. 2016). Benthic estuarine organisms from different estuarine habitats, many of which are of commercial interest such as the mussel *Perna perna,* the swimming crab *Callinectes* spp., and the oyster *Crassostrea,* have been used as biomarkers and are bioaccumulators (Magalhães et al. 2012; Pinheiro et al. 2013; Bordon et al. 2012a, b, 2016; Sousa et al. 2014; Lino et al. 2016). In this case, contaminants can affect organisms at different levels: from genetic to morphological and functional levels and, as a consequence, these impacts can potentially influence ecosystem functions. Additionally, species of commercial interest to the artisanal and industrial fisheries can transfer such contaminants to consumers in the trophic chain, impacting other ecosystems and human populations.

Bioinvasion is an important biological stressor in native benthic estuarine communities. Bumbeer and Rocha (2016) have recently determined the extent to which non-indigenous species (NIS) have spread to natural substrates both inside estuarine areas and in the adjacent open sea. Harbours, always located within estuaries along the SBME, are the main vectors NIS. Several benthic invader species have been reported in the SBME estuaries, such as bivalves, barnacles, swimming crabs, polychaetes, bryozoans, ascidians, sponges, and hydrozoans (e.g., Tavares and Mendonça 1996; Mayer-Pinto and Junqueira 2003; Jara et al. 2006; Julio 2007; Neves et al. 2007).

5.6.4 Estuaries, Climate Change, and Conservation

In addition to all the previously mentioned stressors (loss of habitat, pollutants, contaminants, fisheries, and bioinvasion), it is important to highlight the impacts of climate change on the estuaries of the SBME. Climate change at larger temporal scales could interact synergistically with other stressors, which are forecasted to increase the seasonal stress for the benthic estuarine fauna. As addressed in Sect. 5.2, it is expected that the SBME will suffer from different intensities and frequencies of extreme events that will bring about greater effects on the estuaries. During the summer, intensifications of rainfall and extreme summer storm events are expected, which will increase salinity oscillations and sediment disturbances over short periods. In this scenario, the benthic fauna is expected to experience more stress because of the lower salinities and increased sediment disturbances during the summer. On the other hand, an increase in frontal system events during the winter is expected. During these events, strong southwestern winds blow, developing instabilities with durations of several days that extend for thousands of kilometers (Harari et al. 2008). Such events cause a substantial rise in the sea level (e.g., from −0.4 to +0.5 m on 13–14th March, 2006, Harari et al. 2008), with records of approximately 0.3 m above the annual mean sea level in the SBME (Harari et al. 2008; Mazzuco et al. 2015). Therefore, it is expected that there will be increased stress to the estuarine benthic fauna related to higher salinities during the winter. Such seasonal patterns might affect the distribution of the benthic macrofauna within the salinity gradient, causing an upstream expansion of the intermediate region of disturbance during the winter and a downstream expansion during the summer.

When considering the impacts of climate change under the view of multiple stressors, the importance of integrative studies rather than laboratory controlled single stressor experiments becomes evident. Changes in hydrodynamics and salinity gradients due to an increase in rainfall during the summer or a cold front during the winter will alter not only the distribution of benthic macrofauna but also the resuspension and transport of sediment. Therefore, pollutants and contaminants may be more diluted in estuaries and spread to other coastal areas during the summer, whereas higher concentrations of these pollutants may remain within the estuaries during the winter. The vegetation cover in mangrove and salt marsh habitats might be impacted by such temporal variability. In this case, the boundaries of benthic communities will also be affected. A recent study addressed the possible impacts of the loss of vegetated habitat on the microhabitat selection, behavior and physiological limitations of fiddler crabs. Shading influenced the spatial distribution over time and behavior, with fiddler crabs spending less time outside their burrows as the temperature increased (Sanches 2017).

5.7 Final Considerations

Subtropical estuaries are complex transitional ecosystems under the influence of diverse oceano-meteorological, freshwater, terrestrial, and anthropogenic drivers that promote high dynamic variability along time and space. Such estuarine complexity provides a wide range of goods and services to human populations, including fishery resources and ecosystem processes, such as the transport of organic matter, use of habitats, protection of the coastline, and economic resources and activities. The SBME presents the most continuous area of the remaining Atlantic Rain Forest, which should be supported as a good for humanity. The conservation of such natural areas and their biodiversity could also provide other still underexplored services, including ecotourism.

Understanding the variability of climate events and their impacts on the benthic fauna will be essential to promote better decision-making concerning the management and conservation of estuaries along the SBME. The SBME is most relevant for coastal land use and tourism in Brazil, and this trend will keep growing over the following decades. Besides holding almost half of the Brazilian population in modern cities and industrial centers, the SBME is still home to a number of traditional communities that directly use ecosystem resources for their livelihoods. The area also has a major environmental relevance because it is a transitional zone between tropical and temperate climates and is the distributional limit for many species, many of them endemic. Despite its social and ecological relevance, the whole area is under high pressure from natural (e.g., seasonal cycles and changes in habitat complexity) and anthropogenic disturbances (e.g., climate change, bioinvasion, and habitat loss).

The SBME is the most economically developed and urbanized ecoregion in Brazil and display the most impacted estuaries, but it also holds the greatest number of protected marine and coastal areas in Brazil. Therefore, the SBME may turn out over the next few years as a model region for innovative public policies concerning estuarine conservation even under high disturbance levels by human activities. The conservation of the SBME estuaries will be essential to maintain the coastal biodiversity and ecosystem functions in an area with a variety of human impacts. By strengthening the conservation strategies of the estuaries in the SBME, we may enhance the sustainable development of the region, in accordance with many, if not all, Sustainable Developmental Goals of the United Nations 2030 Agenda. To achieve this goal, more effective strategies are needed to promote the engagement of people towards nature conservation and resource sustainability. The society and decision-makers need to develop novel approaches to promote the management, conservation, and sustainable use of estuaries and thus enhance human well-being for posterity.

References

Abreu IM, Cordeiro RC, Soares-Gomes A et al (2016) Ecological risk evaluation of sediment metals in a tropical euthrophic bay, Guanabara Bay, Southeast Atlantic. Mar Pollut Bull 109:435–445. https://doi.org/10.1016/j.marpolbul.2016.05.030

Abuchacra PFF, Aguiar VMC, Abuchacra RC et al (2015) Assessment of bioavailability and potential toxicity of cu, Zn and Pb, a case study in Jurujuba Sound, Rio de Janeiro, Brazil. Mar Pollut Bull 100:414–425. https://doi.org/10.1016/j.marpolbul.2015.08.012

Aguiar V, de Lima MN, Abuchacra RC et al (2016) Ecological risks of trace metals in Guanabara Bay, Rio de Janeiro, Brazil: an index analysis approach. Ecotoxicol Environ Saf 133:306–315. https://doi.org/10.1016/j.ecoenv.2016.07.012

Alongi DM (1990) The ecology of tropical soft-bottom benthic ecosystems. Oceanogr Mar Biol 28:381–496

Aluizio R (2007) Análise comparativa da fauna associada às linhas de detritos em duas praias estuarinas da Ilha do Mel (Paraná-Brasil). Dissertation, Universidade Federal do Paraná

Alvares CA, Stape JL, Sentelhas PC et al (2014) Köppen's climate classification map for Brazil. Meteorol Zeitschrift 22:711–728. https://doi.org/10.1127/0941-2948/2013/0507

Amaral ACZ, Denadai MR, Turra A, Rizzo AE (2003) Intertidal macrofauna in Brazilian subtropical tide-dominated sandy beaches. J Coast Res 35:446–455

Amaral ACZ, Jablonski S (2005) Conservation of marine and coastal biodiversity in Brazil. Conserv Biol 19:625–631

Andriguetto-Filho JM, Krul R, Feitosa S (2009) Analysis of natural and social dynamics of fishery production systems in Paraná, Brazil: implications for management and sustainability. J Appl Ichthyol 25:277–286. https://doi.org/10.1111/j.1439-0426.2009.01273.x

Angulo RJ, Borzone CA, Noernberg MA et al (2016) The state of Paraná beaches. In: Short AD, Klein AHF (eds) Brazilian beach systems. Springer, Berlin, pp 419–464

Angulo RJ, Lessa GC, Souza MC (2009) The Holocene barrier systems of Paranaguá and northern Santa Catarina coasts, southern Brazil. In: Dillenburg SR, Hesp PA (eds) Geology and geomorphology of Holocene coastal barriers of Brazil, Lecture notes in earth sciences. Springer, Berlin, pp 135–176

Angulo RJ, Lessa GC, Souza MC (2006) A critical review of mid- to late-Holocene Sea-level fluctuations on the eastern Brazilian coastline. Quat Sci Rev 25:486–506

Attolini FDS, Flynn MN, Tararam AS (1997) Influence of *Spartina alterniflora* and tide level on the structure of polychaete associations in an euryhaline salt marsh in Cananéia lagoon estuarine region (SE Brazil). Braz J Oceanogr 45:24–34. https://doi.org/10.1590/S1679-87591997000100003

Barbosa CF, Suguio K (1999) Biosedimentary facies of a subtropical microtidal estuary - an example from southern Brazil. J Sediment Res 69:576–587. https://doi.org/10.2110/jsr.69.576

Barboza CAM (2010) Variabilidade espacial das associações de ofiuróides ao longo de um gradiente de contaminação no Canal da Cotinga, Baía de Paranaguá, Paraná. Dissertation, Universidade Federal do Paraná

Barboza CAM, Martins CC, Lana PC (2015) Dissecting the distribution of brittle stars along a sewage pollution gradient indicated by organic markers. Mar Pollut Bull 100:438–444. https://doi.org/10.1016/j.marpolbul.2015.08.008

Baroni PC, Barrella W et al (2015) Macrofauna bentônica da faixa entremarés da praia de José Menino (Santos-SP). UNISANTA Biosci 4:98–104

Barreiro M, Chang P, Saravanan R (2002) Variability of the South Atlantic convergence zone simulated by an atmospheric general circulation model. J Clim 15:745–763. https://doi.org/10.1175/1520-0442(2002)015<0745:VOTSAC>2.0.CO;2

Bernardino AF, Netto SA, Pagliosa PR et al (2015) Predicting ecological changes on benthic estuarine assemblages through decadal climate trends along Brazilian marine Ecoregions. Estuar Coast Shelf Sci 166:74–82. https://doi.org/10.1016/j.ecss.2015.05.021

Bernardino AF, Pagliosa PR, Christofoletti RA et al (2016) Benthic estuarine communities in Brazil: moving forward to long term studies to assess climate change impacts. Braz J Oceanogr 64:81–96

Bernini E, Rezende CE (2010) Variação estrutural em florestas de mangue do estuário do rio Itabapoana, ES-RJ. Biotemas 23:49–60

Bigarella JJ, Becker RD, Matos DD, Werner A (1978) A Serra do Mar e a porção oriental do Estado do Paraná. Report of the Secretaria de Estado do Planejamento, Governo do Paraná, 248

Blankensteyn A, Moura RS (2002) Lista preliminar das espécies da macrofauna de fundos inconsolidados da Baía de Guaratuba, Paraná, Brasil. Rev Bras Zool 19:715–721. https://doi.org/10.1590/S0101-81752002000300008

Boehs G, Absher TM, Cruz-Kaled A (2004) Composition and distribution of benthic molluscs on intertidal flats of Paranaguá Bay (Paraná, Brazil). Sci Mar 68:537–543. https://doi.org/10.3989/scimar.2004.68n4537

Bordon ICAC, Sarkis JES, Andrade NP et al (2016) An environmental forensic approach for tropical estuaries based on metal bioaccumulation in tissues of *Callinectes danae*. Ecotoxicology 25:91–104. https://doi.org/10.1007/s10646-015-1570-1

Bordon ICAC, Sarkis JES, Gobbato GM et al (2011) Metal concentration in sediments from the Santos estuarine system: a recent assessment. J Braz Chem Soc 22:1858–1865. https://doi.org/10.1590/S0103-50532011001000005

Bordon ICAC, Sarkis JES, Tomás ARG et al (2012a) A preliminary assessment of metal bioaccumulation in the blue crab, *Callinectes danae* S., from the São Vicente channel, São Paulo state, Brazil. Bull Environ Contam Toxicol 88:577–581. https://doi.org/10.1007/s00128-012-0539-5

Bordon ICAC, Sarkis JES, Tomás ARG et al (2012b) Assessment of metal concentrations in muscles of the blue crab, *Callinectes danae* S., from the Santos estuarine system. Bull Environ Contam Toxicol 89:484–488. https://doi.org/10.1007/s00128-012-0721-9

Brandini FP, Thamm C (1994) Variações diárias e sazonais do fitoplâncton e parâmetros ambientais na Baía de Paranaguá. Nerítica 8:55–72

Brauko KM (2008) Efeitos da passagem de sistemas frontais sobre a macrofauna bêntica de praias arenosas do Paraná (Brasil). Dissertation, Universidade Federal do Paraná

Brauko KM, Muniz P, Martins CDC, Lana PC (2016) Assessing the suitability of five benthic indices for environmental health assessment in a large subtropical South American estuary. Ecol Indic 64:258–265. https://doi.org/10.1016/j.ecolind.2016.01.008

Breves-Ramos A, Lavrado HP, de Junqueira AOR, da Silva SHG (2005) Succession in rocky intertidal benthic communities in areas with different pollution levels at Guanabara Bay (RJ-Brazil). Braz Arch Biol Technol 48:951–965. https://doi.org/10.1590/S1516-89132005000800012

Brustolin MC, Thomas MC, Mafra LL, Lana PC (2014) Does *Encope emarginata* (Echinodermata: Echinoidea) affect spatial variation patterns of estuarine subtidal meiofauna and microphytobenthos? J Sea Res 91:70–78. https://doi.org/10.1016/j.seares.2014.03.006

Bumbeer J, Rocha RM (2016) Invading the natural marine substrates: a case study with invertebrates in South Brazil. Fortschr Zool 33(3):e2015021

Caires RA (2014) Biogeografia dos peixes marinhos do Atlântico Sul ocidental: padrões e processos. Arq Zool 45:5–24. https://doi.org/10.11606/issn.2176-7793.v45iespp5-24

Calil M, Soares-Gomes A, Tavares M (2006) Spatial distribution of the major groups of Crustacea from infralitoral soft-bottoms at Sepetiba Bay, RJ, Brazil. J Coast Res ICS 2004:1140–1143

Caliman A, Carneiro LS, José J et al (2013) Biodiversity effects of benthic ecosystem engineers on the spatial patterns of sediment CH 4 concentration in an urban Neotropical coastal lagoon. Acta Limnol Bras 25:302–314. https://doi.org/10.1590/S2179-975X2013000300009

Camargo R, Harari J (1994) Modelagem numérica de ressacas na plataforma sudeste do Brasil a partir de cartas sinóticas de pressão atmosférica na superfície. Bol do Inst Ocean 42:19–34. https://doi.org/10.1590/S0373-55241994000100002

Campos RM, Camargo R, Harari J (2010) Caracterização de eventos extremos do nível do mar em Santos e sua correspondência com as reanálises do modelo do NCEP no Sudoeste do Atlântico Sul. Rev Bras Meteorol 25:175–184. https://doi.org/10.1590/S0102-77862010000200003

Capparelli MV, Abessa DM, McNamara JC (2016) Effects of metal contamination in situ on osmoregulation and oxygen consumption in the mudflat fiddler crab *Uca rapax* (Ocypodidae, Brachyura). Comp Biochem Physiol Part - C Toxicol Pharmacol 185–186:102–111. https://doi.org/10.1016/j.cbpc.2016.03.004

Carcamo A (1980) Observações sobre a fauna bentônica da região de Cananéia (25°00'S-48°00'W), Estado de São Paulo. Dissertation, Universidade de São Paulo

Cardoso C (2010) A macrofauna bêntica da região estuarina-lagunar de Cananéia-SP, com ênfase em Polychaeta. Dissertation, Universidade de São Paulo

Carneiro MER, Marques ANJ, Pereira RC et al (1987) Estudos populacionais de *Ulva fasciata* Delile, indicadora de poluição na Baía de Guanabara, RJ. Nerítica 2:201–212

Castilho-Westphal GG, Dal Pont G, Horodesky A, Ostrensky A (2014) Comunidades ribeirinhas extrativistas e a exploração de bancos de ostras do mangue *Crassostrea* sp., na Baía de Guaratuba-Paraná, litoral sul do Brasil. Biosci J 30:912–923

Checon HH, Corte GN, Silva CF et al (2017) Mangrove vegetation decreases density but does not affect species richness and trophic structure of intertidal polychaete assemblages. Hydrobiologia 795:169–179. https://doi.org/10.1007/s10750-017-3128-0

Christo SW, Ivachuk CS, Ferreira-Júnior AL, Absher TM (2016) Reproductive periods of *Lucina pectinata* (bivalve; Lucinidae) in the Paranaguá estuarine complex, Paraná - Brazil. Braz J Biol 76:300–306. https://doi.org/10.1590/1519-6984.12514

Citadin M, Costa TM, Netto SA (2016) The response of meiofauna and microphytobenthos to engineering effects of fiddler crabs on a subtropical intertidal sandflat. Austral Ecol 41:572–579. https://doi.org/10.1111/aec.12346

Coimbra JC, Geraque EA, Carreño AL, Eichler BB (2007) Ostracodes (Crustacea) from Cananéia-Iguape estuarine/lagoon system and geographical distribution of the mixohaline assemblages in southern and southeastern Brazil. Iheringia Série Zool 97:273–279. https://doi.org/10.1590/S0073-47212007000300010

Colling LA, Bemvenuti CE, Gandra MS (2007) Seasonal variability on the structure of sublittoral macrozoobenthic association in the Patos lagoon estuary, southern Brazil. Iheringia Série Zool 97:257–262. https://doi.org/10.1590/S0073-47212007000300007

Colpo KD, Chacur MM, Guimarães FJ, Negreiros-Fransozo ML (2011) Subtropical Brazilian mangroves as a refuge of crab (Decapoda: Brachyura) diversity. Biodivers Conserv 20:3239–3250. https://doi.org/10.1007/s10531-011-0125-x

Colpo KD, Negreiros-Fransozo ML (2011) Sediment particle selection during feeding by four species of *Uca* (Brachyura, Ocypodidae). Crustaceana 84:721–734. https://doi.org/10.2307/23034320

Copertino MS, Creed JC, Lanari MO et al (2016) Seagrass and submerged aquatic vegetation (VAS) habitats off the coast of Brazil: state of knowledge, conservation and main threats. Braz J Oceanogr 64:53–80. https://doi.org/10.1590/S1679-875920161036064sp2

Corbisier TN (1991) Benthic macrofauna of sandy intertidal zone at Santos estuarlne system, São Paulo, Brazil. Bol do Inst Ocean 39:1–13

Corbisier TN (1994) Macrozoobentos da praia do Codó (Ubatuba, SP) e a presença de *Halodule wrightii* Ascherson. Braz J Oceanogr 42:99–111. https://doi.org/10.1590/S1679-87591994000100008

Corgosinho PHC, Metri R, Baptista C et al (2003) Abundance and diversity of the sublittoral meiofauna on two sand beaches under different hydrodynamic conditions at Ilha do Mel (PR, Brazil). Lundiana 4:89–94

Correia RRS, Guimarães JRD (2016) Impacts of crab bioturbation and local pollution on sulfate reduction, hg distribution andmethylation in mangrove sediments, Rio de Janeiro, Brazil. Mar Pollut Bull 109:453–460. https://doi.org/10.1016/j.marpolbul.2016.05.028

Costa CSB, Davy AJ (1992) Coastal saltmarsh communities of Latin America. In: Coastal plant communities of latin America. Academic Press, New York, pp 179–199

Costa CSB, Marangoni JC, Azevedo AMG (2003) Plant zonation in irregularly flooded salt marshes: relative importance of stress tolerance and biological interactions. J Ecol 91:951–965. https://doi.org/10.1046/j.1365-2745.2003.00821.x

Costa HA, Pereira RMFA, Hoffmann VE (2006) Compreendendo o Espaço Turístico de Balneário Camboriú (SC) como Insumo para o Estudo da Competitividade Local. Tur - Visão e Ação 8:223–234. https://doi.org/10.14210/rtva.v8n2.p223-234

Costa RC, Lopes M, Castilho AL et al (2008) Abundance and distribution of juvenile pink shrimps *Farfantepenaeus* spp. in a mangrove estuary and adjacent bay on the northern shore of são Paulo state, southeastern Brazil. Invertebr Reprod Dev 52:51–58. https://doi.org/10.1080/07924259.2008.9652272

Couto EC (1996) Estrutura espaço-temporal da comunidade macrobêntica da planície intertidal do Saco do Limoeiro - Ilha do Mel (Paraná, Brasil). PhD Thesis, Universidade Federal do Paraná

Couto EC, Almeida M, Lana PC (1995) Diversidade e distribuição da macroinfauna bêntica do Saco do Limoeiro - Ilha do Mel, Paraná. Publicação Espec do Inst Ocean 11:239–247

Creed JC (1997) Morphological variation in the seagrass *Halodule wrightii* near its southern distributional limit. Aquat Bot 59:163–172. https://doi.org/10.1016/S0304-3770(97)00059-4

Creed JC (1999) Distribution, seasonal abundance and shoot size of the seagrass *Halodule wrightii* near its southern limit at Rio de Janeiro state, Brazil. Aquat Bot 65:47–58. https://doi.org/10.1016/S0304-3770(99)00030-3

Creed JC (2000) Epibiosis on cerith shells in a seagrass bed: correlation of shell occupant with epizoite distribution and abundance. Mar Biol 137:775–782. https://doi.org/10.1007/s002270000429

Cunha-Lignon M, Mahiques MM, Schaeffer-Novelli Y et al (2009) Analysis of mangrove forest succession, using sediment cores: a case study in the Cananéia-Iguape coastal system, São Paulo-Brazil. Braz J Oceanogr 57:161–174. https://doi.org/10.1590/S1679-87592009000300001

Grande FR (2016) O conteúdo orgânico do sedimento influência na distribuição intraespecífica de caranguejos-chama-maré. Dissertation, Universidade Estadual Paulista

Domingos AM, Lana PC (2017) Detecting multiple states of trophic connectivity between mangroves and salt marshes. Ecosystems 20:1179–1189. https://doi.org/10.1007/s10021-016-0101-0

Dominguez JML (2006) The coastal zone of Brazil: an overview. J Coast Res 39:16–20

Donnici S, Serandrei-Barbero R, Bonardi M, Sperle M (2012) Benthic foraminifera as proxies of pollution: the case of Guanabara Bay (Brazil). Mar Pollut Bull 64:2015–2028. https://doi.org/10.1016/j.marpolbul.2012.06.024

Santos CMH, Pinheiro MAA, Hattori GY (2009) Orientation and external morphology of burrows of the mangrove crab *Ucides cordatus* (Crustacea: Brachyura: Ucididae). J Mar Biol Assoc United Kingdom 89:1117–1123. https://doi.org/10.1017/S0025315409000502

Duarte LFA, Duran RS, Mendonça JT, Pinheiro MAA (2014) Fishery of the Uçá crab *Ucides cordatus* (Linnaeus, 1763) in a mangrove area in Cananéia, state of São Paulo, Brazil: fishery performance, exploitation patterns and factors affecting the catches. Braz J Oceanogr 62:187–199. https://doi.org/10.1590/S1679-87592014068906203

Echeverría CA, Neves RAF, Pessoa LA, Paiva PC (2010) Spatial and temporal distribution of the gastropod *Heleobia australis* in an eutrophic estuarine system suggests a metapopulation dynamics. Nat Sci 2:860–867. https://doi.org/10.4236/ns.2010.28108

Egres AG, Martins CC, Oliveira VM, Lana PC (2012) Effects of an experimental in situ diesel oil spill on the benthic community of unvegetated tidal flats in a subtropical estuary (Paranaguá Bay, Brazil). Mar Pollut Bull 64(12):2681–2691. https://doi.org/10.1016/j.marpolbul.2012.10.007

Esteves FA, Caliman A, Santangelo JM, Guariento RD, Farjalla VF, Bozelli RL (2008) Neotropical coastal lagoons: an appraisal of their biodiversity, functioning, threats and conservation management. Braz J Biol 68(Suppl. 4):967–981. https://doi.org/10.1590/S1519-69842008000500006

Esteves FA (1998) Ecologia das lagoas costeiras do Parque Nacional da Restinga de Jurubatiba e do Município de Macaé (RJ). NUPEM/UFRJ, Rio de Janeiro

Fistarol GO, Coutinho FH, Moreira APB, Venas T, Cánovas A, Paula SEM Jr, Coutinho R, Moura RL, Valentin JL, Tenenbaum DR, Paranhos R, Valle RAB, Vicente ACP, Amado Filho GM, Pereira RC, Kruger R, Rezende CE, Thompson CC, Salomon P, Thompson F (2015) Environmental and sanitary conditions of Guanabara Bay, Rio de Janeiro. Front Microb 6:1–17. https://doi.org/10.3389/fmicb.2015.01232

Ferreira J (2008) Estudo das associações de anelídeos Polychaeta da Baía de Santos e plataforma continental adjacente (SP, Brasil) e suas interações com parâmetros físicos e geoquímicos estruturadores. PhD Thesis, Universidade de São Paulo

Floeter SR, Soares-Gomes A (1999) Biogeographic and species richness patterns of Gastropoda on the southwestern Atlantic. Rev Bras Biol 59:567–575. https://doi.org/10.1590/S0034-71081999000400006

Flynn MN, Tararam AS, Wakabara Y (1996) Effects of habitat complexity on the structure of macrobenthic association in a *Spartina altemiflora* marsh. Rev Bras Oceanogr 44:09–21. https://doi.org/10.1590/S1413-77391996000100002

Flynn MN, Wakabara Y, Tararam AS (1998) Macrobenthic associations of the lower and upper marshes of a tidal flat colonized by *Spartina alterniflora* in Cananéia lagoon estuarine region (southeastern Brazil). Bull Mar Sci 63:427–442

Fonseca ALDO, Machado EC, Brandini FP, Brandini N (2013) Microphytobenthic biomass on a subtropical intertidal flat of Paranaguá Bay (southern Brazil): spatio-temporal distribution and the influence of environmental conditions. Braz J Oceanogr 61:83–92. https://doi.org/10.1590/S1679-87592013000200001

Galvão TA, Nolasco MC (2013) Urbanization and coral reefs in Guarajuba beach, north coast of Bahia, Brazil. Ocean Coast Manag 77:50–58. https://doi.org/10.1016/j.ocecoaman.2012.03.013

García-Rodríguez F, Brugnoli E, Muniz P et al (2014) Warm-phase ENSO events modulate the continental freshwater input and the trophic state of sediments in a large South American estuary. Mar Freshw Res 65:1–11. https://doi.org/10.1071/MF13077

Gerlach SA (1958) Die Mangroveregion Tropischer Küster als Lebensraum. Z Morphol Okol Tiere 46:636–730. https://doi.org/10.1007/BF00407459

Gern FR, Lana PC (2013) Reciprocal experimental transplantations to assess effects of organic enrichment on the recolonization of benthic macrofauna in a subtropical estuary. Mar Pollut Bull 67:107–120. https://doi.org/10.1016/j.marpolbul.2012.11.026

Gomes FC, Godoy JM, Godoy MLDP et al (2009) Metal concentrations, fluxes, inventories and chronologies in sediments from Sepetiba and Ribeira bays: a comparative study. Mar Pollut Bull 59:123–133. https://doi.org/10.1016/j.marpolbul.2009.03.015

Grimm AM, Tedeschi RG (2009) ENSO and extreme rainfall events in South America. J Clim 22:1589–1609. https://doi.org/10.1175/2008JCLI2429.1

Grohmann PA (2009) Hydroids (Cnidaria, hydrozoa) of the intertidal zone of Governador and Paquetá islands, Guanabara Bay, Rio de Janeiro, Brazil. Iheringia Série Zool 99:291–294. https://doi.org/10.1590/S0073-47212009000300010

Guiss C (1993) Influência da gramínea *Spartina alterniflora* Loiseleur, 1807 sobre a distribuição e densidade de *Anomalocardia brasiliana* (Gmalin, 1791). Dissertation, Universidade Federal do Paraná

Gusmão Jr JB (2017) Sediments and functional traits: applying a functional trait approach to assess marine macrobenthic function. PhD Thesis. Co-tutelle University of Groningen, Universidade Federal do Paraná

Hadlich HL (2010) Variabilidade espacial de associações de poliquetas ao longo de um gradiente de contaminação por esgotos no Canal da Cotinga (Baía de Paranaguá, Brasil). Dissertation, Universidade Federal do Paraná

Hansen A (1882) Recherches sur les Annélides recueillies par M. le professeur Edouard Van Beneden pendant son voyage au Brésil et a la Plata. Mémoires couronnés publiés par l'Académie R des Sci Belgique 44:1–29

Haponiuk R (2007) Variação espacial da macrofauna bentônica e suas relações com variáveis ambientais do sedimento no sublitoral da Baía de Babitonga (Santa Catarina - Brasil). Bachelor Thesis, Universidade Federal do Paraná

Harari J, França C, Camargo R (2008) Climatology and hydrography of Santos estuary. In: Neves R, Baretta J, Mateus M (eds) Perspectives on integrated coastal zone management in South America. IST Press, Lisboa, pp 147–160

Hattori GY, Pinheiro MAA (2003) Fertilidade do caranguejo de mangue *Ucides cordatus* (Linnaeus) (Crustacea, Brachyura, Ocypodidae), em Iguape (São Paulo, Brasil). Rev Bras Zool 20:309–313. https://doi.org/10.1651/C-2334

Henriques MB, Machado IC, Fagundes L (2010) Análise econômica comparativa dos sistemas de cultivo integral e de "engorda" da ostra do mangue *Crassostrea* spp. no estuário de Cananéia, São Paulo, Basil. Bol do Inst Pesca 36:307–316. https://doi.org/10.1590/1519-6984.20713

Hostin LM, Sandrini-Neto L, Oliveira VM (2007) Associações macrofaunais em áreas sujeitas à dragagem do Complexo Estuarino de Paranaguá (CEP). In: Dragagens Portuárias no Brasil. Governo do Estado do Paraná; Sema/PR; Ademadan; Unibem, Antonina, p 312

Isacch JP, Costa CSB, Rodriguez-Gallego L et al (2006) Distribution of saltmarsh plant communities associated with environmental factors along a latitudinal gradient on the south-west Atlantic coast. J Biogeogr 33:888–900. https://doi.org/10.1111/j.1365-2699.2006.01461.x

Jacobi C (1987) Spatial and temporal distribution of Amphipoda associated with mussel beds from the bay of Santos (Brazil). Mar Ecol Prog Ser 35:51–58. https://doi.org/10.3354/meps035051

Jara VC, Miyamoto JHS, Gama BAP et al (2006) Limited evidence of interactive disturbance and nutrient effects on the diversity of macrobenthic assemblages. Mar Ecol Prog Ser 308:37–48. https://doi.org/10.3354/meps308037

Jorcin A (1997) Distribuição espacial (vertical e horizontal) do macrozoobentos na região estuarina da Cananéia (SP), e sua relação com algumas variáveis físicas e químicas. PhD Thesis, Universidade de São Paulo

Julio L (2007) Espécies exóticas e criptogênicas do macrozoobentos de substrato consolidado da região entre-marés da Baía de Guanabara, Rio de Janeiro, RJ. Bachelor Thesis, Universidade Federal do Rio de Janeiro

Junqueira A, Falcão A, Mayer-Pinto M et al (2000) Spatial and temporal variations on intertidal barnacle abundance in a tropical bay. Nauplius 8:195–204

Kassuga A, Masunari S (2015) Spatial distribution of juveniles of the mangrove crab *Ucides cordatus* in Guaratuba Bay, southern Brazil. Panam J Aquat Sci 10:222–229

Kjerfve B, Ribeiro CHA, Dias GTM et al (1997) Oceanographic characteristics of an impacted coastal bay: Baia de Guanahara, Rio de Janeiro, Brazil. Cont Shelf Res 17:1609–1643. https://doi.org/10.1016/S0278-4343(97)00028-9

Knoppers BA, Brandini FP, Thamm C (1987) Ecological studies in the bay of Paranaguá. II. Some physical and chemical characteristics. Nerítica 2:1–36

Kolm H, Lana PC, Soares CR, Lorenzi L, Brogim RA, Odreski L, Siqueira A (1999) Monitoramento das atividades de dragagem (1998/1999) ao longo do canal de acesso aos terminais portuários da Ponta do Félix (Antonina). Unpublished Technical Report. Center for Marine Studies, Universidade Federal do Paraná, p 181

Lana PC (1986) Macrofauna bêntica de fundos sublitorais não consolidados da Baía de Paranaguá. Nerítica 1:79–89

Lana PC (2003) As marismas da Baía de Paranaguá: características gerais, modos de apropriação e implicações para a legislação ambiental. Desenvolv e Meio Ambiente 8:11–23. https://doi.org/10.5380/dma.v8i0.22044

Lana PC, Almeida MVO, Freitas CAF et al (1989) Estrutura espacial das associações macrobênticas sublitorais da gamboa Perequê (Pontal do Sul, Paraná). Nerítica 4:119–136

Lana PC, Angulo RJ, Ribeiro D (1997a) Estudo de Impacto ambiental (EIA) de uma usina termelétrica na Baía de Paranaguá e do Porto de desembarque, subestação e linha de transmissão associados. Copel - Companhia Paranaense de Energia, Curitiba

Lana PC, Camargo MG, Brogim RA, Isaac VJ (1996) O bentos da costa brasileira. Avaliação crítica e levantamento bibliográfico. MMA/CIRM/FEEMA, Rio de Janeiro

Lana PC, Couto EC, Almeida MV (1997b) Polychaete distribution and abundance in intertidal flats of Paranaguá Bay (SE Brazil). Bull Mar Sci 60(2):433–442

Lana PC, Guiss C (1991) Influence of *Spartina alterniflora* on structure and temporal variability of macrobenthic associations in a tidal flat of Paranaguá Bay (southeastern Brazil). Mar Ecol Prog Ser 73:231–244

Lana PC, Guiss C (1992) Macrofauna-plant-biomass interactions in a euhaline salt marsh in Paranaguá Bay (SE Brazil). Mar Ecol Prog Ser 80:57–64. https://doi.org/10.3354/meps080057

Lana PC, Guiss C, Disaró ST (1991) Seasonal variation of biomass and production dynamics for above- and belowground components of a *Spartina alterniflora* marsh in the euhaline sector of Paranaguá Bay (SE Brazil). Estuar Coast Shelf Sci 32:231–241. https://doi.org/10.1016/0272-7714(91)90017-6

Lana PC, Marone E, Lopes RM, Machado EC (2001) The subtropical estuarine complex of Paranaguá Bay, Brazil. In: Seeliger U, Kjerfve B (eds) Coastal marine ecosystem of latin America. Springer, Berlin, pp 131–145

Lana PC, Pagliosa PR, Paiva PC et al (2017) Polychaetes in Brazil: people and places, past, present and future. Boletín del Inst Ocean Venez 1:24

Laurino I (2016) Macrofauna bentonica em estuários protegidos subtropicais: padrões e subsídios para a gestão de ecossistemas costeiros. Dissertation, Universidade Federal de São Paulo

Laut LLM, Martins M, Fontana LF, Silva FS, Mendonça-filho JG, Moreira Clemente IMM, Frontalini F, Raposo D, Belart P, Ballalai J (2016) Ecological status evaluation of Itaipu lagoon (Niterói) based on biochemical composition of organic matter. J Sediment Environ 1(3):297–315. https://doi.org/10.12957/jse.2016.25903

Lavrado HP, Falcão APC, Carvalho-Cunha PC et al (2000) Composition and distribution of decapoda from Guanabara Bay, RJ. Nauplius 8:15–23

Leite DS, Sandrini-Neto L, Camargo MZ et al (2014) Are changes in the structure of nematode assemblages reliable indicators of moderate petroleum contamination? Mar Pollut Bull 83:38–47. https://doi.org/10.1016/j.marpolbul.2014.04.035

Lessa GC, Meyers SR, Marone E (1998) Holocene stratigraphy in the Paranagua Bay estuary, southern Brazil. J Sediment Res 68:1060–1076. https://doi.org/10.2110/jsr.68.1060

Liebmann B, Marengo JA (2001) Interannual variability of the rainy season and rainfall in the Brazilian Amazon Basin. J Clim 14:4308–4318. https://doi.org/10.1175/1520-0442(2001)014<4308:IVOTRS>2.0.CO;2

Lino AS, Galvão PMA, Longo RTL et al (2016) Metal bioaccumulation in consumed marine bivalves in southeast Brazilian coast. J Trace Elem Med Biol 34:50–55. https://doi.org/10.1016/j.jtemb.2015.12.004

Longo LL, Amado Filho GM (2014) Knowledge of Brazilian benthic marine fauna throughout time. História, Ciências, Saúde 21:995–1010. https://doi.org/10.1590/S0104-59702014000300011

López-Barrera EA, Longo GO, Monteiro-Filho ELA (2012) Incidental capture of green turtle (*Chelonia mydas*) in gillnets of small-scale fisheries in the Paranaguá Bay, southern Brazil. Ocean Coast Manag 60:11–18. https://doi.org/10.1016/j.ocecoaman.2011.12.023

Lorenzi L (1998) Composição e distribuição da macrofauna bêntica em gamboas da Baía de Paranaguá (Paraná, Brasil). Dissertation, Universidade Federal do Paraná

Lotufo T (1997) Ecologia das ascídias da Baía de Santos (SP): período reprodutivo, crescimento e aspectos sucessionais. Dissertation, Universidade de São Paulo

Lupo AR, Nocera JJ, Bosart LF et al (2001) South American cold surges: types, composites and case studies. Mon Weather Rev 129:1021–1041. https://doi.org/10.1175/1520-0493(2001)129<1021:SACSTC>2.0.CO;2

Machado E, Daniel C, Brandini N, Queiroz R (1997) Temporal and spatial dynamics of nutrients and particulate suspended matter in Paranaguá Bay, PR, Brazil. Nerítica 11:15–34

Machado GBO, Gusmão JB Jr, Costa TM (2013) Burrow morphology of *Uca uruguayensis* and *Uca leptodactylus* (Decapoda: Ocypodidae) from a subtropical mangrove forest in the western Atlantic. Integr Zool 8:307–314. https://doi.org/10.1111/j.1749-4877.2012.00297.x

Machado IC, Fagundes L, Henriques M (2015) Multidimensional assessment of sustainability extractivism of mangrove oyster *Crassostrea* spp. in the estuary of Cananéia. Braz J Biol 75:670–678. https://doi.org/10.1590/1519-6984.20713

Magalhães CA, Taniguchi S, Cascaes MJ, Montone RC (2012) PCBs, PBDEs and organochlorine pesticides in crabs *Hepatus pudibundus* and *Callinectes danae* from Santos Bay, state of São Paulo, Brazil. Mar Pollut Bull 64:662–667. https://doi.org/10.1016/j.marpolbul.2011.12.020

Mantovanelli A, Marone E, Silva ET et al (2004) Combined tidal velocity and duration asymmetries as a determinant of water transport and residual flow in Paranaguá Bay estuary. Estuar Coast Shelf Sci 59:523–537. https://doi.org/10.1016/j.ecss.2003.09.001

Marengo JA (2001) Mudanças climáticas globais e regionais: avaliação do clima atual do Brasil e projeções de cenários climáticos do futuro. Rev Bras Meteorol 16:1–18

Marengo JA, Ambrizzi T, Rocha RP et al (2010) Future change of climate in South America in the late twenty-first century: intercomparison of scenarios from three regional climate models. Clim Dyn 35:1089–1113. https://doi.org/10.1007/s00382-009-0721-6

Maria TF, Paiva P, Vanreusel A, Esteves AM (2013) The relationship between sandy beach nematodes and environmental characteristics in two Brazilian sandy beaches (Guanabara Bay, Rio de Janeiro). An Acad Bras Cienc 85:257–270. https://doi.org/10.1590/S0001-37652013005000019

Maria TF, Wandeness AP, Esteves AM (2016) State of the art of the meiofauna of Brazilian sandy beaches. Braz J Oceanogr 64:17–26

Marone E, Camargo R (1994) Efeitos da maré meteorológica na Baía de Paranaguá, PR. Nerítica 8:73–85

Marques AN, Monna F, da Silva Filho EV et al (2006) Apparent discrepancy in contamination history of a sub-tropical estuary evaluated through 210Pb profile and chronostratigraphical markers. Mar Pollut Bull 52:532–539. https://doi.org/10.1016/j.marpolbul.2005.09.048

Marques LV, Creed JC (2008) Biologia e ecologia das fanerógamas marinhas do Brasil. Oecologia Aust 12:315–331. https://doi.org/10.4257/oeco.2008.1202.12

Marques MCM, Silva SM, Liebsch D (2015) Coastal plain forests in southern and southeastern Brazil: ecological drivers, floristic patterns and conservation status. Rev Bras Bot 38:1–18

Martins CC, Bícego MC, Mahiques MM et al (2011) Polycyclic aromatic hydrocarbons (PAHs) in a large south American industrial coastal area (Santos estuary, southeastern Brazil): sources and depositional history. Mar Pollut Bull 63:452–458. https://doi.org/10.1016/j.marpolbul.2011.03.017

Mattos G, Cardoso RS, Santos AS (2013) Environmental effects on the structure of polychaete feeding guilds on the beaches of Sepetiba Bay, South-Eastern Brazil. J Mar Biol Assoc UK 93:973–980. https://doi.org/10.1017/S0025315412000707

Mattos LA, Mendes LMD, Bede LM et al (2014) Influence of environmental variables on the distribution of intertidal porcellanid crabs in Sepetiba Bay, Rio de Janeiro, Brazil. Lat Am J Aquat Res 42:980–988. https://doi.org/10.3856/vol42-issue5-fulltext-5

Mayer-Pinto M, Junqueira AOR (2003) Effects of organic pollution on the initial development of fouling communities in a tropical bay, Brazil. Mar Pollut Bull 46:1495–1503. https://doi.org/10.1016/S0025-326X(03)00249-2

Mazzuco ACA, Christofoletti RA, Pineda J et al (2015) Temporal variation in intertidal community recruitment and its relationships to physical forcings, chlorophyll-a concentration and sea surface temperature. Mar Biol 162:1705–1725. https://doi.org/10.1007/s00227-015-2689-6

Mello GH (2008) Expansão e estrutura urbana de Santos (SP): aspectos da periferização, da deterioração, da intervenção urbana, da verticalização e da sociabilidade. Dissertation, Universidade de São Paulo

Mendes CLT, Soares-Gomes A, Tavares M (2006) Seasonal and spatial distribution of sublittoral soft-bottom mollusks assemblages at Guanabara Bay, Rio de Janeiro, Brazil. J Coast Res 39:136–140

Mendonça JT, Machado IC (2010) Mangrove oyster (*Crassostrea* spp.) (Sacco, 1897) extractivism in Cananéia estuary (São Paulo, Brazil) from 1999 to 2006: capture and management evaluation. Braz J Biol 70:65–73. https://doi.org/10.1590/S1519-69842010000100010

Miloslavich P, Klein E, Díaz JM et al (2011) Marine biodiversity in the Atlantic and Pacific coasts of South America: knowledge and gaps. PLoS One 6:e14631. https://doi.org/10.1371/journal.pone.0014631

Molisani MM, Marins RV, Machado W et al (2004) Environmental changes in Sepetiba Bay, SE Brazil. Reg Environ Chang 4:17–27. https://doi.org/10.1007/s10113-003-0060-9

Monteiro L (2007) Variação espaço-temporal da macro e meiofauna bênticas localizadas sob um cultivo de ostras na Baía de Guaratuba, Paraná – Brasil. Bachelor Thesis, Universidade Federal do Paraná

Morais GC (2015) Variabilidade em distintas escalas espaço-temporais da macrofauna bêntica de um estuário subtropical. PhD Thesis, Universidade Federal do Paraná

Morais GC, Camargo MG, Lana PC (2016) Intertidal assemblage variation across a subtropical estuarine gradient: how good conceptual and empirical models are? Estuar Coast Shelf Sci 170:91–101. https://doi.org/10.1016/j.ecss.2015.12.020

Murolo PPDA, Carvalho PVVDC, Carvalho MLB et al (2006) Spatio-temporal variations of microphytobenthos in the Botafogo and Siri estuaries. BrazJ Oceanogr 54:19–30

Nascimento ER, Salamuni E, Queiroz GL et al (2013) Evidências de determinação morfotectônica e neotectônica no relevo da Serra do Mar no estado do Paraná. Rev Bras Geomorfol 14:287–299. https://doi.org/10.20502/rbg.v14i3.402

Natálio LF, Pardo JCF, Machado GBO et al (2017) Potential effect of fiddler crabs on organic matter distribution: a combined laboratory and field experimental approach. Estuar Coast Shelf Sci 184:158–165. https://doi.org/10.1016/j.ecss.2016.11.007

Natividade C, Pereira M, Andriguetto J (2006) Small-scale fishing landings on the coast of the state of Parana, Brazil, from 1975 to 2000, with emphasis on shrimp data. J Coast Res ICS 2004:1272–1275

Negrello-Filho OA, Lana PC (2013) Short-term stability of estuarine benthic assemblages: are storms pattern-defining events? Fortschr Zool 30:266–272

Netto SA, Lana PC (1994) Effects of sediment disturbance on the structure of benthic fauna in a subtropical tidal creek of southeastern Brazil. Mar Ecol Prog Ser 106:239–248. https://doi.org/10.3354/meps106239

Netto SA, Lana PC (1995) Zonação e estratificação da macrofauna bêntica em um banco areno-lodoso do setor euhalino de alta energia da Baía de Paranaguá (Paraná, Brasil). Iheringia Ser Zool 79:27–37

Netto SA, Lana PC (1997) Intertidal zonation of benthic macrofauna in a subtropical salt marsh and nearby unvegetated flat (SE Brazil). Hydrobiologia 353:171–180. https://doi.org/10.1023/A:1003090701675

Neves CS, Rocha RM, Pitombo FB, Roper JJ (2007) Use of artificial substrata by introduced and cryptogenic marine species in Paranaguá Bay, southern Brazil. Biofouling 23:319–330. https://doi.org/10.1080/08927010701399174

Neves F, Echeverria A, Pessoa LA et al (2013) Factors influencing spatial patterns of molluscs in a eutrophic tropical bay. J Mar Biol Assoc United Kingdom 93:1–13. https://doi.org/10.1017/S0025315412001105

Neves RAF, Valentin JL (2011) Revisão bibliográfica sobre a macrofauna bentônica de fundos não-consolidados, em áreas costeiras prioritárias para conservação no Brasil. Labomar Arq para Ciências do Mar 44:59–80

Neves R, Echeverría CA, Pessoa LA (2011) Resposta da espécie *Heleobia australis* (Gastropoda: Hydrobiidae) a variações de salinidade e exposição a hidrocarbonetos. Bol do Laboratório Hidrobiol 24:19–25

Nieto-Ferreira R, Rickenbach TM, Wright EA (2011) The role of cold fronts in the onset of the monsoon season in the South Atlantic convergence zone. Q J R Meteorol Soc 137:908–922. https://doi.org/10.1002/qj.810

Noernberg MA, Lautert LFC, Araujo A et al (2006) Remote sensing and GIS integration for modelling the Paranaguá estuarine complex - Brazil. J Coast Res 3:1627–1631. 01/2004; 39:1627-1631

Noernberg MA, Mizerkowski BD, Paloschi NG, Bento JP (2014) Hydrodynamics and bio-optical assessment of two pristine subtropical estuaries in southern Brazil. Braz J Oceanogr 62:265–278. https://doi.org/10.1590/S1679-87592014069006204

Nonato EF (1964) Anelídeos poliquetas. In: Vanzolini PE (ed) História natural dos organismos aquáticos do Brasil. Bibliografia comentada. FAPESP, São Paulo, pp 133–138

Oigman S, Omena EP (1999) Efeito da grama *Halodule wrightii* na distribuição de anelídeos poliquetas na região entremares da Ilha do Japonês, Cabo Frio-RJ. In: XII semana nacional de oceanografia. UERJ, Rio de Janeiro, pp 101–103

Oliveira-Filho EC, Pirani J, Giulietti AM (1983) The Brazilian seagrasses. Aquat Bot 16:251–267. https://doi.org/10.1016/0304-3770(83)90036-0

Oliveira-Neto J, Batista E, Metri R, Metri C (2014) Local distribution and abundance of *Cardisoma guanhumi* Latreille, 1928 (Brachyura: Gecarcinidae) in southern Brazil. Braz J Biol 74:1–7. https://doi.org/10.1590/1519-6984.02912

Omena E, Creed JC (2004) Polychaete fauna of seagrass beds (*Halodule wrightii* Ascherson) along the coast of Rio de Janeiro (Southeast Brazil). Mar Ecol 25:273–288. https://doi.org/10.1111/j.1439-0485.2004.00031.x

Omena EP, Lavrado HP, Paranhos R, Silva TA (2012) Spatial distribution of intertidal sandy beach polychaeta along an estuarine and morphodynamic gradient in an eutrophic tropical bay. Mar Pollut Bull 64:1861–1873. https://doi.org/10.1016/j.marpolbul.2012.06.009

Ovalle ARC, Rezende CE, Lacerda LD, Silva CAR (1990) Factors affecting the hydrochemistry of a mangrove tidal creek, Spetib Bay, Brazil. Estuar Coast Shelf Sci 31:639–650. https://doi.org/10.1016/0272-7714(90)90017-L

Pagliosa PR (2006) Distribuição da macrofauna bêntica do entremarés ao sublitoral em uma praia estuarina da Baía da Babitonga, Sul do Brasil. Rev Biotemas 19:25–33

Pagliosa PR, Lana PC (2000) Population dynamics and secondary production of *Nereis oligohalina* (Nereididae: Polychaeta) from a subtropical marsh in Southeast Brazil. Bull Mar Sci 67:259–268

Pagliosa PR, Lana PC (2005) Impact of plant cover removal on macrobenthic community structure of a subtropical salt marsh. Bull Mar Sci 77:1–17

Palacio FJ (1982) Revisión zoogeográfica marina del sur del Brasil. Boletim do Instituto Oceanográfico 31:69–92

Pereira CDS, Abessa DMS, Choueri RB et al (2014) Ecological relevance of sentinels' biomarker responses: a multi-level approach. Mar Environ Res 96:118–126. https://doi.org/10.1016/j.marenvres.2013.11.002

Pereira CVP, Pessoa LA, Aurélio M et al (2013) Recovery process on soft-bottom macrobenthic communities after artificial disturbance in tropical polluted estuary (Guanabara Bay, Rio de Janeiro, Brazil). J Mar Sci 3:161–166. https://doi.org/10.4236/ojms.2013.34018

Pieper WW (2007) Distribuição espacial da macrofauna bêntica na região sublitoral do complexo estuarino de Paranaguá, eixo norte-sul - Baía das Laranjeiras. Bachelor Thesis, Universidade Federal do Paraná

Pinheiro MAA, Duarte LFA, Toledo TR et al (2013) Habitat monitoring and genotoxicity in *Ucides cordatus* (Crustacea: Ucididae), as tools to manage a mangrove reserve in southeastern Brazil. Environ Monit Assess 185:8273–8285. https://doi.org/10.1007/s10661-013-3172-9

Por F, Shimizu G, Almeida Prado-Por MS et al (1984) The Blackwater river estuary of rio Una do Prelado (Sao Paulo,Brazil): preliminary hydrobiological data. Rev d'Hydrobiol Trop 17:245–258

Prantoni AL, Lana PC, Belmonte-Lopes R, Erseus C (2017) Genetic diversity of marine oligochaetous clitellates in selected areas of the South Atlantic as revealed by DNA barcoding. Invertebr Syst. in press

Ribeiro V, Junqueira A (1999) Abundância e diversidade dos moluscos bentônicos de fundos vegetados e não vegetados (*Halodule wrightii*) da Ilha do Japonês, Cabo Frio - R.J. In: XII semana nacional de oceanografia. UERJ, Rio de Janeiro, pp 23–25

Ribeiro MC, Metzger JP, Martensen AC, Ponzoni FJ, Hirota MM (2009) The Brazilian Atlantic Forest: how much is left, and how is the remaining forest distributed? Implications for conservation. Biol Conserv 142:1141–1153

Rocha B (2010) Ophiuroidea (Echinodermata) em substratos consolidados naturais e artificiais da Baía de Sepetiba - RJ. Bachelor Thesis, Universidade Federal do Rio de Janeiro

Rocha DC (2006) O uso da suficiência taxonômica para avaliar efeitos de um gradiente de pouição sobre o macrobentos sublitoral na Baía de Paranaguá (Paraná - Brasil). Bachelor Thesis, Universidade Federal do Paraná

Rocha M, Silva E, Riascos N (2013) Avaliação da influência da oxigenação e da qualidade do sedimento sobre a sobrevivência de *Scolelepis chilensis* (Spionidae: Polychaeta) da Baía de Guanabara,RiodeJaneiro.Biotemas26:85–92.https://doi.org/10.5007/2175-7925.2013v26n4p85

Rodrigues SK, Abessa DMS, de Rodrigues APC et al (2017) Sediment quality in a metal-contaminated tropical bay assessed with a multiple lines of evidence approach. Environ Pollut 228:265–276. https://doi.org/10.1016/j.envpol.2017.05.045

Rosa LC (2009) As praias estuarinas da Baía de Paranaguá (Região sul do Brasil): aspectos morfodinâmicos e ecológicos. PhD Thesis, Universidade Federal do Rio Grande

Rosa LC, Borzone CA (2008) Uma abordagem morfodinâmica na caracterização fisica das praias estuarinas da Baia de Paranaguá, sul do Brasil. Rev Bras Geoci 38:237–245

Rumjaneck R, Lavrado H (1999) Efeito da grama marinha Halodule wrightii sobre a comunidade de crustáceos da região entremarés da Ilha do Japonês, Cabo Frio, R.J. In: XII semana nacional de oceanografia. UERJ, Rio de Janeiro, pp 98–100

Sanches FHC (2017) Distribuição espacial de caranguejos-chama-maré: efeitos do sombreamento, competição interespecífica e seleção sexual. PhD Thesis, Universidade Estadual Paulista

Sandrini-Neto L, Lana PC (2014) Does mollusc shell debris determine patterns of macrofaunal recolonisation on a tidal flat? Experimental evidence from reciprocal transplantations. J Exp Mar Bio Ecol 452:9–21

Sandrini-Neto L, Pereira L, Martins CC et al (2016) Antioxidant responses in estuarine invertebrates exposed to repeated oil spills: effects of frequency and dosage in a field manipulative experiment. Aquat Toxicol 177:237–249. https://doi.org/10.1016/j.aquatox.2016.05.028

Santi L, Tavares M (2009) Polychaete assemblage of an impacted estuary, Guanabara Bay, Rio de Janeiro, Brazil. Braz J Oceanogr 57:287–303

Santi L, Tavares M, Omena EP (2006) Patterns of species richness and species density of sublittoral soft-bottom polychaetes in a grossly polluted urban bay: Guanabara Bay, Rio de Janeiro, Brazil. J Coast Res 2004:1127–1131

Santos HF (2004) Composição e estrutura da comunidade bêntica do costão rochoso da Ilha da Galheta, Baía de Paranaguá, Paraná. Bachelor Thesis, Universidade Federal do Paraná

Santos J, Gomes R, Vasconcellos R et al (2014) Effects of morphodynamics and across-shore physical gradients on benthic macroinfauna in two sandy beaches in southeastern Brazil. J Mar Biol Assoc UK 94:671–680

Sayão-Aguiar B, Pinheiro MAA, Colpo KD (2012) Sediment bioturbation potential of *Uca rapax* and *Uca uruguayensis* as a result of their feeding activity. J Crustac Biol 32:223–229. https://doi.org/10.1163/193724011X615451

Schaeffer-Novelli Y, de Mesquita HSL, Cintrón-Molero G (1990) The Cananéia Lagoon estuarine system, São Paulo, Brazil. Estuaries 13:193–203. https://doi.org/10.1007/BF02689853

Severino-Rodrigues E, Pita JB, Graça-Lopes R (2001) Pesca artesanal de siris (Crustacea, Decapoda, Portunidae) na região estuarina de Santos e São Vicente (SP), Brasil. Bol do Inst Pesca 27:7–19

Short AD (2006) Australian beach systems — nature and distribution. J Coast Res 221:11–27. https://doi.org/10.2112/05A-0002.1

Silva L (2008) Variação espacial da macrofauna bêntica no sub-litoral da Baía de Guaratuba, Paraná. Bachelor Thesis, Universidade Federal do Paraná

Silva-Camacho DDS, Gomes RDS, Santos JNS, Araújo FG (2017) Distribution of benthic fauna in sediment grains and prop roots of a mangrove channel in south-eastern Brazil. J Mar Biol Assoc UK 97:1–9. https://doi.org/10.1017/S0025315416000485

Soares HC, Gherardi DFM, Pezzi LP et al (2014) Patterns of interannual climate variability in large marine ecosystems. J Mar Syst 134:57–68. https://doi.org/10.1016/j.jmarsys.2014.03.004

Soares MLG, Estrada GCD, Fernandez V, Tognella MMP (2012) Southern limit of the western South Atlantic mangroves: assessment of the potential effects of global warming from a biogeographical perspective. Estuar Coast Shelf Sci 101:44–53. https://doi.org/10.1016/j.ecss.2012.02.018

Soares-Gomes A, Gama BAP, Baptista Neto JA et al (2016) An environmental overview of Guanabara Bay, Rio de Janeiro. Reg Stud Mar Sci 8:319–330. https://doi.org/10.1016/j.rsma.2016.01.009

Soares-Gomes A, Mendes CLT, Tavares M, Santi L (2012) Taxonomic sufficiency of polychaete taxocenes for estuary monitoring. Ecol Indic 15:149–156. https://doi.org/10.1016/j.ecolind.2011.09.030

Sordo L (2008) Alterações na estrutura e funcionamento de um banco de *Halodule wrightii* (Cymodoceaceae) durante um florescimento massivo de epífitas na baía de Paranaguá (Paraná, Brasil). Dissertation, Universidade Federal do Paraná

Sordo L, Fournier J, Oliveira VM et al (2011) Temporal variations in morphology and biomass of vulnerable *Halodule wrightii* meadows at their southernmost distribution limit in the southwestern Atlantic. Bot Mar 54:13–21. https://doi.org/10.1515/BOT.2011.007

Sousa ECPM, Zaroni LP, Gasparro MR, Pereira CDS (2014) Review of ecotoxicological studies of the marine and estuarine environments of the Baixada Santista (São Paulo, Brazil). Braz J Oceanogr 62:133–147. https://doi.org/10.1590/S1679-87592014063006202

Sousa PHGO, Siegle E, Tessler MG (2013) Vulnerability assessment of Massaguaçú Beach (SE Brazil). Ocean Coast Manag 77:24–30. https://doi.org/10.1016/j.ocecoaman.2012.03.003

Souza F (2009) Distribuição espacial da macrofauna bêntica do sublitoral da Baía de Trapandé, complexo estuarino lagunar de Cananéia-Iguape – SP. Bachelor Thesis, Universidade Federal do Paraná

Souza FM, Brauko KM, Gilbert ER et al (2016) Complex spatial and temporal variation of subtropical benthic macrofauna under sewage impact. Mar Environ Res 116:61–70. https://doi.org/10.1016/j.marenvres.2016.02.008

Souza FM, Gilbert ER, Camargo MG, Pieper WW (2013) The spatial distribution of the subtidal benthic macrofauna and its relationship with environmental factors using geostatistical tools: a case study in Trapandé Bay, southern Brazil. ZOOLOGIA 30:55–65

Spalding MD, Fox HE, Allen GR et al (2007) Marine ecoregions of the world: a bioregionalization of coastal and shelf areas. Bioscience 57:573–583. https://doi.org/10.1641/B570707

Speranzini BT (2017) On physical oceanography of the Santos estuary. Dissertation, Delft University of Technology

Spier D, Gerum HLN, Noernberg MA, Lana PC (2016) Flood regime as a driver of the distribution of mangrove and salt marsh species in a subtropical estuary. J Mar Syst 161:11–25. https://doi.org/10.1016/j.jmarsys.2016.05.004

Stanski G, Mantelatto FL, Leão-Castilho A (2016) Hermit crab bycatch fauna (Decapoda, Anomura) off the coast of Santa Catarina state, Brazil: diversity and spatial-temporal distribution. Lat Am J Aquat Res 44:546–556. https://doi.org/10.3856/vol44-issue3-fulltext-13

Suzuki KN, Machado EC, Machado W et al (2012) Selenium, chromium and cobalt diffusion into mangrove sediments: radiotracer experiment evidence of coupled effects of bioturbation and rhizosphere. Water Air Soil Pollut 223:3887–3892. https://doi.org/10.1007/s11270-012-1157-x

Tararam AS, Wakabara Y (1987) Benthic fauna living on *Spartina alterniflora* of Cananéia estuarine region (25°02'S - 47°56'W). Bol do Inst Ocean 35:103–113. https://doi.org/10.1590/S0373-55241987000200001

Tavares M, Mendonça J (1996) *Charybdis hellerii* (a. Milne Edwards, 1867) (Brachyura: Portunidae), eighth nonindigenous marine decapod recorded from Brazil. Crustac Res 25:151–157

Teixeira VL, Almeida SSS, Fleury BG et al (1984a) Comunidades de algas bentônicas e fauna acompanhante de poças de maré de Niterói (RJ). In: Anais do IV Congresso da Sociedade Brasileira de Botânica. Taubaté, pp 93–97

Teixeira VL, Pereira RC, Almeida SAS, Marques ANJ (1984b) Estudo comparativo de populações de *Codium decorticatum* (Woodward) Howe em Niterói (RJ), em diferentes condições ambientais. In: Anais do IV Congresso da Sociedade Brasileira de Botânica. Taubaté, pp 87–91

Teixeira VL, Pereira RC, Marques AN et al (1987) Seasonal variations in infralitoral seaweed communities under a pollution gradient in Baia de Guanabara, Rio de Janeiro (Brazil). Cienc Cult 39:423–428

Thomas MC (2011) Dispersão e fluxo gênico de metapopulações de nematoides marinhos de vida livre. PhD Thesis, Universidade Federal do Paraná

Thomas MC, Lana PC (2011) A new look into the small-scale dispersal of free-living marine nematodes. Zool 28:449–456. https://doi.org/10.1590/S1984-46702011000400006

Tommasi LR (1970) Observações sobre a fauna bêntica do Complexo Estuarino-Lagunar de Cananéia (SP). Bol Do Inst Ocean 19:43–56

Tommasi LR (1971) Equinodermes do Brasil: II. equinodermes da baía do Trapandé, situada no complexo estuarino-lagunar de Cananéia, SP. Bol do Inst Ocean 20:23–26. https://doi.org/10.1590/S0373-55241971000100002

Tschá MK, Bachmann L, Abilhoa V, Boeger WA (2017a) Past connection and isolation of catchments: the sea-level changes affect the distribution and genetic variability of coastal freshwater fishes. Estuar Coast Shelf Sci 190:31–39. https://doi.org/10.1016/j.ecss.2017.02.030

Tschá MK, Baggio RA, Marteleto FM et al (2017b) Sea-level variations have influenced the demographic history of estuarine and freshwater fishes of the coastal plain of Paraná, Brazil. J Fish Biol 90:968–979. https://doi.org/10.1111/jfb.13211

Van Der Ven PH (2005) Distribuição espacial e temporal de crustáceos decápodes de substratos não consolidados sublitorais da Baía de Guanabara, Rio de Janeiro, RJ, (Brasil). Dissertation, Universidade Federal Fluminense

Varoli FMF (1990) Avaliação da macrofauna bentônica da zona do entremarés de dois baixios do Sistema Estuarino-Lagunar de Iguape-Cananéia. Bioikos 4:24–39

Vicente MMDR (2008) A meiofauna sublitoral do complexo estuarino de Paranaguá (Paraná, Brasil): composição, distribuição e variabilidade temporal. Dissertation, Universidade Federal do Paraná

Vieira J (2007) Variação espacial da macrofauna bentônica no sublitoral inconsolidado da Baía da Babitonga (Santa Catarina, Brasil). Bachelor Thesis, Universidade da Região de Joinville

Westphal GGC, Ostrensky A (2016) Use of side-scan sonar for estimations of *Crassostrea brasiliana* (Lamarck, 1819) stocks in subtidal banks on the south coast of Brazil. Braz J Oceanogr 64:49–56. https://doi.org/10.1590/S1679-87592016104506401

Wunderlich AC, Pinheiro MAA, Rodrigues AMT (2008) Biology of the mangrove Uça crab, *Ucides cordatus* (Crustacea: Decapoda: Brachyura), in Babitonga Bay, Santa Catarina, Brazil. Rev Bras Zool 25:188–198. https://doi.org/10.1590/S0101-81752008000200005

Xavier E, Gama B, Porto T et al (2008) Effects of disturbance area on fouling communities from a tropical environment: Guanabara Bay, Rio de janeiro, Brazil. Braz J Oceanogr 56:73–84

Yoneshigue-Valentin Y, Fernandes DRP, De-Paula JC (2012) Macroalgas marinhas. In: Meniconi MFG, Silva TA, Fonseca ML et al (eds) Baía de guanabara: sintese do conhecimento ambiental, 1st edn. PETROBRAS, Rio de Janeiro, pp 1–479

Chapter 6
Benthic Estuarine Assemblages from the Southern Brazilian Marine Ecoregion

Sérgio A. Netto, Paulo R. Pagliosa, André Colling, Alessandra L. Fonseca, and Kalina M. Brauko

Abstract This chapter examines the estuaries within the Southern Brazilian Marine Ecoregion (SBE), a coastline of some 850 km in the tropical-temperate transition zone of the extreme southern portion of Brazil. A particular attention is given to the ecology of benthic communities, considering their structure, functioning, impacts, and conservation issues. In a microtidal and wave dominated environment, there are tens of small coastal plain estuaries and lagoons along the SBE. Although there is much overlapping of characteristics between small estuaries and lagoons, differences in the freshwater influence, spatial organization, and variability of gradients constitute important drivers since they directly affect species composition, abundance, and dominance of ecological guilds. This is particularly true for estuaries with intermittently open inlets, which are dominant in SBE. El Nino Southern Oscillation (ENSO) has a strong influence on the regional climate, affecting interannual estuarine hydrodynamics, benthic fauna distribution and recruitment. The environmental quality status of 27 SBE estuarine systems according to the AZTI Marine Biotic Index (AMBI index) was largely classified as moderately disturbed (80%),

S. A. Netto (✉)
Laboratório de Ciências Marinhas, Universidade do Sul de Santa Catarina, Tubarão, SC, Brazil
e-mail: sergio.netto@unisul.br

P. R. Pagliosa · K. M. Brauko
Laboratório de Biodiversidade e Conservação Marinha, Núcleo de Estudos do Mar, Universidade Federal de Santa Catarina, Florianópolis, SC, Brazil
e-mail: paulo.pagliosa@ufsc.br

A. Colling
Laboratório de Ecologia de Invertebrados Bentônicos, Instituto de Oceanografia, Universidade Federal do Rio Grande, Rio Grande, RS, Brazil
e-mail: lcolling@furg.br

A. L. Fonseca
Laboratório de Oceanografia Química, Coordenação Especial de Oceanografia, Universidade Federal de Santa Catarina, Florianópolis, SC, Brazil
e-mail: alessandra.larissa@ufsc.br

P. C. Lana, A. F. Bernardino (eds.), *Brazilian Estuaries*, Brazilian Marine Biodiversity, https://doi.org/10.1007/978-3-319-77779-5_6

whereas 20% of the sites were of slightly or undisturbed classifications. Different categories of marine protected areas (MPAs) are implemented in the estuaries in the SBE. Although MPAs, the Brazilian Coastal Management Plan as well as other policies could foster the sustainable use of coastal resources and ecosystem services, difficulties in coastal planning and in policy implementation still allow unsustainable practices, impacts, and habitat loss over all SBE estuaries.

Keywords Coastal lagoons · Small coastal plain estuaries · Benthic fauna · Santa Catarina · Rio Grande do Sul

6.1 Introduction

The Southern Brazilian Marine Ecoregion (SBE) or Rio Grande Ecoregion (Spalding et al. 2007) encompasses about 850 km from the North of the Island of Santa Catarina (26°50′S), State of Santa Catarina, to the Chuí (33°44′S), on the extreme southern portion of Brazil, at the State of Rio Grande do Sul. The SBE has a typical microtidal environment along a highly embayed coast with predominance of rocky headlands in the north and extensive sand barriers in the south.

Estuaries of SBE typically have small catchments (except Patos Lagoon on the southern region) and are particularly well-endowed with small coastal plain estuaries and coastal lagoons. The small coastal plain estuaries typically have a funnel-shaped morphology and arise from marine flooding of the lower course of a myriad of small catchment rivers. Different from the sizeable estuaries of the East-Southeast Sector of Brazilian Coast, the small coastal plain estuaries have no granitic rocks at the entrance, but are fluvial valleys usually flourished by mangrove forests northward, and by saltmarshes southward of the SBE.

On the other hand, the southern Brazilian coastal lagoons are an evolving coastal landform that may go through a cycle from an open embayment to a partially back-barrier lagoon with progressive infilling, to a segmentation into small lagoons with unstable inlets, and then coastal lakes (Cooper 1994). Natural and gradual shifts from lagoons to temporarily open/closed lagoons and lakes (and vice versa) are long-term processes that result from large-scale (e.g., sea-level and climate changes) and local processes (e.g., sediment supply, alongshore drift, and coastal morphology). However, the environmental shift may also be hastened by anthropogenic activities at the ecological scale, such as hydrological management (Schock et al. 2014), artificially opening of lagoons (Netto et al. 2012), or modifications as a result of climate change (Chapman 2012). The combination of sediment accretion rates and sea level rise will determine the temporal volumetric capacity of the lagoon, its import/export status, and the resultant evolution. The relative importance of a particular process in a lagoon depends upon the local environmental setting in which the lagoon is located, and the evolutionary path followed by a lagoon depends upon the magnitude and relative importance of each of the operative processes. The dynamism of these forces promotes both long-term and short-term changes. In the long-term (seasons and years), it influences the connectivity with the sea, while in the short-term (tidal cycles), it affects the amount of seawater inflow.

Along the SBE there are 32 small coastal plain estuaries, 16 with mangrove forests occupying the quaternary fluvial facies, and 16 with no transitional land-sea vegetation or just sparse mangrove trees and patches of small beds of *Spartina alterniflora* saltmarsh limited to the mouth of the system (Fig. 6.1, Table 6.1). According to the present connectivity with the sea, the local coastal lagoons water bodies can roughly be divided into three major types: open (permanently connected to the sea), the intermittently open/closed (which includes seasonally or non-seasonally closed or those normally closed), and the closed (presently without connection with the sea, i.e., the coastal lakes). Today, along the SBE there are six lagoons permanently open to the sea (Barra Velha, Conceição, Laguna, Camacho, Tramandai-Armazém, and Patos), and seven with temporarily open/closed bars (Acaraí, Lagoinha do Leste, Ibiraquera, Urussanga-Velha, Garopaba, Sombrio, Peixe) (Fig. 6.1, Table 6.2), and tens of closed lagoons.

Lagoons and small coastal plain estuaries share many common features including vegetation and fauna. However, differences in the freshwater influence and spatial organization of estuarine gradients may introduce major local variability and directly affect species composition, abundance, and dominance of ecological guilds (Pérez-Ruzafa et al. 2011). While coastal plain estuaries are drowned river valleys with hydrodynamics mainly conducted by river flux, with low residence time and clear gradients from freshwater to marine conditions, coastal lagoons are inland shallow and shore-parallel water bodies relatively more isolated from the open sea by low-lying sand barriers. This is particularly true for temporarily open/closed lagoons. Coastal lagoons present more mixed waters and their hydrodynamics is mainly controlled by the wind, higher residence time, and intra-lagoon gradients hidden by complex patterns in substrate properties and freshwater/marine inputs.

Contrary to the small coastal plain estuaries, the southern Brazilian coastal lagoons vary widely in their shape, size, ecological processes, and in the quantity and detail of research as well. In common they share social values and problems. Research on SBE lagoons largely focused at permanently open lagoons (e.g., Sierra de Ledo and Soriano-Sierra 1999; Seeliger et al. 1997; Odebrecht et al. 2017) as they are under more pressure due to larger urban developments. However, more than a half of the existing lagoons along southern Brazil are temporary open/closed estuaries which recently come under growing pressures, particularly artificial inlet managing (e.g., Netto et al. 2012; Crippa et al. 2013; Netto and Fonseca 2017). The value of the natural goods and services provided by coastal lagoons are among the highest of natural ecosystems, and can be categorized as suggested by Anthony et al. (2009) as pragmatic, scholarly, inspirational, and tacit. These provisions include: (a) ecosystem services that indirectly support human uses, such as nursery grounds for species that support commercial fisheries and protection from storms; (b) scientific inquiry and historical study seeking to extend knowledge, as to improve management, sustainable use, and tourism revenues; (c) inspiration to different sorts of artistic expressions; (d) recreation, visual and sonorous enjoyment of scenery,

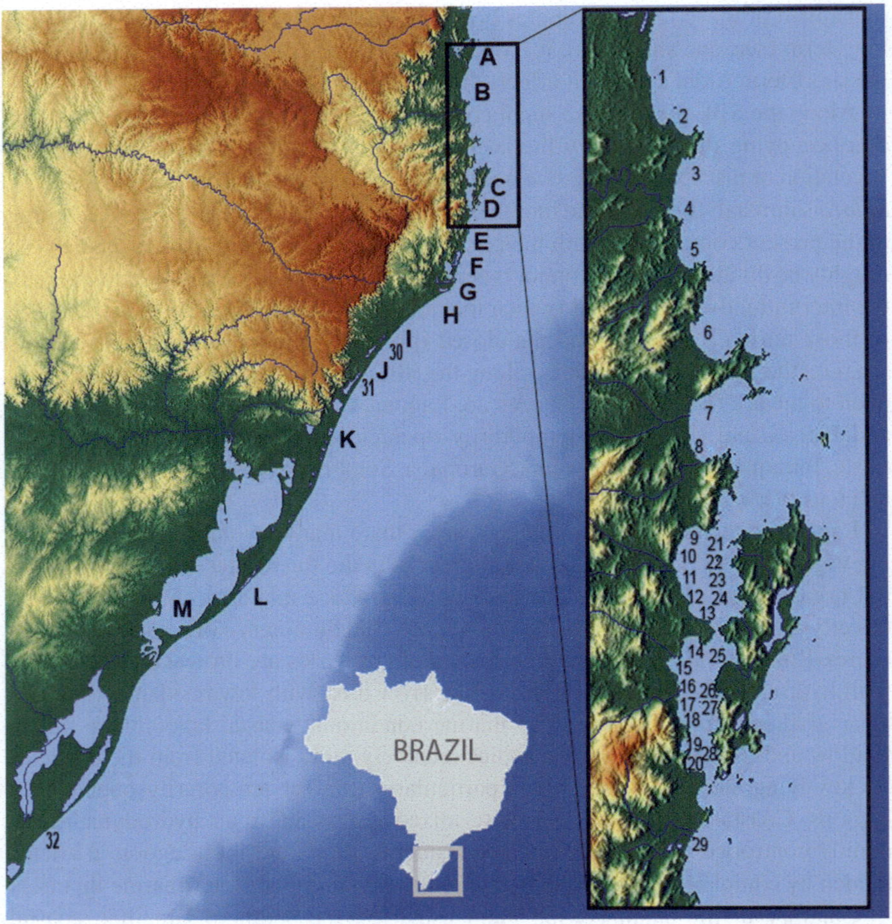

Fig. 6.1 Estuaries of the Southern Brazilian Ecoregion (SBE). Capital letters are coastal lagoons and numbers are coastal plain estuaries. Coastal lagoons = A Acaraí, B Barra Velha, C Conceição, D Lagoinha do Leste, E Garopaba, F Ibiraquera, G Laguna, H Camacho, I Urussanga-Velha, J Sombrio, K Tramandaí-Armazém, L Peixe, M Patos. Coastal plain estuaries = 1 Itajubá, 2 Piçarras, 3 Gravatá, 4 Itajaí, 5 Camboriú, 6 Perequê, 7 Tijucas, 8 Inferninho, 9 Cachoeira-Camarão, 10 Biguaçu, 11 Serraria, 12 Três Henriques, 13 Büchele, 14 Maruím, 15 Curtume, 16 Aririu, 17 Cubatão Sul, 18 Praia do Pontal, 19 Praia de Fora, 20 Massiambu, 21 Ratones-Papaquara, 22 Veríssimo, 23 Saco Grande, 24 Itacorubi, 25 Tavares, 26 Tapera, 27 Ribeirão, 28 Tapera da Caicanga-Açu, 29 Madre-Paulo Lopes, 30 Araranguá, 31 Mampituba, 32 Barra do Chuí. See Tables 6.1 and 6.2 for details

and sense of place. Despite their ecological and social relevance, SBE coastal lagoons are seriously threatened by pollution, eutrophication, inlet hydrological modifications, urbanization, aquaculture, and a variety of modification in their margins and watersheds, caused by fast and unplanned occupation practices within the lagoons' floodplains.

Table 6.1 Coastal plain estuaries of the Brazilian Southern Ecoregion (SBE) and marginal vegetation cover

Id.	Coastal plain estuary	Brazilian state	Vegetation cover
1	Itajubá	SC	Sparse mangrove trees and saltmarshes
2	Piçarras	SC	Sparse mangrove trees and saltmarshes
3	Gravatá	SC	Sparse mangrove trees and saltmarshes
4	Itajaí	SC	Sparse mangrove trees and saltmarshes
5	Camboriú	SC	Mangrove forest and saltmarshes
6	Perequê	SC	Mangrove forest and saltmarshes
7	Tijucas	SC	Currently with no vegetation
8	Inferninho	SC	Currently with no vegetation
9	Cachoeira-Camarão	SC	Sparse mangrove trees and saltmarshes
10	Biguaçu	SC	Sparse mangrove trees and saltmarshes
11	Serraria	SC	Currently with no vegetation
12	Três Henriques	SC	Currently with no vegetation
13	Büchele	SC	Currently with no vegetation
14	Maruím	SC	Currently with no vegetation
15	Curtume	SC	Mangrove forest and saltmarshes
16	Aririu	SC	Mangrove forest and saltmarshes
17	Cubatão Sul	SC	Mangrove forest and saltmarshes
18	Praia do Pontal	SC	Mangrove forest and saltmarshes
19	Praia de Fora	SC	Mangrove forest and saltmarshes
20	Massiambu	SC	Mangrove forest and saltmarshes
21	Ratones-Papaquara	SC	Mangrove forest and saltmarshes
22	Veríssimo	SC	Mangrove forest and saltmarshes
23	Saco Grande	SC	Mangrove forest and saltmarshes
24	Itacorubi	SC	Mangrove forest and saltmarshes
25	Tavares	SC	Mangrove forest and saltmarshes
26	Tapera	SC	Mangrove forest and saltmarshes
27	Ribeirão	SC	Mangrove forest and saltmarshes
28	Tapera da Caicanga-Açu	SC	Mangrove forest and saltmarshes
29	Madre-Paulo Lopes	SC	Sparse mangrove trees and saltmarshes
30	Araranguá	SC	Sand bar with no vegetation
31	Mampituba	SC-RS	Sparse saltmarshes
32	Barra do Chuí	RS	Sparse saltmarshes

Refer to Fig. 6.1 for location of the lagoons along SBE
SC Santa Catarina, *RS* Rio Grande do Sul

6.2 Environmental Settings

The SBE is a transitional climatic zone between the South subtropical to humid temperate, or according to updated Brazilian climate map of Köppen (Alvares et al. 2014), it has Cfa type climate (temperate, without dry season, with hot summer). During the spring and summer months, the South Atlantic Tropical Anticyclone, a

Table 6.2 Coastal lagoons of the Brazilian Southern Ecoregion (SBE), indicating the type (open or intermittent), vegetation cover, and area (km^2)

Id.	Lagoon	Brazilian state	Type	Vegetation cover	Area (km^2)
A	Acaraí	SC	Intermittent	Mangrove forest and saltmarshes	4.3
B	Barra Velha	SC	Open	Mangrove forest and saltmarshes	2.15
C	Conceição	SC	Open	Sparse mangrove trees and saltmarshes	20.6
D	Lagoinha do Leste	SC	Intermittent	Sparse saltmarshes	0.1
E	Garopaba	SC	Intermittent	Sparse saltmarshes	4.36
F	Ibiraquera	SC	Intermittent	Sparse saltmarshes	7.92
G	Laguna	SC	Open	Sparse mangrove trees and saltmarshes	182
H	Camacho	SC	Open	Saltmarshes	25.4
I	Urussanga-Velha	SC	Intermittent	Saltmarshes	1.84
J	Sombrio	SC	Intermittent	Saltmarshes	44
K	Tramandaí-Armazém	RS	Open	Sparse saltmarshes	12.8
L	Peixe	RS	Intermittent	Sparse saltmarshes	50
M	Patos	RS	Open	Saltmarshes	10,144

Refer to Fig. 6.1 for location of the lagoons along SBE
SC Santa Catarina, *RS* Rio Grande do Sul

high-pressure center that produces a wet and warm air mass, predominates at the region, generating NE winds. During autumn and winter, the passage of cold fronts associated with the displacement of the Polar Migratory Anticyclone generates S/SE winds. The total annual precipitation in the ecoregion ranges from 1000 to 1500 mm and it is, in general, evenly distributed throughout the year. The mean air temperatures are lower than 18 °C in the winter, and in the summer lower than 27 °C (Bernardino et al. 2015). El Nino Southern Oscillation (ENSO) has a strong influence on the regional climate, with significant differences in precipitation rates, and on the intensity and direction of prevailing winds between years with positive and negative southern oscillation indices (Grimm et al. 2003). During El Nino events (positive ENSO phase), rainfall is often above normal and the water flow from the estuary towards the coast increases. On contrary, during La Nina (negative ENSO phase) rainfall is often below normal, favoring the input of salt water into the lagoons (Odebrecht et al. 2010). Thus, ENSO strongly affects interannual estuarine hydrodynamics (Fernandes et al. 2002), variability of planktonic and benthic primary producers (Odebrecht et al. 2010), benthic fauna distribution and recruitment (Colling et al. 2007), and many ecologically and commercially important estuarine species (Seeliger 2001; Garcia et al. 2003; Vieira et al. 2008).

Diverse barrier types and associated surficial landforms occur along the SBE (Hesp et al. 2009; Dillenburg et al. 2009). At the northern portion, from north of the Island of Santa Catarina to Santa Marta Cape, the coastline is characterized by a

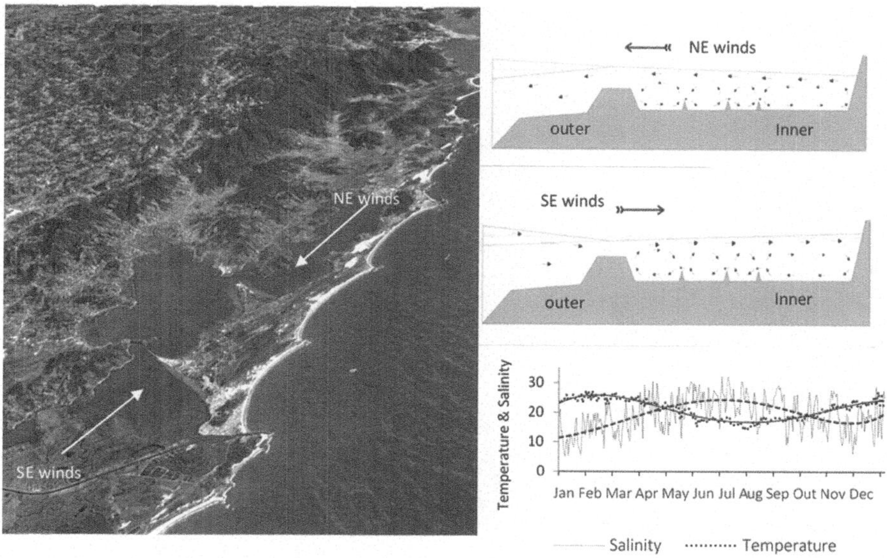

Fig. 6.2 Schematic representation of wind influence on lagoon circulation, and influence on temperature and salinity values (modified from Castelão and Moller 2003)

diversity of formations, such as small coastal bays and coves, lagoon inlets, large promontories, long and arched beaches, rocky shores and barriers dominated by transgressive dune fields. Southwards, the coastline changes direction and it is mostly a straight-line SW-NE orientated, characterized by long sand beaches and barriers dominated, in large scale, by active transgressive dune fields (see Hesp et al. 2009 and Short and Klein 2016 for details).

Shallowness and microtidal regime determine that the wind and precipitation exert key roles in the circulation and sediment movement of coastal plain estuaries and coastal lagoons. Sediment infill of some elongated and shore-parallel lagoons, such as Conceição, Laguna, and Patos, involves the development of a series of cuspate divisions (septation) due to wind waves that build spits isolating the lagoon into separate basin (Woodroffe 2002). At Patos lagoon, in the southern part of SBE, large catchments and seasonal precipitation regulate variations of freshwater discharge, with higher values during the austral winter/spring (July to October) and lower during the summer/autumn (November to June) (Garcia 1997). Wind effects on circulation is particularly important for the small and medium sized lagoons located at the north of the SBE, characterized by small catchments, and where there is a lack in rainfall differences over the year. Predominant NE winds during summer and spring force lagoon water masses towards the southern margins (seaward), resulting in a decreased salinity and increase of suspended materials from continental origin. In contrast, during winter and autumn, periods of strong S-SE winds favor marine water intrusion into the lagoons resulting in lower turbidity and higher salinity values (Fig. 6.2; Eichler et al. 2006; Meurer and Netto 2007).

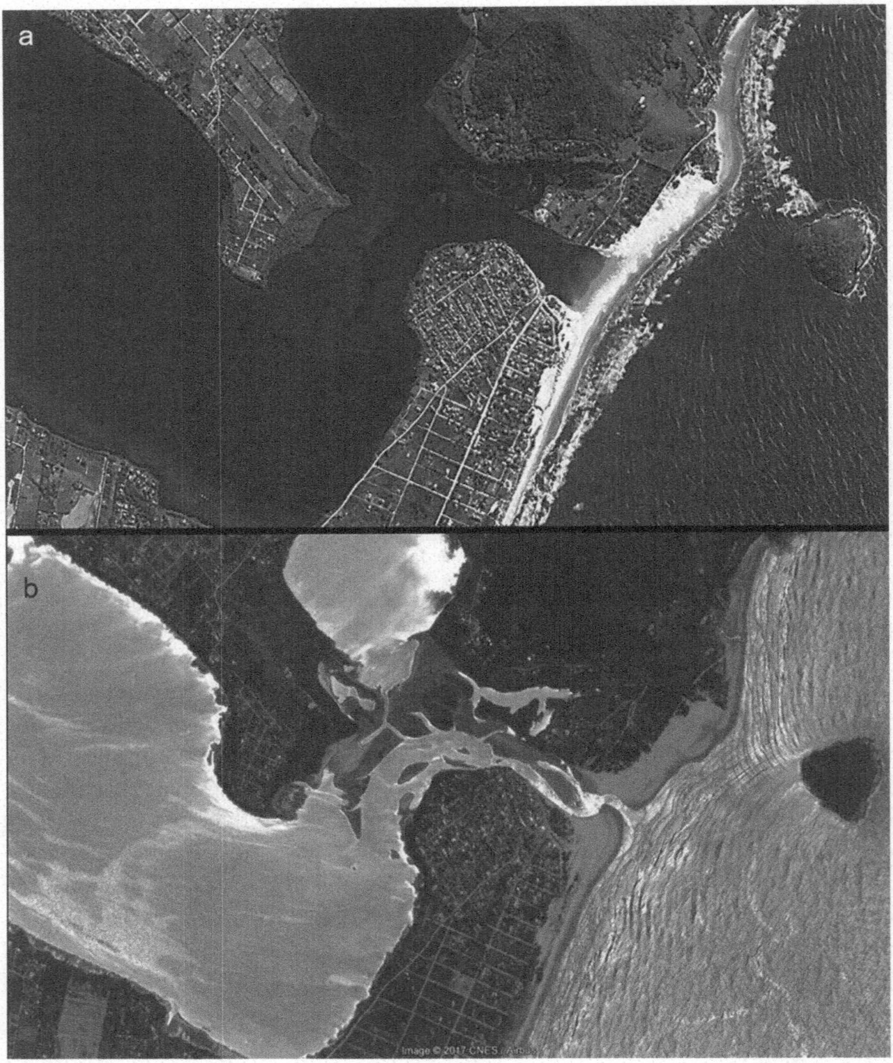

Fig. 6.3 Ibiraquera lagoon, a typical temporarily open/closed coastal lagoon of Brazilian Southern Ecoregion (SBE): (**a**) during isolation from the sea, lagoon water level increases and disturbance from wind and water generated turbulence are minimal, offering a fairly homogeneous and stable environment; (**b**) connectivity during open phase permits the exchange of matter and energy between the lagoons and the sea, exposes sand banks and generates strong physical-chemical gradients that directly affect the biota

The connectivity dynamics between the lagoons and the sea play a key role in their overall functioning, as intermittent closing (Fig. 6.3). The openness of the lagoon mouth is determined by the balance between scouring forces (primarily catchment runoff and tidal prism) and blocking forces (primarily onshore and longshore deposition of sediments; Whitfield and Bate 2007; McSweeney et al. 2017).

Permanently open lagoons, such as Patos, Laguna, and Conceição, are primarily characterized by a wide spatiotemporal range in environmental conditions (e.g., salinity, temperature, oxygen), biological productivity, and movement of resources and consumers with other adjacent marine areas (Pérez-Ruzafa et al. 2011; Abreu et al. 2017). Temporarily closed lagoons (Fig. 6.3) are intermittently isolated from the sea by the formation of a sand berm across the estuary mouth, and typically present a small river catchment and high rates of longshore and onshore sediment transport (Cooper et al. 1999; Schallenberg et al. 2010). Once closed, depending on the freshwater inputs and the time of closure, the system may become gradually fresher or even more saline due to evaporation rates. An intermittent open lagoon may lead to remarkable changes in its physical-chemical environment over short time periods, which in turn triggers major biological responses in both pelagic and benthic compartments (Niekerk et al. 2005; Netto et al. 2012).

The mineralization of organic matter usually is higher in the inner sector (oligohaline) of the estuaries, where most of the materials are retained. In permanently stratified coastal lagoons, as the Conceição lagoon, or stratified systems, as the Madre-Paulo Lopes estuary, haloclines coincide with oxyclines due to the elevated rates of allochthonous and autochthonous organic matter mineralization in bottom waters (Fonseca and Braga 2006; Fontes and Abreu 2009). Heterotrophic conditions (e.g., hypoxic and anoxic waters) associated to high macrophyte biomass and high turbidity have been described to coastal plain with stratified (Madre-Paulo Lopes) and homogeneous (Ratones-Papaquara) water column. Anoxic and hypoxic events were observed in low hydrodynamics and deep regions due to the geomorphology isolation. Those events were enhanced during periods of low rainfall and high water residence time. High precipitation events boosted the exportation of eutrophic estuarine waters to the ocean, which improves the water quality of the SBE systems. The estuaries became dominated by continental freshwaters and it is expected an increase of nutrients in shallow continental shelf, that might relate with toxic algal blooms (e.g., red tide) (Fernandes and Brandini 2010; Proença 2004; Omachi et al. 2007). On the other hand, meteorological tides caused by the southerly winds can increase water residence time of coastal systems, enhancing physical and chemical stratification of the water column and the eutrophication process (Abreu et al. 2010).

The changes in nutrient concentrations are related to the uptake by primary producers, retention in sediment, recycling times and sources (river discharge, domestic and industrial effluents), as observed in Patos and Conceição lagoons (Fonseca 2004; Haraguchi et al. 2015; Silva et al. 2017). Denitrification and phosphorus adsorption by the benthic compartment are important biogeochemical processes, both removing dissolved nutrients from water column and regulating eutrophication in these coastal lagoons (Fonseca 2004). In a temporal perspective, some lagoons, such as Patos and Conceição, show an increase in the concentrations of nitrogen and phosphorous (Niencheski et al. 2006; Haraguchi et al. 2015; Lanari and Copertino 2017), which suggests an eutrophication trend. A significant N:P ratio increase was also shown, suggesting that changes in land use and increased urbanization in the watershed are altering the nutrient delivery, which may impact phytoplankton composition (Haraguchi et al. 2015; Silva et al. 2017). Although the phytoplankton biomass (chlorophyll *a*) do not present clear temporal changes (Abreu et al. 2010; Silva et al. 2017), significant trends of increasing biomass of cyanobacteria and dinoflagellates are observed at interannual scale (Haraguchi et al. 2015; Fontes et al. 2011; Fonseca and Braga 2006). At Patos lagoon, diatoms are the main phytoplankton group and in the seasonal scale, the chlorophyll *a* variation is closely related to these microalgae, influenced by the short water retention time and low light availability during winter (Haraguchi et al. 2015). The combined effects of climate (ENSO and rainfall) and hydrology (freshwater discharge and water physical-chemical parameters) drive the onset and magnitude of drift macroalgal blooms in the Patos Lagoon estuary, where the wind action further controls the persistence and possible occurrence of large green tides (Lanari and Copertino 2017).

Shoreline vegetation is dominated by plants of the genera *Spartina*, *Scirpus*, and *Juncus* in all SBE lagoons (Soriano-Sierra 1999; Costa and Davy 1992, Valgas 2009). Well-structured mangrove forests are found only in the coastal plain estuaries. The typical species from mangroves are *Avicennia schaueriana*, *Laguncularia racemosa*, and *Rhizophora mangle*. Isolated species typical from mangroves (*A. schaueriana* and *L. racemosa*) can be found at the inlet of Conceição Lagoon (Soriano-Sierra 1999) and at Laguna, the southernmost distribution of mangroves vegetation in America. *A. schaueriana* is the dominant mangrove species in SBE, with up to 4333 stems/ha, basal area ranging from 3.8 to 6.3 m^2/ha, and height from 6 to 9 m in old-growth stands (Rovai et al. 2012). In the Itacorubi coastal plain estuary, the mangrove forests and saltmarsh productivity are 415 gC/m^2/year and 711 gC/m^2/year, respectively (Panitz 1986). In the southernmost lagoon, the mean net above- and belowground productivity of *Spartina alterniflora* ranges from 669 to 1928 gC/m^2/year, respectively, while the aboveground *Spartina densiflora* values reach about 2260 gC/m^2/year (Peixoto and Costa 2004; Cunha et al. 2005).

The euphotic zone of most SBE coastal lagoons reach the sediment as they are usually shallow (<2 m) with low particulate suspended material in water column, favoring the primary and secondary benthic production. Blooms of macroalgae *Ulva* sp. and *Enteromorpha* sp. are common along SBE lagoons. In Conceição lagoon, the productivity of those species may reach values up to 7000 gC/m^2/year, followed by the macrophyte *Ruppia* sp. (3042 gC/m^2/year) and the microphytobenthos (50 gC/m^2/year) (dos Santos et al. 2017; Fonseca 2004). In the central region of Conceição lagoon, where the halocline is permanent, it was described high rates of phytoplankton production (1080 gC/m^2/year) in deeper water. In this system, significative primary productions were measured under low solar radiance (438 gC/m^2/year) and oxygen concentrations (360 gC/m^2/year) which demonstrates the importance of carbon fixation by Cyanophyceae and anaerobic anoxygenic bacteria, respectively (Fontes et al. 2011; 2015). The phytoplankton production in the Patos Lagoon varies from 0.004 and 0.320 gC/m^2/h (or 18 gC/m^2/year and 1401 gC/m^2/year; Abreu et al. 1994) and are largely determined by seasonality and the contribution of flagellates (Seeliger et al. 1997). In this lagoon, estimates on *Ruppia maritima* productivity indicate net annual values ranging from 39.6 to 43.2 gC/m^2/year, and its variability is determined by species-specific seasonal growth cycles.

6.3 Benthic Fauna

Protected from the physical harshness of the open ocean, within calm, diverse, and plenty of food habitats, the benthic fauna thrives in estuaries. Large and productive populations of producers and primary consumers, maintained by the ability of estuaries to trap nutrients and food particles, support high biomass of benthic invertebrates. In contrast, changes in chemical conditions caused by local, marine, and riverine inputs have a relevant role in the structure of the estuarine benthic fauna. As natural stress levels increase with confinement, but also with the frequent oscillation between closeness and openness, estuarine benthic invertebrates are typically species poor when compared to adjacent seas.

Benthic invertebrates are an important component of any aquatic ecosystem, but shallowness amplifies their pivotal role in estuaries. They significantly contribute to energy flow and as diet of many other estuarine animals. The activity of some benthic organisms also increases the removal of fixed nitrogen from aquatic ecosystems by fostering nitrification and denitrification in the oxic–anoxic transition zone of the marine sediment (Bonaglia et al. 2014). Benthic organisms are largely used as biological indicators because they can provide information on environmental conditions either due to the sensitivity of assemblages or single species, or because of some general feature that makes them integrate environmental signals over a long period of time (Teixeira et al. 2010; Schratzberger and Ingels 2017).

6.3.1 Meiofauna

The meiobenthos or meiofauna comprises a particular group of microscopic invertebrates ranging from 0.03 mm to 0.5 mm in size. They represent the most abundant group of benthic metazoans. Along SBE estuaries, on average, their abundance ranges from 10^5 to 10^6 individuals per square meter, corresponding to a biomass of 1 to 2 g dry weight per square meter. The meiofauna is also more diverse than any other component of the marine biota with 24 phyla (out of a total of 35 known invertebrate phyla) with meiobenthic representatives. Despite of its importance, the current knowledge of the estuarine meiofauna along SBE is, so far, scarce and heterogeneous. Data on meiofauna with taxonomic resolution higher than phylum or class are only available for Santa Catarina state, northern portion of SBE.

Netto and Fonseca (2017) made the only attempt to comparatively examine the meiofauna of several coastal lagoons along the Brazilian coast. A large-scale sampling program was carried out from Florianopolis (Santa Catarina State) to Torres (Rio Grande do Sul State). The study investigated to what extent the differences in openness of coastal lagoons structure meiofauna communities. Using particular lagoon status (open or closed) as replicates of their evolution over time, they showed, using free-living nematodes as surrogate, that open and closed lagoons are mutually exclusive alternative states of equilibrium, and that temporarily open/closed lagoons are an intermediate or transition phase between them. The gradient in the structural connectivity between lagoons and the sea, due to their regime shifts, changes the movement of resources and consumers, and the internal physical-chemical gradients that directly affect the meiofaunal species diversity, abundance, and trophic status.

The number of nematode genera and taxa diversity are generally higher in the open lagoons and coastal plain estuaries, intermediate in temporarily closed lagoons, and lower at permanently closed ones (Table 6.3 and Fig. 6.4). Density is normally higher in lagoons (open or temporally closed), decreasing in coastal plain estuaries and lower in closed lagoons. Differences within each estuarine typology increase with openness. These patterns are a result of combined effects of the input of marine species, the presence of environmental gradients, and higher environmental heterogeneity (Netto and Fonseca 2017). Overall, freshwater nematode assemblages are impoverished when compared to marine and brackish systems. In turn, the transitional structure of the nematode assemblages of temporarily open/closed lagoons reflects the intermediate pattern of isolation compared to lagoons and coastal lakes.

Most of the nematode genera recorded in open lagoons are those typically found in coastal plain estuaries. The meiofauna of and the small coastal plain river estuaries (e.g., Ratones estuary; Netto and Gallucci 2003) shows relative similarity with SBE open lagoons, though the percentage of marine species is higher (Table 6.3). The proportion of brackish or marine species is reduced in open/closed lagoons with an increasing number of brackish/freshwater or freshwater genera. In closed lagoons, freshwater or brackish/freshwater genera largely dominate. Along the coastal lagoons of Santa Catarina, only five genera occurred in all the three types of lagoons, namely *Anonchus* (Aphanolaimidae), *Anoplostoma* (Anoplostomatidae), *Desmodora* (Desmodoridae), *Dichromadora,* and *Hypodontolaimus* (Chromadoridae) (Netto and Fonseca 2017). Besides, the percentage of exclusive genera (those found exclusively in only one type of lagoon) decreased with increasing connectivity (Table 6.3).

Changes in the structural connectivity between coastal lagoons and the sea affect their trophic status (Fig. 6.5). The impoverishment of the nematode assemblages and the substitution of brackish water species by freshwater species due to the decrease in the connectivity promotes a decrease of trophic diversity (an index based on the proportion of each feeding type; Heip et al. 1985). While closed lagoons were largely dominated by predator/omnivores, more opened lagoons were numerically dominated by nonselective deposit-feeders and epigrowth feeders (Fonseca and Netto 2006).

Similarities of the nematode assemblages within and between lagoons of SBE also change according to the stable state (Netto and Fonseca 2017). While habitat connection and faunal exchange by open inlets increased similarity between more connected lagoons, with variations in the composition controlled by gradients, isolation increased variability of nematode assemblages among lagoons (β-diversity) with an increasing dominance of species replacement over richness (Fig. 6.5). At the same time, internal faunal variability (e.g., differences between outer and inner

Table 6.3 Summary characteristics of nematode assemblages in closed lagoons, intermittently open/closed lagoons, open lagoons, and coastal plain estuaries along the Brazilian Southern Ecoregion (SBE)

	Closed lagoon	Temporarily open/ closed lagoon	Open lagoon	Coastal plain estuary
Number of genera	20	73	69	86
Freshwater genera (%)	65	22	1	1
Freshwater/brackish genera (%)	30	30	28	12
Brackish/marine genera (%)	5	48	71	87
Exclusive genera (%)	65	35	30	30
Most frequent genera (%)	*Semitobrilus* (60) *Trischistoma* (48)	*Desmodora* (81) *Theristus* (80)	*Desmodora* (78) *Theristus* (77)	*Haiplectus* (95) *Anoplostoma* (95)
Total density (inds.10/cm^2)	3–407 (57)	9–5474 (678)	6–5283 (674)	2–2563 (115)
Trischistoma (inds.10/cm^2)	29	-	-	-
Semitobrilus (inds.10/cm^2)	21	-	-	-
Ironus (inds.10/cm^2)	5.2	0.07	-	-
Microlaimus (inds.10/cm^2)	-	143	80	2
Spirinia (inds.10/cm^2)	-	108	185	2
Desmodora (inds.10/cm^2)	0.04	73	37	12
Terchellingia (inds.10/cm^2)	-	8	8	61
Haliplectus (inds.10/cm^2)	-	-	1	149
Anoplostoma (inds.10/cm^2)	0.07	51	44	124

Exclusive genera percentages genera found exclusively in lagoon types or estuary; frequent genera minimum–maximum densities of nematodes (and average inds.10/cm^2); the most abundant genera (mean inds.10/cm^2); "-" no occurrence. Data from Netto and Fonseca (2017) and Netto and Gallucci (2003)

portions of lagoons) is higher within open lagoons than in closed lagoons, with temporarily open/closed lagoons assuming an intermediate position. This pattern may emerge because of the connectivity that modulates the degree to which the inlet state facilitates or impedes the exchange of matter, energy, and specimens among landscape elements. Besides, differences in structural connectivity can lead to internal homogeneity or strong physical-chemical gradients that directly affect species

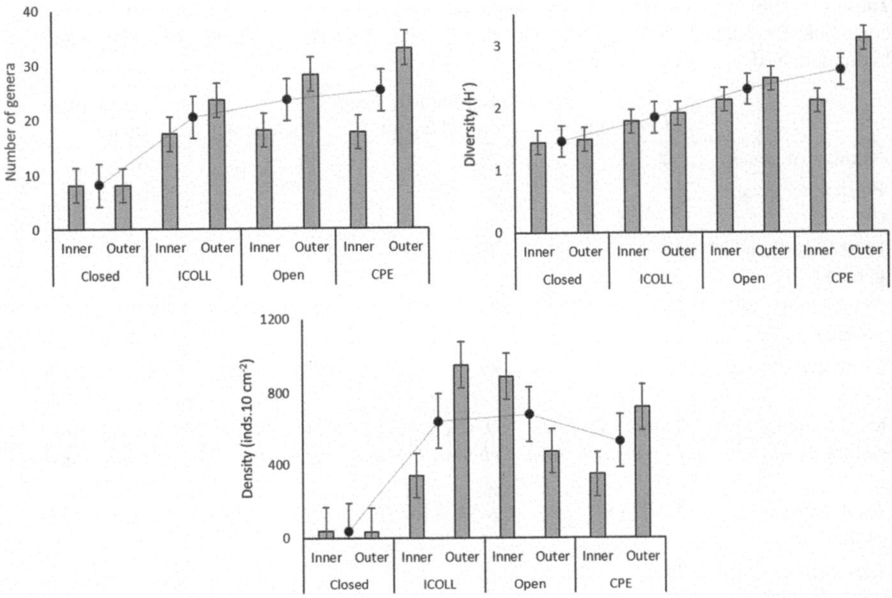

Fig. 6.4 Mean (±SE) of nematode number of genera, diversity (H′), and density (inds.10/cm^2) in inner and outer portions (columns) within closed lagoons, intermittently open/closed lagoons (ICOLL), open lagoons, and coastal plain estuary (CPE) in the Brazilian Southern Ecoregion (SBE). *Black dots* represent the total mean value (±SE) of the community descriptors in each type of estuary

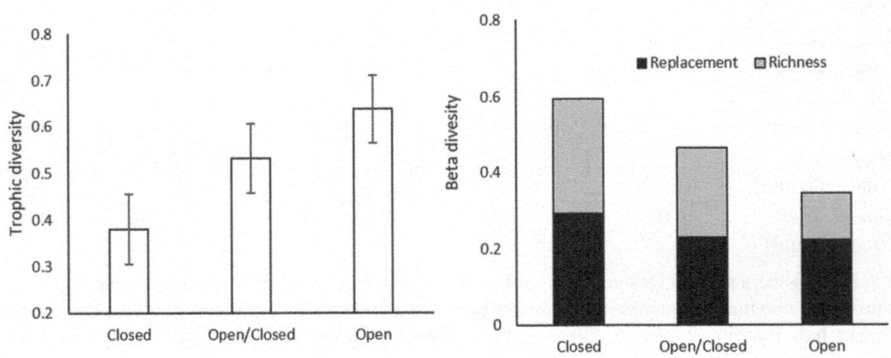

Fig. 6.5 Mean value (±SE) of the index of trophic diversity and nematode variation in genera composition (total mean β-diversity, contribution of replacement and richness differences) in closed lagoons, intermittently open/closed lagoons, open lagoons, and coastal plain estuaries in the Brazilian Southern Ecoregion (SBE) (reproduced from Netto and Fonseca 2017)

composition. While the low variability of nematode assemblages among lagoons is likely to be a result of faunal transport due to their physical link, the high dissimilarities of the assemblages between coastal lakes might be consequence of their spatially disconnection and exposure to discrete and variable surroundings. Thus, as the lagoons loose connectivity, gradually shifting the state, local processes become increasingly more important in structuring these communities than large-scale drivers.

Punctual studies focused on spatial variability of meiofaunal assemblages showed that the most important structuring variables on an individual lagoon scale are salinity and sediment characteristics, which are in turn largely modulated by the hydrodynamic conditions (Kapusta et al. 2002, 2004, 2005; Fonseca and Netto 2006). Besides, the amount of mangrove-derived detritus positively affects the diversity and density of the meiofauna in small coastal plain estuaries bordered by mangroves, indicating that nematodes may show a high degree of specificity in the microhabitat choice and high efficiency in exploiting these microhabitats originated by mangrove litter decomposition (Netto and Gallucci 2003).

6.3.2 Macrofauna

Connectivity of benthic macrofaunal communities (invertebrates retained by ≥0.5 mm mesh size) within and between lagoons and small coastal plain estuaries in SBE suggests that these organisms are exposed to similar structuring drivers. The dispersion and colonization of the macrofauna occurs primarily on the water column with a major passive component throughout large distances. The connections among the many small systems via adjacent ocean facilitate the exchange of individuals among them. As a result, benthic macrofaunal communities of local estuarine systems show similar species composition and richness, differing somewhat in both the spatiotemporal distribution of abundances and from the present closed lagoons.

The macrobenthic community at the inner portion of coastal plain estuaries is generally more abundant than at outer portions of estuaries (Rosa Filho and Bemvenuti 1998; Netto and Gallucci 2003; Pagliosa and Barbosa 2006). Species composition is therefore typically transitional, meaning that these regions have a key role in exchanges between the watershed and the marine system. In fact, the muddy sediments at inner estuarine portions represent a shifting gradient from the gross particles in freshwater habitats to finer sandy sediments in outward estuaries. Notwithstanding, the outer estuaries present the higher species richness, a pattern more related with the greater water renovation at the entrance of the system. Contrary to coastal plain estuaries, differences in species numbers and abundances along the main axis of coastal lagoons may present a more complex pattern, with intra-lagoon gradients (Bemvenuti and Netto 1998; Bemvenuti and Rosa Filho 2000; Fonseca and Netto 2006; Netto et al. 2012). Still, although such general patterns are variable among lagoons, many open and intermittently open/closed lagoons show higher species richness and/or densities at the outermost portions (i.e., Patos, Ibiraquera, Conceição, Laguna, and Lagoinha do Leste). Those spatial patterns are probably related to the site-specific inputs of salt water, presence of aquatic vegetation, and the heterogeneity of sedimentary habitats found within lagoons.

Polychaete annelids and oligochaetes are the richest taxonomical groups in SBE estuarine systems, followed by bivalves, gastropods, crustaceans, and insects. Altogether, 52 species of polychaetes, 19 of crustaceans, 14 of bivalves, 11 of oligochaetes, and 7 of gastropods, as well as 50 species of insects, have been reported for local lagoons and coastal plain estuaries. These numbers should yet be seen with caution because they represent a consensus among local experts in which just some of the previous taxa were well studied in terms of taxonomy, biology, and ecology. Species richness ranged from 41 to 74, but the mean richness by sample is very similar. The highest densities occur in the open lagoons while the lowest in the coastal plain estuaries, where the species diversity is high (Table 6.4, Fig. 6.6). The numerically dominant and most frequent macrofaunal species show similar patterns among open/closed lagoons, open lagoons, and coastal plain estuaries.

Table 6.4 Summary characteristics of macrobenthic communities in closed lagoons, intermittently open/closed lagoons, open lagoons, and coastal plain estuaries along the Brazilian Southern Ecoregion (SBE)

	Closed lagoon	Temporarily open/closed lagoon	Open lagoon	Coastal plain estuary
Number of species	67	41	74	53
Exclusive species (%)	51	02	34	14
Most frequent species	*Caladomya ortoni* *Fissimentum* sp. *Polypedilum* sp.	*Heteromastus similis* *Laeonereis acuta* *Nephtys fluviatilis*	*Heleobia australis* *Heteromastus similis* *Monocorophium acherusicum*	*Heteromastus similis* *Monokalliapseudes schubarti* Tubificidae
Mean total density (ind/m^2)	5813	8326	20,770	1960
Annelida				
Oligochaeta				
Bothrioneurum sp. (ind/m^2)	522	-	-	-
Polychaeta				
Nephtys fluviatilis (ind/m^2)	-	730	998	178
Laeonereis acuta (ind/m^2)	-	599	865	25

(continued)

Table 6.4 (continued)

	Closed lagoon	Temporarily open/closed lagoon	Open lagoon	Coastal plain estuary
Capitella spp. (ind/m²)	-	555	294	11
Heteromastus similis (ind/m²)	-	553	707	512
Arthropoda				
Crustacea				
Sinelobus stanfordi (ind/m²)	1553	-	-	-
Monokalliapseudes schubarti (ind/m²)	-	658	777	778
Monocorophium acherusicum (ind/m²)	-	585	371	01
Insecta				
Aedokritus sp. (ind/m²)	1653	-	-	-
Goeldichironomus sp. (ind/m²)	744	-	-	-
Mollusca				
Bivalvia				
Erodona mactroides (ind/m²)	-	493	1083	-
Gastropoda				
Heleobia australis (ind/m²)	57	2922	13,484	01

"-" no occurrence

The gastropod *Heleobia australis*, a small opportunistic species (around 5 mm), occurs in all estuaries of SBE, including the freshwater closed lagoons. *H. australis* may be particularly abundant in vegetated sites, including drifting algae. This species displays efficient morphophysiological strategies of dispersion and occupation of non-colonized areas (i.e., fluctuation by retaining air inside their light shells) that allow them to numerically dominate most of the SBE lagoons. This species can be considered tolerant to pollution, reaching up to 300,000 ind/m² near urban centers (Danulat et al. 2002).

In the three typical estuarine systems of SBE (coastal plain estuaries, open/closed lagoons, and permanently open lagoons), the polychaetes *Heteromastus similis*, *Laeonereis acuta*, and *Nephtys fluviatilis* are often recorded in high numbers. They inhabit the subtidal and intertidal zones in mangroves, salt marshes, and tidal flats. At the mangrove, the tentacle-crowned suspension-feeder polychaete *Manayunkia brasiliensis* is an endemic species in southern and southeastern Brazil with a highly patched distribution (Pagliosa et al. 2016). The polychaete *Nephtys fluviatilis*, usually found in less saline waters, is an active carnivorous and deposit feeder, predating

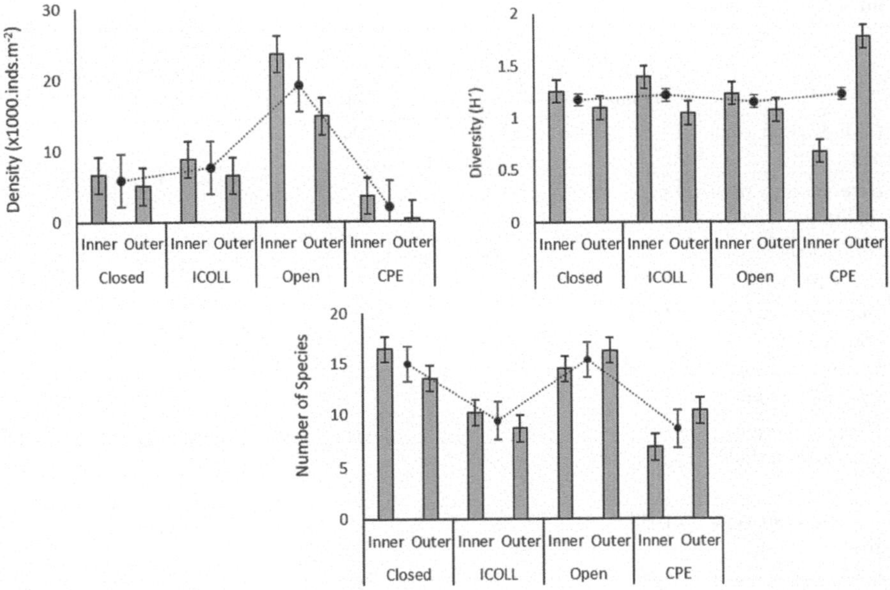

Fig. 6.6 Mean (± SE) macrofaunal species richness, Shannon diversity (H′), and density (ind/m^2) in inner and outer portions (columns) within closed lagoons, Intermittently Closed/Open Lagoons (ICOLL), open lagoons, and Coastal Plain Estuaries (CPE) in the Brazilian Southern Ecoregion (SBE). *Black dots* represent the total mean value (±SE) of the descriptors in each type of estuary

on *L. acuta* and *H. similis* (Bemvenuti 1987a), meiofaunal ostracods and nematodes. The infaunal *L. acuta* and *H. similis* escape predation in deeper (approximately 20 cm) sediment strata. In superficial sediments, the juveniles of *L. acuta* are under stronger predation pressure, which controls species abundance (Bemvenuti 1992). The major predators of infaunal species in shallow mudflats are the small crab, *Rhithropanopeus harrisii* and juvenile crabs *Cyrtograpsus angulatus* and *Callinectes sapidus* and the shrimp *Farfantepenaeus paulensis*, that use shallow areas during warm months as nursery grounds (Bemvenuti 1987b; Bemvenuti 1997a).

The amphipod *Monocorophium acherusicum* and the tanaid *Monokalliapseudes schubarti* are very abundant in open lagoons and coastal plain estuaries. The tanaid shows at least six annual cohorts (Fonseca and D'Incao 2003), forming U-shaped tubes up to 15 cm deep in intertidal and shallow water muddy sands. It is a deposit and filter-feeder (Montagnolli et al. 2004; Freitas-Júnior et al. 2013), being the main prey for fishes (*Micropogonias furnieri* and *Odontesthes bonarienses*) and other crustaceans (*F. paulensis, C. sapidus,* and *Neohelice granulata*). The species also plays an essential role in estuarine trophic webs by transferring energy from the detritus compartment to higher trophic levels.

The bivalve *Erodona mactroides* occurs in coastal lagoons, from Laguna to southernmost of the SBE (Bemvenuti 1997b; Netto and Pereira 2008). They occur in subsurface sediments and can reach densities of more than 20,000 ind/m^2 in shallow sublittoral areas. Their adult stocks are usually located at inner lagoon (or near

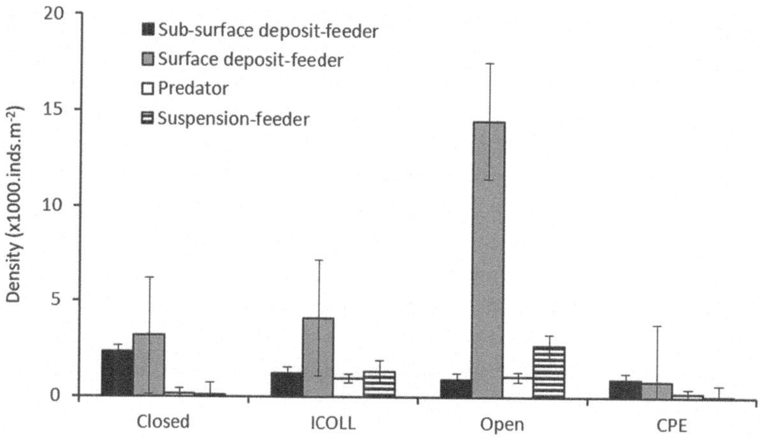

Fig. 6.7 Mean value (±SE) of the trophic diversity and composition of macrofaunal assemblages in closed lagoons, intermittently open/closed lagoons, open lagoons, and coastal plain estuaries in the Brazilian Southern Ecoregion (SBE)

sites with freshwater input) but the larvae can be carried following freshwater discharges, reaching the sheltered mudflats where they successfully recruit. In the upper estuary, all size classes are present (larger individuals reaching up to 30 mm) but at lower densities (461 ind/m^2), reaching an average biomass of 281 g/m^2, while in the lower estuary, despite the higher densities, the average biomass rarely exceeds 105 g/m^2 (Bemvenuti et al. 1978). The key ecological role of this species is linked to its importance as one of the main prey for decapods such as *Callinectes sapidus*. Moreover, *E. mactroides* also influence the occurrence of the estuarine cirriped *Amphibalanus improvisus*, which in the absence of any other hard substrate, attaches to their shells (Bemvenuti 1997b).

The chironomids *Caladomya ortoni, Fissimentum* sp., and *Polypedilum* are the most frequent species, and along with *Aedokritus*, Goeldichironomus, the crustacean *Sinelobus stanfordi* and the oligochaete *Bothrioneurum* numerically dominate the bottoms of closed lagoons. In general, chironomid species in coastal lakes presents a patchy distribution related with the food availability (Lemes-Silva et al. 2014, 2016). Macrofaunal predators are highly dependent on variations of prey abundance (like the ostracod species *Stenocypris major* and *Cytheridella ilosvayi*, other chironomids and copepods), while macrofaunal herbivores usually forage on the abundant resources found in soft-bottoms throughout the year (wood detritus, coarse and fine particulate organic matter, and microphytobenthos).

The macrofauna from SBE estuaries is predominantly composed of microphages (subsurface deposit-feeders, surface deposit-feeders, and suspension-feeders). Microphages handle food particles in the bulk using only part of the feeding apparatus, while macrophages are nimble organisms that move through sediments in search of their prey. Among the microphages, surface deposit-feeders and suspension-feeders are relatively more abundant in open lagoons, whereas subsurface deposit-feeders dominate in closed lagoons (Fig. 6.7). The occurrence patterns of feeding groups among estuarine types in SBE seem to be related with a fine scale composition and distribution of the resources in sediments, and with the organisms' ability to handle food (Pagliosa et al. 2005). Surface deposit-feeders and suspension-feeders conduct food towards the mouth using their grooved palps or tentacular crowns with branched and ciliated lobes. In contrast, subsurface deposit-feeders swallow bulk particles with less handling of food. Microphages that extensively handle food tend to inhabit fine sands, feeding sites with high phytoplankton and microphytobenthos productivity. However, these patches of high food quality may be ephemeral, that is, environments where the nutritional status changes quickly due to continuous material flux at the sediment surface. On the other hand, microphages would occur in more protected sites. They consist of species that ingest sediments without much effort in sorting and only digest organic matter along with microorganisms associated to the sediment matrix.

6.4 Temporal Trends

Knowledge of seasonal fluctuation of benthic fauna is scanty along the SBE. In open lagoons, simultaneous comparison of seasonal variability between estuarine meiofauna and macrofauna indicates a clear temporal asynchrony in their variation. High abundance and richness of the meiofauna is evident during autumn and winter, whereas the macrofauna is more diversified and abundant during summer and spring periods (Fonseca and Netto 2006; Meurer and Netto 2007). Such divergent temporal trend exhibited by the meiofauna and macrofauna, also observed in some North American estuaries (e.g., Montagna and Kalke 1992), might result from a complex array of variables and processes, such as competition for food sources, predatory pressures, and differential response to disturbances, all ultimately linked to the contrasting mechanisms for diversity maintenance.

The reproduction and recruitment of macrobenthic species in SBE open lagoons are related to increases in temperature, benthic production, and sediment organic content during summer periods, which results in increased macrobenthic abundance and species number (Bemvenuti 1987a; Rosa and Bemvenuti 2006; Fonseca and Netto 2006; Meurer and Netto 2007). In the permanently open Patos and Laguna lagoons, for example, extreme values in total macrofauna density between winter and summer are 9000 and 17,000 ind/m^2 in Patos and 10,600 and 18,900 ind/m^2 in Laguna, respectively (Bemvenuti 1987a; Fonseca and Netto 2006). In contrast to permanently open lagoons, temporal oscillations of benthic communities in temporarily open/closed lagoons are primarily modulated by the inlet dynamics. Intermittent breaching of lagoon mouth leads to remarkable changes in its physical-chemical environment over short time periods (hours/days), which in turn triggers major biological responses in the benthic compartment. The barrier breaching results in strong advection and flushing into the sea. Together with sediment scouring from the lagoon, microphytobenthic biomass, sediment organic content, and benthic fauna numbers decrease. Following the shock produced by breaching, a new benthic fauna association slowly emerges (Netto et al. 2012). Many lagoons in South Africa and Australia experience seasonal opening and closure of inlets, with clear seasonal variations of streamflow and onshore sediment transport (Allanson and Baird 2008; Ranasinghe and Pattiaratchi 2003). This is not the case of SBE lagoons, even in the southern portion where rainy periods are more pronounced (Sbruzzi 2015). Despite species similarities among lagoons and coastal plain estuaries, the temporal dynamics of numerically dominant species seems to be quite different. At coastal plain systems, the population peaks in abundance (and greatest spatial variation of data within each estuary) of major species occurs in autumn-winter, while at lagoon systems they occur in summer-spring (Rosa Filho and Bemvenuti 1998).

Macrofaunal communities respond to environmental fluctuations at different time scales by changing their composition and structure (Morrisey et al. 1992). Seasonal cycles of benthic macrofauna (Alden et al. 1997; Ysebaert et al. 2005) and their interannual fluctuations are broadly associated with temperature, rainfall, and substrate alterations (Herman et al. 2001; Teske and Wooldridge 2003). El Niño Southern Oscillation (ENSO) has been extensively studied on the southern portion of SBE (e.g., Odebrecht et al. 2017 and references herein), where it has a major role in controlling the long-term variability of the composition and abundance of benthic invertebrates. In the last 20 years, large interannual variations of freshwater discharge (198 m^3/s to 4021 m^3/s) and four *El Niño* events (1997–1998; 2002–2003; 2009–2010; 2014–2015) affected the species composition by sustaining oligohaline and decreasing estuarine species numbers. During this same period, species numbers presented large interannual variability, with higher values (>28,000 ind/m^2) observed during the polyhaline summer seasons of 1997, 1999, 2000, 2008, and 2013, and lower (<6000 ind/m^2) related to the oligohaline periods influenced by El Niño events. However, low density values might also be related to extremely dry conditions (high salinity and low freshwater discharge) as observed in summer of 2005 (Fig. 6.8; Colling et al. 2010). In the main channel of the Patos Lagoon (>10 m deep), variations of abundance and species composition can also be attributed to a natural macrobenthic response related to environmental variations. Along the year 2000, for example, high salinity values decreased through seasons to almost a limnetic scenario in spring, reducing diversity and abundance of marine species. A contrasting situation was observed in 2006 when the maintenance of brackish waters throughout the year favored the input and survival of marine species from coastal adjacent areas inside the estuary, increasing macrobenthic diversity mainly by marine mollusks and polychaete worms (Pinotti et al. 2011).

The response of individual macrobenthic species to ENSO may differ. The *El Niño* events affected negatively the numerically dominant *Monokalliapseudes schubarti*, whose density values drop from a mean of higher than 5000 ind/m^2 to less than 400 ind/m^2 (Rosa and Bemvenuti 2006; Bemvenuti and Colling 2010) (Fig. 6.8). In contrast, intense recruitment of the bivalve *Erodona mactroides* recurrently occurs associated to *El Niño* events and pos-*El Niño* periods. Depending on synchronicity of reproduction and estuarine discharge, the larvae of *E. mactroides* generated by the adult reproductive stocks in the upper Patos Lagoon are carried by

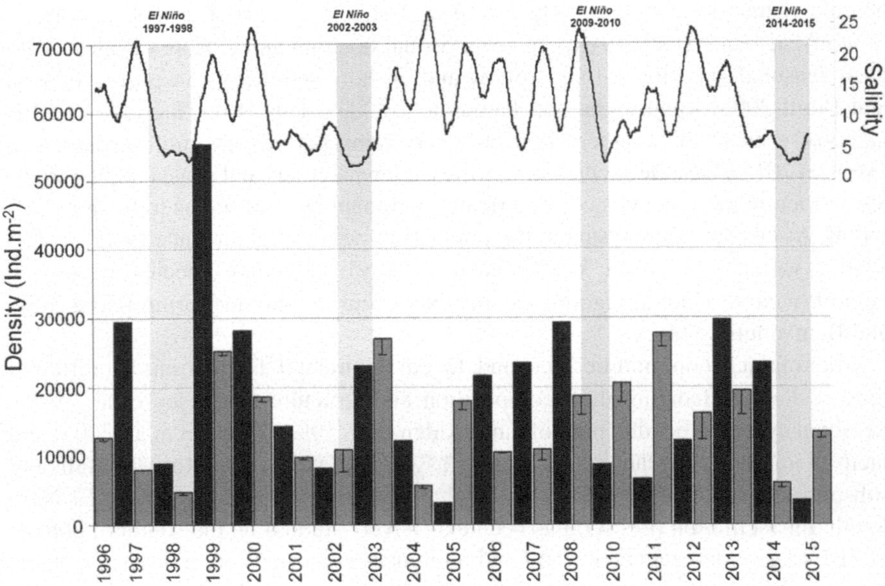

Fig. 6.8 Average densities (ind/m^2) of total macrobenthic fauna for the winter (*gray bars*) and summer (*black bars*) seasons between 1996 and 2015. Seasonal average values (*line*) of salinity between winter 1996 and summer 2015. Periods influenced by ENSO-El Niño are indicated with *light gray bars*

ebb tides during spring and summer, successfully recruiting in sheltered embayments within low estuarine regions (Bemvenuti 1997b; Colling et al. 2010).

6.5 Conservation, Impacts, and Management Issues

Four different categories of marine protected areas (MPAs) are present in estuaries from SBE: the Marine Extractive Reserve (Resex) of Pirajubaé and the Carijós Ecological Station, in the north, with small coastal plain estuaries bordered by mangroves; the Southern Right Whale Environmental Protection Area, in the center, that includes the inlet region of Garopaba and Ibiraquera temporarily open/closed lagoons, fragments of saltmarshes of Laguna, and the Camacho and Urussanga lagoons; and the Lagoa do Peixe National Park, in the south, that encompass total area of this temporarily open/closed lagoon. MPAs together with the Brazilian National Plan of Coastal Management (PNGC) and supported state and municipal plans, as well as several other legal, planning, and conservation measures could lead to a more sustainable management of coastal ecosystems and resources. However, despite almost three decades of establishing the PNGC, the overlap between laws, policies, plans, and provisions relevant to coastal planning and decision-making has not avoided unsustainable practices, impacts, and habitat loss all over the SBE estuaries. Even within the MPAs, SBE estuaries are under constant environmental pressure (e.g., artificial opening of Peixe and Ibiraquera lagoons, construction of roads in Resex Pirajubaé) that lead to losses of habitat and taxonomic and functional biodiversity. Although coastal management efforts in SBE have stimulated sound technical analysis, information gaps and inadequate exchange and coordination of information within and between government, civil society, private sector, and academia are commonplace.

Interactions between humans and SBE estuaries occur at least since the middle Holocene. Signs of pre-ceramic coastal populations can still be seen by some of the world's largest shell mounds or "sambaquis" (in some cases reaching 30 m high and several hundred meters in diameter) predominantly composed by the estuarine carib pointed-venus or *berbigão Anomalocardia flexuosa* (DeBlasis et al. 2007; Colonese et al. 2017). Over the time, the lagoons mouths, particularly Patos Lagoon, became important sites of commerce, which magnified environmental changes (e.g., Seeliger and Odebrecht 2010). Human interactions with SBE estuaries have continued and accelerated over the past decades, so that at present no estuarine system is in its natural state.

A survey in the current literature of SBE accounted for the main causes of changes on sediment characteristics, the associated environmental impacts, and their consequences for the benthic fauna (Table 6.5):

1. The growing urbanization in estuaries is primarily expressed by filling activities over the margins for the construction of roads, ports, commerce and households, promoting physical suffocation on the sediment surface, causing anoxia and habitat suppression. These practices usually result in the death of benthic organisms, losses of species, and assemblage functional diversity. A recent long-term local study showed that despite the potential for faunal colonization provided by the relative improvement of the conditions surrounding impacted areas by landfilling, the effective restoration did not occur once the stressors were not removed (i.e., pollutants in sediments and changes due to sediment refilling) (Pagliosa et al. 2016).
2. The construction of infrastructure and the removal of coastal vegetation (Caruso 1990; Rovai et al. 2012) are among the causes of changes on the average grain size of sediments, which affect sorting, rates of sediment transport, and the local biogeochemistry cycles (Pagliosa et al. 2005, 2006b). The changes directly affect estuarine functions through alterations in the trophic guilds of benthic meiofauna and macrofauna (Pagliosa et al. 2012; Felix et al. 2015), mainly affecting filter-feeder, surface deposit-feeder, and burrower organisms. Those activities have caused the reduction of stocks of the bivalve *Anomalocardia flexuosa*, the white shrimp *Litopenaeus schmitti,* and the pink shrimp *Farfantepenaeus brasiliensis*, resources used by local fishing communities (Pezzuto and Echternacht 1999; Spínola et al. 2014; Pezzuto and Souza 2015). The coastal construction has attracted hard-bottom organism (Pagliosa et al. 2012a).
3. The dredging and sediment deposition activities with overloading suspended material result in change of depth, coarser sediment deposits, increased turbidity

Table 6.5 Causes, main environmental impacts and the consequences for benthic fauna recorded in estuaries of the Southern Brazil Ecoregion (SBE)

Cause	Environmental impact	Consequence for benthic fauna
1. Filling of coastal areas	Anoxia Habitat suppression	Death of individuals Losses of taxonomic and functional biodiversity
2. Coastal construction with removal of vegetation	Changes in sorting of particles Changes in rates of particles transport Changes in biogeochemistry	Changes in filter-feeders, surface deposit-feeders, and burrowing organisms Attraction of hard-bottom organisms
3. Dredging and deposition of sediments	Reduction of primary productivity Change of depth Change in sediment size Remobilization of metals	Removal and reduction of surface deposit-feeders and filter-feeding organisms Changes in diversity and abundance
4. Marine spill incidents and harbor/marina activities	Changes in water and sediment biogeochemistry Death of coastal vegetation Habitat suppression Sink of pollutants	Increase in invasive species Losses of taxonomic and functional biodiversity Changes in abundance, size, biomass, and secondary productivity Endocrine disruption DNA damage
5. Artificial breaching of temporarily open/closed lagoons	Reduction of primary productivity Increase of sediment infill Death of coastal vegetation	Losses of taxonomic and functional biodiversity
6. Aquaculture	Anoxia Eutrophication Change in sediment size and organic content	Increase in invasive species Losses of taxonomic and functional biodiversity Alterations in species evenness
7. Fishing	Decrease of benthic primary production Change sediment cohesion and erosion and deposition rates	Overexploitation and collapse of target populations Changes in diversity and abundance Loss of low mobility and long-lived species
8. Lack of adequate sanitation	Anoxia Changes in quality of food for benthic fauna Increase of fine sediments Release of emerging contaminants	Losses of taxonomic and functional biodiversity Changes in size, biomass, and secondary productivity Endocrine disruption DNA damage

due to resuspension of silt and clay (Schettini et al. 2000; Bemvenuti et al. 2005), and remobilization of metals. The increasing turbidity has implications for surface deposit-feeder and filter-feeder organisms, as well as for primary productivity. Changes in water compounds and in sediments have altered both nutrient dynamics (Pagliosa et al. 2005) and the relationships among several types of chemicals (Pagliosa et al. 2006a,b). Dredged areas can experience drastic decrease of both species richness and density, which requires efficient strategies of resilience of the dominant species to minimize the effects of these impact sources (Bemvenuti et al. 2005). In disposal sites at coastal marine zone, the absence of detrimental effects on benthic assemblages suggests that the resident biota can be adapted to dynamic sedimentary conditions, and that the dispersion of dredge material through the water column might minimize sediment deposition and damage to the benthic fauna (Angonesi et al. 2006).

4. A recent study indicated that mangrove and saltmarshes are the most sensible areas to marine spills (Silva et al. 2008). Marine spill incidents and harbor/marina activities cause sink of pollutants, changes in coastal vegetation, habitat suppression, and changes in water and sediment biogeochemistry. A sulfuric acid spill within Patos Lagoon altered the water quality, causing physical damages in zooplankton (Montú and Gloeden 1998) and the decline of macrofaunal richness and abundance (Bemvenuti et al. 2003). Another impact related to harbor activities is the larvae transport by ship's ballast water, which plays an important role in the dispersion of nonindigenous species. The occurrence of the decapod *Rhithropanopeus harrisii* (D'incao and Martins 1998) and the bivalve *Limnoperna fortunei* in SBE are attributed to this source of larval dispersion, being *L. fortunei* highly dependent of limnetic scenarios to their dispersion through estuaries (Capítoli et al. 2008).
5. Along SBE, more than half of the coastal lagoons' entrance channels become temporarily blocked by the buildup of marine sandbars. Artificial breaching of open/closed lagoons involves dredging and/or bulldozing the lagoon inlet at a level lower than natural breakout and has occurred for various reasons. While alleviating actual or perceived water quality is often cited as a trigger for artificial openings, the opening alone is not likely to significantly improve water quality. The limited tidal flushing and exchange efficiencies means that pollutants (particularly those entering from tributaries furthest from the entrance) may be moved around within the lagoon but may not be removed (e.g., Spurway et al. 2000). Artificial opening aiming to enhance fish and prawn recruitment and subsequent production is another reason always cited for dredging/bulldozing the lagoon, but its efficiency is largely unclear and virtually impossible to address without a detailed sampling and analysis of offshore and coastal larval populations. Finally, unplanned and irregular occupation within lagoon floodplains can further result in pressure for intervention to artificially breach lagoon inlet to avoid damage or inconvenience to low-lying properties. As inlet dynamics play a key role in the overall functioning coastal lagoons, intervention in the behavior of lagoon entrances is generally accompanied by negative environmental impacts (e.g., crash of the population of macrobenthic invertebrates, fish and vegetation and increase of sediment infill), potentially reducing lagoon resilience (Whitfield and Bate 2007; Crippa et al. 2013; Netto et al. 2012; Netto and Fonseca 2017). Monitoring, establishment of local estuarine management plan with short- and long-term goals, and permanent policy review would ensure that the most ecologically appropriate and cost-effective options are being implemented at any given location.

6. The Center-North of SBE ecoregion is the main Brazilian area for aquaculture, being responsible for ~21,000 tons/year of molluscs and shrimps. The seabed enrichment as a by-product of aquaculture has large impacts on the availability of food for benthic fauna. Aquaculture activities reduce benthic diversity, increase anaerobic metabolism in the sediment, and increase both the rate of sedimentation and organic matter content in the sediment (Bonetti et al. 2007; Netto and Meurer 2007; Netto and Valgas 2007, 2010). Aquaculture in general is one of the ways by which cultured and associated species can be dispersed outward from their native regions. Particularly, the commercial oyster industry is responsible for accidental transport of many harmful shell-borer polychaetes as *Polydora haswelli, P. ecuadoriana, P. carinhosa* (Radashevsky et al. 2006), and *Boccardiella bihamata* (Junqueira et al. 2009). These worms bore into shells of live oysters *Crassostrea brasiliana, C. gigas, C. rhizophorae*, the barnacle *Megabalanus* and empty shells of the gastropods *Pugilina morio, Stramonita haemastoma, Strombus pugilis*, and *Tegula viridula* inhabited by hermit crabs *Clibanarius vittatus, Paguristes tortugae* and *Pagurus brevidactylus*.
7. Fishery along SBE estuaries is a small-scale or artisanal activity. The most important estuarine fishery, the shrimp *Penaeus paulensis*, is overexploited or collapsed in the main fishery grounds, Patos and Laguna Lagoons (Sunye et al. 2014; Haimovici and Cardoso 2017). Local shrimp bottom tending gears may be either active (1—a hand light mini trawl, known as "berimbau"; 2—bag nets, known as coca; 3—motorized otter trawl, only in Patos Lagoon) or passive (fyke nets, normally used in groups, known as "aviãozinho") (Netto and Perereira 2008; Haimovici and Cardoso 2017). Experimental evaluations of trawling on macrofauna of mud-bottoms (Angonesi and Bemvenuti 2004) and on meiofauna and macrofauna of sandy vegetated and unvegetated bottoms (Costa and Netto 2014) showed negative impacts on superficial benthic forms, such as the gastropod *H. australis* and the ostracod *Cyprideis multidentate.* The passive shrimp fishing gear "aviãozinho," a very unselective gear (Vieira et al. 1996), is largely the actual dominant gear for shrimp in SBE lagoons. It is composed by a group of five to seven fyke nets (25 mm mesh size) set in contact to the bottom, fixed with stakes forming a cage-like structure (around 30 m^2). The nets, kept in the same place during months, change sediment composition, decrease microphytobenthic biomass, and affect the trophic structure of the meiofauna (Netto and Pereira 2008).
8. The rapid and unplanned urbanization without the appropriate treatment of residuals have changed the water and sediment estuarine properties (Pagliosa et al. 2005, Pagliosa et al. 2006a, 2006b), even promoting the release of emerging contaminants such as synthetic hormones and the herbicide glyphosate. Estuarine macrobenthic communities and populations, in turn, have shown altered responses in abundance, richness, size of individuals, biomass and secondary productivity, followed by DNA damage in the polychaete *Laeonereis culveri* (Pagliosa and Barbosa 2006; Weis et al. 2017).

The measurements of what might threaten estuarine outputs—namely impacts to the ecosystem itself as well as the social and economic systems to which they are linked—can be greatly facilitated by the use of efficient chemical markers of eutrophication and bioindicator tools. In this sense, the delivery of more integrated policy and management strategies is intrinsically linked to the conservation of estuarine functions, towards the preservation of complex mechanisms and processes that generate goods and services, even if not directly benefiting people at a first moment (Mahoney and Bishop 2017). Thus, assessments of estuarine quality might be highly improved when populations, species, or species guilds, with distinguishing properties and capabilities to support particular functions within an ecosystem, are used as real-world indicators. Productivity is also an important ecosystem function to consider because while it may not often be a direct service, it underpins many other kinds of output.

The environmental quality status along 27 SBE estuarine systems was analyzed based on the AZTI Marine Biotic Index (AMBI; Borja et al. 2000) and Trophic Index (TRIX; Vollenweider et al. 1998). According to the AMBI index, the environmental quality was largely classified as moderately disturbed (80%), whereas 20% of the sites were of slightly or undisturbed classifications (Fig. 6.9a). As estuaries are constantly subjected to natural sources of stress, they tend to be dominated by more tolerant taxa, and the indices based on macrofaunal indicator species in general tend to underestimate habitat quality assessments (Tweedley et al. 2016). However, regardless of natural background variability, consistent AMBI responses to contamination gradients have been found in Brazilian estuaries (Brauko et al. 2015, 2016), meaning that the AMBI levels of disturbance in the SBE estuaries might be even informative to some extent. Despite maintaining similar moderate ecological qualities, the proportion of ecological groups of fauna composing the AMBI values suggests community responses linked to salinity and hydrodynamic changes along the estuaries (see bar colors in Fig. 6.9). Freshwater communities of closed lagoons were dominated by ecological group III or tolerant species, followed by the absence of group I, composed of species more sensitive to disturbance. As the estuaries change with increasing salinity inputs from temporarily open/closed lagoons to open and coastal plain estuaries, the proportion of ecological groups I (of sensitive species) and IV (of opportunist species), in general, increases, while the relative contribution of group III decreases.

Such natural subsequent changes in the physical-chemical settings and transitional communities along the estuarine systems have not yet led to inconsistencies in the habitat quality diagnosis shown by both AMBI and TRIX. These indices are based on distinct biological compartments, but agreed on their diagnosis of environmental stress despite the additional influence of natural background variability. Similarly, average values of TRIX indicated that the SBE estuarine waters are mesotrophic in general (Fig. 6.9b). However, TRIX values indicated that nearly 40% of open and closed estuarine waters were eutrophic, mostly driven by high nitrate and low oxygen concentrations within those systems. The proximity to urbanized areas and the lower resident water time (hours to weeks) also favor loads of served water.

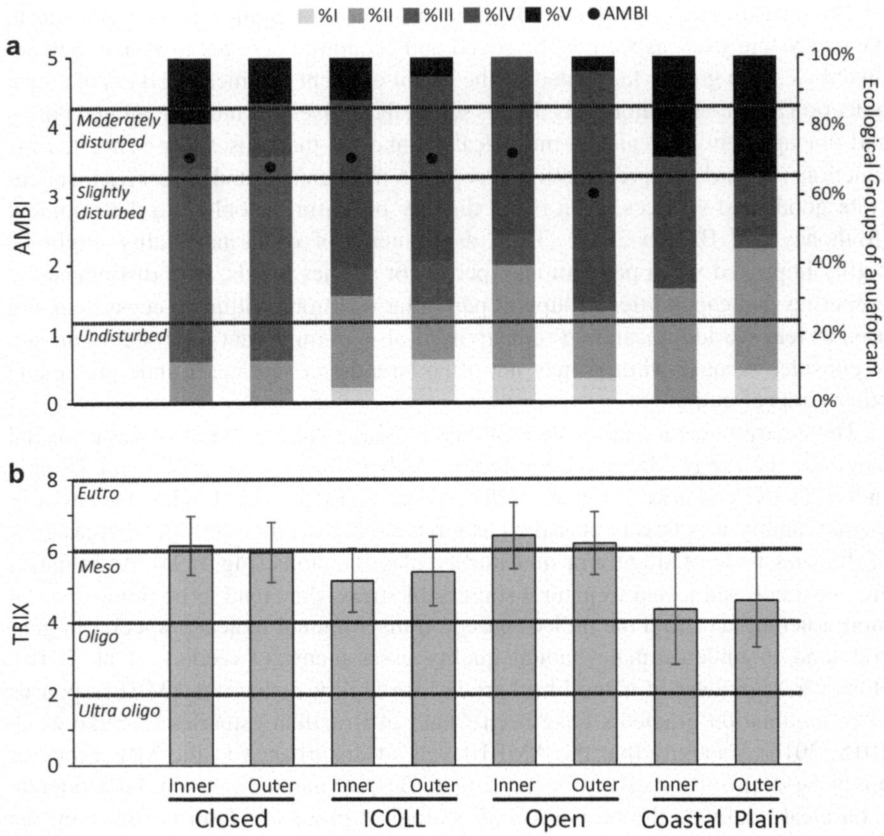

Fig. 6.9 Environmental quality shown by: (**a**) AZTI Marine Biotic Index (AMBI) and percentage of ecological groups of macrofauna (I—sensitive; II—indifferent; III—tolerant; IV—second order opportunistic fauna; V—first order opportunistic fauna). Index values were calculated and their ecological status was therefore attributed from undisturbed to slightly disturbed and moderately disturbed. (**b**) Water trophic state estimated by trophic Index (TRIX) (mean ± SE, $n = 160$), from ultra-oligotrophic to eutrophic conditions in SBE estuaries. Indices were applied over inner and outer stations within closed, intermittently closed/open (ICOLL), and open lagoons, and coastal plain estuaries

Despite relatively optimistic macrofaunal health scenarios of slight and moderate disturbance, previous evidence of diffuse pollution impacts highlight important local constrains for estuarine conservation. In particular, sediment and water chemical markers as well as benthic community parameters have responded to the increasing urbanization in estuaries of the SBE within the last decade (Pagliosa et al. 2005, Pagliosa and Barbosa 2006; Felix et al. 2015). In addition, more recent findings in Weis et al. (2017) showed that some of the more urbanized estuaries are experiencing loss of environmental quality linked to changes in a bioindicator estuarine species, the polychaete *Laeonereis acuta* (see details of comparisons between Urban and Non-urban estuaries in Fig. 6.10). There is very convincing evidence of molecu-

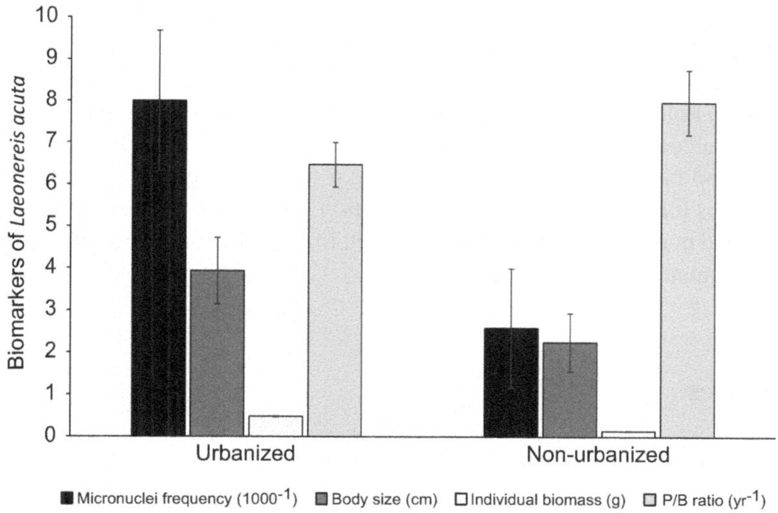

Fig. 6.10 Biomarkers of different levels of biological organization (micronuclei frequency, body size, individual biomass and P/B ratio) in *Laeonereis acuta* (mean ± SE) along urbanized and non-urbanized estuaries of the Southern Brazil Ecoregion (SBE). Adapted from Weis et al. (2017)

lar damage with increased micronuclei frequency, higher individual biomass and size, as well as losses in population production (P/B ratios) associated to a set of contaminants including total nitrogen, aluminum, copper, and lead contents (Fig. 6.10). The rapid sub-cellular or molecular changes provide the detection of early stages of pollution (from hours to days), and as responses progress to the higher biological levels of populations (e.g., biomass or body size) and communities (e.g., indicator species indexes) the responses take longer periods to emerge (from weeks to years). Eventual biomonitoring protocols of anthropic impacts could therefore benefit from the combination of rapid sub-cellular indicators along with biotic indices based on community responses of higher ecological and functional relevance.

6.6 Final Considerations

Despite sensitive to anthropic impacts, the estuarine systems included in the Southern Brazilian Marine Ecoregion still present satisfactory conditions that sustain benthic biodiversity and ecosystem goods and services. Notwithstanding, the growing coastal urbanization intensified by agricultural and industrial activities may change the relatively optimistic scenario observed in the last decades.

As other impacts, climate change is expected to intensify environmental risks and amplify extreme weather events and coastal hazards. Although the slow-onset nature of climate changes, there is an urgent need to build institutional capacity to better understand and address climate change impacts in estuaries and chart adaptive pathways. Thus, the systematic application of long-term monitoring protocols (such as the Brazilian Monitoring Network for Coastal Benthic Habitats, ReBentos) within estuaries is essential to understand and predict changes. Besides, recommendations (see Polette et al. 2015) for building adaptive capacity, resilience, and sustainability in coastal communities should include: (1) raise public awareness about the coast and climate change through active social learning processes; (2) create opportunities for meaningful public participation in coastal management efforts; (3) integrate and mainstream coastal management, from pollution to disaster risk reduction, and climate change adaptation efforts.

References

Abreu PC, Odebrecht C, González A (1994) Particulate and dissolved phytoplankton production of the Patos lagoon estuary, southern Brazil: comparison of methods and influencing factors. J Plankton Res 16(7):737–753

Abreu PC, Bergesch M, Proença LA et al (2010) Short- and long-term chlorophyll *a* variability in the shallow microtidal Patos lagoon estuary, Southern Brazil. Estuar Coasts 33:554–569

Abreu PC, Marangoni JC, Odebrecht C (2017) So close, so far: differences in long-term chlorophyll a variability in three nearby estuarine-coastal stations. Mar Biol Res 13:9–21

Alden RW, Weisberg SB, Ranasinghe JA (1997) Optimizing temporal sampling strategies for benthic environmental monitoring programs. Mar Pollut Bull 34(11):913–922

Allanson BR, Baird D (eds) (2008) Estuaries of South Africa. Cambridge University Press, Cambridge

Alvares CA, Stape JL, Sentelhas PC et al (2014) Köppen's climate classification map for Brazil. Meteorol Z 22(6):711–728

Angonesi LG, Bemvenuti CE (2004) Effects of shrimp trawling upon macrobenthic fauna in Patos lagoon estuary. J Coast Res SI 39:1368–1372

Angonesi LG, Bemvenuti CE, Gandra MS (2006) Effects of dredged sediment disposal on the coastal marine macrobenthic assemblage in Southern Brazil. Braz J Biol 66(2A):413–420

Anthony A, Atwood J, August P (2009) Coastal lagoons and climate change: ecological and social ramifications in U.S. Atlantic and Gulf coast ecosystems. Ecol Soc 14(1):8

Bemvenuti CE (1987a) Predation effects on a benthic community in estuarine soft sediments. Atlantica 9(1):5–32

Bemvenuti CE (1987b) Macrofauna bentônica da região estuarial da Lagoa dos Patos, RS – Brasil. Publ Acad Ciências do Estado de São Paulo 54(1):428–459

Bemvenuti CE (1992). Interações biológicas da macrofauna bentônica numa enseada estuarina da Lagoa dos Patos, RS, Brasil. Doctoral thesis, Universidade de São Paulo, São Paulo

Bemvenuti CE (1997a) Benthic invertebrates. In: Seeliger U, Odebrecht C, Castello JP (eds) Subtropical convergence environments: the coast and sea in the Southwestern Atlantic. Springer, Berlin, pp 43–47

Bemvenuti CE (1997b) Unvegetated intertidal flats and subtidal bottoms. In: Seeliger U, Odebrecht C, Castello JP (eds) Subtropical convergence environments: the coast and sea in the Southwestern Atlantic. Springer, Berlin, pp 78–85

Bemvenuti CE, Colling LA (2010) As comunidades de macroinvertebrados bentônicos. In: Seeliger U, Odebrecht C (eds) O estuário da Lagoa dos Patos, um século de transformações. Editora da FURG, Rio Grande, pp 101–114

Bemvenuti CE, Netto GA (1998) Distribution and seasonal patterns of the sublittoral benthic macrofauna of Patos lagoon (South Brazil). Rev Bras Biol 58:211–221

Bemvenuti CE, Rosa Filho JS (2000) Estrutura e dinâmica das associações de macroinvertebrados bentônicos dos ambientes estuarinos do Rio Grande do Sul: um estudo de caso. In: Anais do Workshop Avaliação e Ações Prioritárias para a Zona Costeira e Marinha, PROBIO, Federal University of Rio Grande, Rio Grande

Bemvenuti CE, Capítoli RR, Gianuca NM (1978) Estudos de ecologia bentônica na região estuarial da Lagoa dos Patos. II - Distribuição quantitativa do macrobentos infralitoral. Atlântica 3:23–32

Bemvenuti CE, Rosa Filho JS, Elliott M (2003) Changes in soft-bottom macrobenthic assemblages after a sulphuric acid spill in the Rio Grande Harbor (RS, Brazil). Braz J Biol 63(2):183–194

Bemvenuti CE, Angonesi LG, Gandra MS (2005) Effects of dredging operations on soft bottom macrofauna in a harbor in the Patos lagoon estuarine region of Southern Brazil. Braz J Oceanogr 65(4):573–581

Bernardino AF, Netto AS, Pagliosa PR et al (2015) Predicting ecological changes on benthic estuarine assemblages through decadal climate trends along Brazilian marine ecoregions. Estuar Coast Shelf Sci 166:74–82

Bonaglia S, Nascimento FJA, Bartoli M et al (2014) Meiofauna increases bacterial denitrification in marine sediments. Nat Commun 16:1–9

Bonetti C, Bonetti J, Barcellos RL (2007) Caracterização sedimentar e geoquímica de sistemas costeiros com ênfase na avaliação da influência de sítios de cultivo de moluscos. In: Barroso GF, Poersch LHS, Cavalli RO (eds) Sistemas de cultivos aquícolas na zona costeira do Brasil: recursos, tecnologias, aspectos ambientais e sócio econômicos. Museu Nacional, Rio de Janeiro, pp 139–150

Borja A, Franco J, Pérez V (2000) A marine biotic index to establish the ecological quality of soft-bottom benthos within European estuarine and coastal environments. Mar Pollut Bull 40:1100–1114

Brauko KM, Souza FM, Muniz P et al (2015) Spatial variability of three benthic indices for marine quality assessment in a subtropical estuary of Southern Brazil. Mar Pollut Bull 91:454–460

Brauko KM, Muniz P, Martins CC et al (2016) Assessing the suitability of five benthic indices for environmental health assessment in a large subtropical South American estuary. Ecol Indic 64:258–265

Capítoli RR, Colling LA, Bemvenuti CE (2008) Cenários de distribuição do mexilhão dourado *Limnoperna fortunei* (Mollusca – Bivalvia) sob distintas condições de salinidade no complexo lagunar Patos-Mirim, RS – Brasil. Atlantica 30(1):35–44

Caruso MML (1990) O desmatamento da Ilha de Santa Catarina de 1500 aos dias atuais. Editora da UFSC, Florianópolis

Castelão RM, Möller OO Jr (2003) Sobre a circulação tridimensional forçada por ventos na Lagoa dos Patos. Atlantica 25(2):91–106

Chapman PM (2012) Management of coastal lagoons under climate change. Estuar Coast Shelf Sci 110:32–35

Colling LA, Bemvenuti CE, Gandra MS (2007) Seasonal variability on the structure of sublittoral macrozoobenthic association in the Patos lagoon estuary, southern Brazil. Iheringia Ser Zool 97:257–263

Colling LA, Bemvenuti CE, Pinotti RM (2010) Temporal variability of the bivalve *Erodona mactroides* BOSC, 1802 during and after El Niño phenomenon (2002/2003) in a subtropical lagoon, Southern Brazil. Acta Limnol Bras 22(4):410–423

Colonese AC, Netto SA, Francisco AS et al (2017) Shell sclerochronology and stable isotopes of the bivalve *Anomalocardia flexuosa* (Linnaeus, 1767) from southern Brazil: implications for environmental and archaeological studies. Palaeogeogr Palaeoclimatol Palaeoecol 484:7. https://doi.org/10.1016/j.palaeo.2017.01.006

Cooper JAG (1994) Lagoons and microtidal coasts. In: Carter RWG, Woodroffe CD (eds) Coastal evolution. Late quaternary shoreline morphodynamics. Cambridge University Press, Cambridge, pp 219–265

Cooper JAG, Wright CI, Mason TR (1999) Geomorphology and sedimentology of South African estuaries. In: Allanson BR, Baird D (eds) Estuaries of South Africa. Cambridge University Press, Cambridge, pp 5–25
Costa CSB, Davy AJ (1992) Coastal saltmarsh communities of Latin America. In: Seeliger U (ed) Coastal Plant Communities of Latin America. Academic Press, New York, pp 179–212
Costa KG, Netto SA (2014) Effects of small-scale trawling on benthic communities of estuarine vegetated and non-vegetated habitats. Biodivers Conserv 23:1041–1055
Crippa LB, Stenert C, Maltchik L (2013) Does the management of sandbar openings influence the macroinvertebrate communities in southern Brazil wetlands? A case study at Lagoa do Peixe National Park - Ramsar site. Ocean Coast Manag 71:26–32
Cunha SR, Asmus M, Costa CSB (2005) Production dynamics of *Spartina alterniflora* saltmarshes in the estuary of Patos lagoon (RS, Brazil): a simulation model approach. Brazil J Aquat Sci Technol 9:75–85
D'Incao F, Martins STS (1998) Occurrence of *Rhithropanopeus harrisii* (Gouls, 1841) in the southern coast of Brazil (Decapoda, Xanthidae). Nauplius 6:191–194
Danulat E, Muniz P, García-Alonso J et al (2002) First assessment of the highly contaminated harbour of Montevideo, Uruguay. Mar Pollut Bull 44(6):554–565
DeBlasis P, Kneip A, Scheel-Ybert R et al (2007) Sambaquis e paisagem: dinâmica natural e arqueologia regional no litoral do sul do Brasil. Arqueol Sudamer/Arqueol Sul-Amer 3(1):29–61
Dillenburg S, Barboza E, Tomazelli L et al (2009) The Holocene Coastal Barriers of Rio Grande do Sul. In: Dillenburg SR, Hesp PA (eds) Geology and Geomorphology of Holocene Coastal Barriers of Brazil. Springer, Berlin, pp 53–91
Dos Santos AC, Fonseca ALDO, Koerich G (2017) Caracterização do estado trófico e da produção primária dos sistemas estuarinos e lagunares de Santa Catarina (Brasil) e sua relação com a atividade antrópica. XVII Congresso Latino-Americano de Ciências do Mar, Camboriú, SC, Brasil, p 1465–1468
Eichler PPB, Casteão GP, Pimenta FM, Eichler BB (2006) Avaliação da saúde ecológica do Sistema Estuarino de Laguna (SC) baseado nas espécies de foraminíferos e tecamebas. Pesq Geocien 33:101–115
Felix G, Cândido TF, Silveira CB et al (2015) Resposta de associações bênticas ao lançamento de resíduos de drenagem urbana. Brazil J Aquat Sci Technol 19(2):47–57
Fernandes LF, Brandini FP (2010) The potentially toxic diatom *Pseudo-nitzschia* H. Peragallo in the Paraná and Santa Catarina states, southern Brazil. Iheringia Ser Bot 65:47–62
Fernandes EHL, Dyer KR, Möller OO et al (2002) The Patos lagoon hydrodynamics during an El Nino event (1998). Cont Shelf Res 22(11):1699–1713
Fonseca ALO (2004) Variação sazonal e espacial das características hidroquímicas, dos fluxos de nutrientes e do metabolismo na interface água-sedimento da Lagoa da Conceição (SC, Brasil). Doctoral thesis, University of São Paulo, São Paulo
Fonseca A, Braga ES (2006) Spatial and seasonal variation of dissolved inorganic nutrients and phytoplankton biomass in the pelagic system of the Conceição lagoon, Southern Brazil. J Coast Res 39:1229–1233
Fonseca DB, D'Incao F (2003) Growth and reproductive parameters of Kalliapseudes schubartii in the estuarine region of the Lagoa dos Patos (southern Brazil). J Mar Biol Assoc UK 83:931–935
Fonseca G, Netto SA (2006) Shallow sublittoral benthic communities of the Laguna estuarine system, South Brazil. Braz J Oceanogr 54(1):41–54
Fontes MLS, Abreu PA (2009) Spatiotemporal variation of bacterial assemblages in a shallow subtropical coastal lagoon in Southern Brazil. Microb Ecol 58(1):140–153
Fontes MLS, Suzuki MT, Cottrell MT et al (2011) Primary production in a subtropical stratified coastal lagoon, contribution of anoxygenic phototrophic bacteria. Microb Ecol 61:223–237
Freitas-Júnior F, Christoffersen ML, Araújo JP et al (2013) Spatiotemporal distribution and population structure of *Monokalliapseudes schubarti* (Tanaidacea: Kalliapseudidae) in an estuary in Southern Brazil. Sci World J 2013:1. https://doi.org/10.1155/2013/363187

Garcia CAE (1997) Hydrographic characteristics. In: Seeliger U, Odebrecht C, Castello JP (eds) Subtropical convergence environments: the coast and sea in the Southwestern Atlantic. Springer, Berlin, pp 18–20

Garcia AM, Vieira JP, Winemiller KO (2003) Effects of 1997-1998 El Niño on the dynamics of the shallow-water fish assemblage of the Patos lagoon estuary (Brazil). Estuar Coast Shelf Sci 57(3):489–500

Grimm AM, Barros VR, Doyle ME (2003) Climate variability in Southern South America associated with El Niño and La Niña events. J Clim 1:35–38

Haimovici M, Cardoso LG (2017) Long-term changes in the fisheries in the Patos lagoon estuary and adjacent coastal waters in Southern Brazil. Mar Biol Res 13:135–150

Haraguchi L, Carstensen J, Abreu PC et al (2015) Long-term changes of the phytoplankton community and biomass in the subtropical shallow Patos lagoon estuary, Brazil. Estuar Coast Shelf Sci 162:76–87

Heip C, Vincx M, Vranken G (1985) The ecology of marine nematodes. Oceanogr Mar Biol Ann Rev 23:399–489

Herman PM, Middelburg JJ, Heip CH (2001) Benthic community structure and sediment processes on an intertidal flat: results from the ECOFLAT project. Cont Shelf Res 21(18):2055–2071

Hesp PA, Giannini P, Martinho C et al (2009) The Holocene barrier systems of the Santa Catarina coast, southern Brazil. In: Dillenburg SR, Hesp PA (eds) Geology and geomorphology of Holocene coastal barriers of Brazil. Springer-Verlag, Berlin Heidelberg, pp 93–133

Junqueira AOR, Tavares MDS, Schaeffer-Noveli I et al (2009) Zoobentos. In: Lopes RM, Coradin L, Pombo VB et al (eds) Informe Sobre as Espécies Exóticas Invasoras Marinhas no Brasil. Ministério do Meio Ambiente, Brasília, pp 145–371

Kapusta SC, Würdig NL, Bemvenuti CE (2002) Distribuição vertical da meiofauna, inverno e verão, no estuário de Tramandaí- Armazém, RS – Brasil. Acta Limnol Bras 14(1):81–90

Kapusta SC, Bemvenuti CE, Würdig NL (2004) Meiofauna spatial-temporal distribution in a subtropical estuary of southern coast Brazil. J Coastal Res SI 39:1238–1242

Kapusta SC, Würdig NL, Bemvenuti CE et al (2005) Meiofauna structure in Tramandaí-Armazém estuary (South Brazil). Acta Limnol Bras 17(4):349–359

Lanari M, Copertino MS (2017) Drift macroalgae in the Patos lagoon estuary (Southern Brazil): effects of climate, hydrology and wind action on the onset and magnitude of blooms. Mar Biol Res 13:1–12

Lemes-Silva AL, Pagiosa PR, Petrucio MM (2014) Inter- and intra-guild patterns of food resource utilization by chironomid larvae in a subtropical coastal lagoon. Limnology 15:1–12

Lemes-Silva AL, Pires JR, Pagiosa PR et al (2016) Distribution of aquatic macroinvertebrate assemblages in a subtropical coastal lake: response to environmental parameters. Fund Appl Limnol 188:113–127

Mahoney PC, Bishop MJ (2017) Assessing risk of estuarine ecosystem collapse. Ocean Coast Manag 140:46–58

McSweeney SL, Kennedy DM, Rutherfurd ID et al (2017) Intermittently closed/open lakes and lagoons: their global distribution and boundary conditions. Geomorphology 292:142. https://doi.org/10.1016/j.geomorph.2017.04.022

Meurer AZ, Netto SA (2007) Seasonal dynamics of benthic communities in a shallow sublitoral site of Laguna estuarine system (south, Brazil). Braz J Aquat Sci Technol 1(2):53–62

Montagna PA, Kalke RD (1992) The effect of freshwater inflow on meiofaunal and macrofaunal populations in the Guadalupe and Nueces estuaries, Texas. Estuaries 15:307–326

Montagnolli W, Zamboni A, Luvizotto-Santos R et al (2004) Acute effects of *Microcystis aeruginosa* from the Patos lagoon estuary, Southern Brazil, on the microcrustacean *Kalliapseudes schubartii* (Crustacea: Tanaidacea). Arch Environ Cont Toxicology 46(4):463–469

Montú M, Gloeden IM (1998) O derramamento ácido do navio NT Bahamas e as primeiras consequências sobre o zooplâncton do estuário da Lagoa dos Patos (Rio Grande, RS). Nauplius 6:195–197

Morrisey DJ, Underwood AJ, Howitt L et al (1992) Temporal variation in soft-sediment benthos. J Exp Mar Biol Ecol 164:233–245

Netto SA, Fonseca GF (2017) Regime shifts in coastal lagoons: evidence from free-living marine nematodes. PLoS One 12(2):e0172366

Netto SA, Gallucci F (2003) Meiofauna and macrofauna communities in a mangrove from the Island of Santa Catarina, South Brazil. Hydrobiologia 505:159–170

Netto SA, Meurer AZ (2007) Influência de efluentes de cultivo de camarão na estrutura das comunidades bênticas do Sistema Estuarino de Laguna (SC, Brasil). In: Barroso GF, Poersch LHS, Cavalli RO (eds) Sistemas de Cultivos aquícolas na zona costeira do Brasil: recursos, tenologias, aspectos ambientais e sócio-economicos. Museu Nacional, Rio de Janeiro, pp 129–138

Netto SA, Pereira TJ (2008) Benthic community response to a passive fishing gear in a coastal lagoon (South Brazil). Aquat Ecol 43(2):521–538

Netto SA, Valgas IS (2007) Efeitos de cultivos de mexilhões sobre a estrutura da meiofauna. In: Barroso GF, Poersch LHS, Cavalli RO (eds) Sistemas de Cultivos aquícolas na zona costeira do Brasil: recursos, tenologias, aspectos ambientais e sócio-economcos. Museu Nacional, Rio de Janeiro, pp 119–128

Netto SA, Valgas IS (2010) The response of nematode assemblages to intensive mussel farming in coastal sediments (Southern Brazil). Environ Monit Assess 162:81–93

Netto SA, Domingos AM, Kurtz M (2012) Effects of artificial breaching of a temporarily open/closed estuary on benthic macroinvertebrates (Camacho lagoon, southern Brazil). Estuar Coasts 35:1069–1081

Niekerk L, Merwe JH, Huizinga P (2005) The hydrodynamics of the Bot River Estuary revisited. Water SA 31(1):73–85

Niencheski LFH, Zepka-Baumgarten MG, Cabrera L et al (2006) Patos lagoon: indicators of organic pollution. J Coast Res 39:1356–1359

Odebrecht C, Abreu PC, Bemvenuti CE et al (2010) The Patos lagoon estuary, Southern Brazil. Biotic responses to natural and anthropogenic impacts in the last decades (1979-2008). In: Kennish MJ, Paerl HW (eds) Coastal Lagoons, Critical Habitats of Environmental Change. CRC Press, Boca Raton, pp 433–455

Odebrecht C, Secchi ER, Abreu PC et al (2017) Biota of the Patos lagoon estuary and adjacent marine coast: long-term changes induced by natural and human-related factors. Mar Biol Res 13:3–8

Omachi CY, Tamanaha MS, Proença LAO (2007) Bloom of *Alexandrium fraterculus* in coastal waters of Itajaí, SC, Southern Brazil. Braz J Oceanogr 55(1):57–61

Pagliosa PR, Barbosa FAR (2006) Assessing the environment–benthic fauna coupling in protected and urban areas of southern Brazil. Biol Conserv 129:408–417

Pagliosa PR, Fonseca A, Bosquilha GE et al (2005) Phosphorus dynamics in water and sediments in urbanized and non-urbanized rivers in Southern Brazil. Mar Pollut Bull 50:965–974

Pagliosa PR, Fonseca A, Barbosa FAR (2006a) Evidence of systemic changes in trace metal concentrations in subtropical estuarine sediments as a result of urbanization. J Coast Res 39:1078–1083

Pagliosa PR, Fonseca A, Barbosa FAR et al (2006b) Urbanization impact on subtropical estuaries: a comparative study of water properties in urban areas and in protected areas. J Coast Res 39:731–735

Pagliosa PR, Cantor M, Scherner F et al (2012) Influence of piers on functional groups of benthic primary producers and consumers in the channel of a subtropical coastal lagoon. Braz J Oceanogr 60(1):65–73

Pagliosa PR, Oortman MS, Rovai AS et al (2016) Is mangrove planting insufficient for benthic macrofaunal recovery when environmental stress is persistent? Ecol Eng 95:290–301

Panitz CMN (1986) Produção e decomposição de serapilheira no mangue do rio Itacorubi, Ilha de Santa Catarina, Florianópolis, Brasil. Doctoral thesis, Federal University of São Carlos, São Carlos

Peixoto AE, Costa CSB (2004) Produção primária líquida aérea de *Spartina densiflora* Brong (Poaceae) no estuário da laguna dos Patos, Rio Grande do Sul, Brasil. Iheringia Ser Bot 59:27–34

Pérez-Ruzafa A, Marcos C, Pérez-Ruzafa IM, Pérez-Marco M (2011) Coastal lagoons: "transitional ecosystems" between transitional and coastal waters. J Coast Conserv 15:369–392
Pezzuto PR, Echternacht AM (1999) Avaliação de impactos da construção da via expressa SC-sul sobre o berbigão *Anomalocardia brasiliana* (Gmelin, 1791) (Mollusca: Pelecypoda) na reserva extrativista marinha do Pirajubaé (Florianópolis, SC - Brasil). Atlantica 21:105–119
Pezzuto PR, Souza DS (2015) A pesca e o manejo do berbigão (*Anomalocardia brasiliana*) (Bivalvia: Veneridae) na Reserva Extrativista Marinha do Pirajubaé, SC, Brasil. Desenv Meio Amb 34:169–189
Pinotti RM, Colling LA, Bemvenuti CE (2011) Temporal dynamics of deep inffralittoral macrobenthic fauna inside a subtropical estuarine environment. Braz J Aquat Sci Technol 15(1):226–241
Polette M, Muehe D, Soares MLG, Glavovic BC (2015) The promise of coastal management in Brazil in times of global climate change. In: Glavovic BC, Kelly M, Kay R, Travers A (eds) Climate change and the coast: building resilient communities. CRC Press, Boca Raton, pp 379–413
Proença LAO (2004) A red water caused by *Mesodinium rubrum* on the coast of Santa Catarina, Southern Brazil. Braz J Oceanogr 52(2):153–161
Radashevsky VI, Lana PC, Nalesso RC (2006) Morphology and biology of *Polydora* species (Polychaeta: Spionidae) boring into oyster shells in South America, with the description of a new species. Zootaxa 1353:1–37
Ranasinghe R, Pattiaratchi C (2003) The seasonal closure of tidal inlets: causes and effects. Coast Eng J 45(04):601–627
Rosa LC, Bemvenuti CE (2006) Temporal variability of the estuarine macrofauna of the Patos lagoon, Brazil. Rev Biol Mar Ocean 41(1):1–9
Rosa Filho JS, Bemvenuti CE (1998) Caracterización de las comunidades macrobentónicas de fondos blandos en regiones estuarinas de Rio Grande do Sul (Brasil). Thalassas 18:43–56
Rovai AS, Soriano-Sierra EJ, Pagliosa PR et al (2012) Secondary succession impairment inrestored mangroves. Wetl Ecol Manag 20:447–459
Sbruzzi JB (2015) Análise da dinâmica da embocadura da lagoa do Peixe– RS utilizando dados de sensoriamento remoto orbital. Dissertação de mestrado, Universidade Federal do Rio Grande do Sul, Porto Alegre
Schallenberg M, Larned S, Hayward S et al (2010) Contrasting effects of managed opening regimes on water quality in two intermittently closed and open coastal lakes. Estuar Coast Shelf Sci 86:587–597
Schettini CAF, Pereira FOJ, Spillere L (2000) Caracterização oceanográfica e biogeoquímica dos estuários dos rios Tavares e Defuntos, Reserva Extrativista de Pirajubaé, Florianópolis, SC. Not Técn Fac Ciênc Mar 4:11–28
Schock NT, Murry BA, Uzarski DG (2014) Impacts of agricultural drainage outlets on Great Lakes coastal wetlands. Wetlands 24:297–307
Schratzberger M, Ingels J (2017) Meiofauna matters: the roles of meiofauna in benthic ecosystems. J Exp Mar Biol Ecol. https://doi.org/10.1016/j.jembe.2017.01.007
Seeliger U (2001) The Patos lagoon estuary. In: Seeliger U, Kjerfve B (eds) Coastal marine ecosystems of Latin America. Springer, New York, pp 167–183
Seeliger U, Odebrecht C (2010) O Estuário da Lagoa dos Patos: um século de transformações. Editora da FURG, Rio Grande
Seeliger U, Odebrecht C, Castello JP (1997) Subtropical convergence environments: the coast and sea in the Southwestern Atlantic. Springer, Berlin
Short AD, Klein AHF (eds) (2016) Brazilian beach systems. Springer, Berlin
Sierra de Ledo B, Soriano-Sierra EJ (eds) (1999) O ecossitema da lagoa da Conceição. SDM-FEPEMA, Florianópolis
Silva AF, Klein AHF, Petermann RM et al (2008) Índice de sensibilidade do litoral (ISL) ao derramamento de óleo, para a Ilha de Santa Catarina e áreas do entorno. Braz J Aquat Sci Technol 12:73–89

Silva VEC, Fonseca A, Franco D et al (2017) Space time evolution of the trophic state of a subtropical lagoon, Lagoa da Conceição, Florianópolis Island of Santa Catarina. Brazil Rev Bras Rec Hid. https://doi.org/10.1590/2318-0331.011716027

Soriano-Sierra EJ (1999) Ecossistema de marismas da lagoa da Conceição. II. A fitocenosis. In: Sierra de Ledo B, Soriano-Sierra EJ (eds) O ecossitema da lagoa da Conceição. SDM-FEPEMA, Florianópolis, pp 171–183

Spalding MD, Fox HE, Allen GR et al (2007) Marine ecoregions of the world: a bioregionalization of coastal and shelf areas. Bioscience 57:573e583

Spínola JL, Teixeira CF, Andriguetto Filho JM (2014) Desafios à cogestão: os impactos da Via Expressa Sul sobre o extrativismo na RESEX Marinha do Pirajubaé. Desenv Meio Amb 32:139–150

Spurway P, Roper T, Stephens K (2000) Community consultation, conservation, crustaceans, calculations and consents – Entrance Management Policies for ICOLLs. In: Proceedings 10th Annual NSW Coastal Conference, Yamba, 2000

Sunye PS, Pereira TJ, Russo A, Netto SA (2014) A pesca do camarão-rosa no Sistema Estuarino de Laguna, SC: história e acasos das políticas de manejo. In: Haimovici M, Andriguetto JM, Sunye PS (eds) A pesca marinha e estuarino no Brasil: estudos de caso multidisciplinares. Editora da Furg, Rio Grande, pp 75–85

Teixeira H, Borja A, Weisberg SB et al (2010) Assessing coastal benthic macrofauna community condition using best professional judgement - developing consensus across North America and Europe. Mar Pollut Bull 60:589–600

Teske P, Wooldridge T (2003) What limits the distribution of subtidal macrobenthos in permanently open and temporally open/closed south African estuaries? Salinity vs. sediment particle size. Estuar Coast Shelf Sci 57:225–238

Tweedley JR, Hallett CS, Warwick RM et al (2016) The hypoxia that developed in a microtidal estuary following an extreme storm produced dramatic changes in the benthos. Mar Fresh Res 67(3):327–341

Valgas, IS (2009) As marismas de *Spartina alterniflora* e os taboais de *Typha domingensis* do Sistema Estuarino de Laguna (Santa Catarina, Brasil): distribuição espacial, estrutura e macrofauna associada. Master thesis, Universidade Federal do Paraná, Pontal do Paraná

Vieira JP, Vasconcellos MCA, Silva RES, Fischer LGF (1996) Rejeição da pesca do camarão-rosa (*Penaeus paulensis*) no estuário da Lagoa dos Patos, RS, Brasil. Atlantica 18:123–142

Vieira JP, Garcia AM, Grimm AM (2008) Preliminary evidences of El Niño effects on the mullet fishery of Patos lagoon estuary (Brazil). Braz Arch Biol Technol 51:433–440

Vollenweider RA, Giovanardi F, Montanari G, Rinaldi A (1998) Characterization of the trophic conditions of marine coastal waters with special reference to the NW Adriatic Sea: proposal for a trophic scale, turbidity and generalized water quality index. Environmetrics 9:329–357

Weis WA, Soares CHL, Quadros DPC et al (2017) Urbanization effects on different biological organization levels of an estuarine polychaete tolerant to pollution. Ecol Indic 73:698–707

Whitfield A, Bate G (2007) A review of information on temporarily open/closed estuaries in the warm and cool temperate biogeographic regions of South Africa, with particular emphasis on the influence of river flow on these systems. Water Research Commission Report, Pretoria

Woodroffe CD (2002) Coasts: form, process and evolution. Cambridge University Press, Cambridge

Ysebaert T, Fettweis M, Meire P et al (2005) Benthic variability in intertidal soft-sediments in the mesohaline part of the Schelde estuary. Hydrobiologia 540:197–216

The manufacturer's authorised representative in the EU is Springer Nature Customer Service Centre GmbH, Europaplatz 3, 69115 Heidelberg, Germany. If you have any concerns regarding our products, please contact ProductSafety@springernature.com

Printed and bound by CPI Group (UK) Ltd, Croydon, CR0 4YY
22/09/2026
02221059-0005